Automation, Control and Complexity

AN INTEGRATED APPROACH

Edited by

Tariq Samad and **John Weyrauch**
Honeywell Technology Center, Minneapolis, USA

JOHN WILEY & SONS, LTD
Chichester • Weinheim • New York • Brisbane • Singapore • Toronto

Copyright © 2000 by John Wiley & Sons Ltd
Baffins Lane, Chichester,
West Sussex, PO 19 1UD, England

National 01243 779777
International (+44) 1243 779777

e-mail (for orders and customer service enquiries): cs-books@wiley.co.uk

Visit our Home Page on http://www.wiley.co.uk
or
http://www.wiley.com

All Rights Reserved. No part of this publication may be reproduced, stored in a retrieval system, or transmitted, in any form or by any means, electronic, mechanical, photocopying, recording, scanning or otherwise, except under the terms of the Copyright Designs and Patents Act 1988 or under the terms of a licence issued by the Copyright Licensing Agency, 90 Tottenham Court Road, London, W1P 9HE, UK, without the permission in writing of the Publisher, with the exception of any material supplied specifically for the purpose of being entered and executed on a computer system, for exclusive use by the purchaser of the publication.

Neither the authors nor John Wiley & Sons Ltd accept any responsibility or liability for loss or damage occasioned to any person or property through using the material, instructions, methods or ideas contained herein, or acting or refraining from acting as a result of such use. The authors and Publisher expressly disclaim all implied warranties, including merchantability of fitness for any particular purpose. There will be no duty on the authors of Publisher to correct any errors or defects in the software.

Designations used by companies to distinguish their products are often claimed as trademarks. In all instances where John Wiley & Sons is aware of a claim, the product names appear in initial capital or capital letters. Readers, however, should contact the appropriate companies for more complete information regarding trademarks and registration.

Other Wiley Editorial Offices

John Wiley & Sons, Inc., 605 Third Avenue,
New York, NY 10158-0012, USA

Wiley-VCH Verlag GmbH
Pappelallee 3, D-69469 Weinheim, Germany

Jacaranda Wiley Ltd, 33 Park Road, Milton,
Queensland 4064, Australia

John Wiley & Sons (Canada) Ltd, 22 Worcester Road
Rexdale, Ontario, M9W 1L1, Canada

John Wiley & Sons (Asia) Pte Ltd, 2 Clementi Loop #02-01,
Jin Xing Distripark, Singapore 129809

Library of Congress Cataloguing in Publication Data

Automation, control and complexity: an integrated approach/edited by Tariq Samad and John Weyrauch.
 p. cm
 includes bibliographical references and index.
 ISBN 0 471 81654 X
 1. Automatic control. 2. Technological complexiity. I. Samad, Tariq. II. Weyrauch, John.
 TJ213. A834 2000-04-10
 629.8—dc21

British Library Cataloguing in Publication Data

A catalogue record for this book is available from the British Library

ISBN 0 471 81654 X

Produced from PostScript files supplied by the authors
Printed and bound in Great Britain by Bookcraft (Bath) Ltd
This book is printed on acid-free paper responsibly manufactured from sustainable forestry, in which at least two trees are planted for each one used for paper production.

Contents

Preface	**vii**
1. Introduction: Complexity Management for Automation and Control	**1**
Tariq Samad	
1.1 Domain Knowledge and Solution Technologies	1
1.2 Complexity Management – Motivation	2
1.3 Objectives for Automation	4
1.4 Managing Complexity	8
1.5 Complexity and Automation – Technologies and Applications	13
References	15
PART 1 AUTOMATION AND PEOPLE	**17**
2. Advanced Technology in Complex Systems: Automation, People, Culture	**19**
Edward L. Cochran and Peter Bullemer	
2.1 Introduction	19
2.2 What Makes a System Complex	20
2.3 The Role of Individual Humans in Complex Systems	22
2.4 Operations Teams and Complex Systems	25
2.5 Lessons for Managing Automation in Complex Process	32
2.6 Conclusion	33
Acknowledgments	33
3. The Human Factor in Complexity	**35**
Chris Miller	
3.1 Introduction	35
3.2 Definitional Issues	36
3.3 Human Impact of System Complexity	39
3.4 Strategies for Mitigating the Human Impacts of Complexity	41
3.5 New Approaches to Managing Perceived Complexity	53
3.6 Conclusion	55
References	57
4. Perceived Complexity and Mental Models in Human-Computer Interaction	**59**
Victor Riley	
4.1 Introduction	59
4.2 Mental Models and the Basis of Consistency	59
4.3 Metaphors as a Means of Reducing Perceived Complexity	62
4.4 Airplanes à la Mode	64

	4.5 Human-Factored Systems vs. Human-Centered Systems	70
	4.6 Conclusions	72
	References	73

PART 2 SENSING AND CONTROL   75

5. Active Multimodeling for Autonomous Systems   77
Tariq Samad

	5.1 Introduction	77
	5.2 Active Multimodels	78
	5.3 Uses of Models	79
	5.4 Types of Models	81
	5.5 Multimodels	85
	5.6 An Example: Active Multimodel Control of UAVs	89
	5.7 Conclusions	92
	5.8 Acknowledgments	94
	References	95

6. Randomized Algorithms for Control and Optimization   97
Rudolf Kulhavý

	6.1 Introduction	97
	6.2 Statistical Simulation	98
	6.3 Managing Complex Models	99
	6.4 Managing Complex Designs	105
	6.5 More Applications	109
	6.6 Practical Aspects	110
	6.7 Conclusion	113
	References	113

7. Complexity Management via Biology   115
Blaise Morton and Tariq Samad

	7.1 Introduction	115
	7.2 The Central Nervous System	116
	7.3 The CNS as an Intelligent Control System	119
	7.4 Lessons Learned from Biological Brains	122
	7.5 Architectural Outlines	124
	7.6 A Philosophical Controversy	127
	7.7 Conclusion	128
	References	129

8. Sensors in Control Systems   131
J. David Zook, Ulrich Bonne, and Tariq Samad

	8.1 Introduction	131
	8.2 Sensor Fundamentals and Classifications	132
	8.3 Sensors in Control Systems	134
	8.4 Sensor Technology Developments	138
	8.5 Biological Systems, Chemical Sensors, and Biosensors	145
	8.6 Sensor-Enabled Visions for the Future	148
	References	149

PART 3   SOFTWARE AND COMPLEX SYSTEMS — 151

9. Managing the Complexity of Software — 153
Jonathan W. Krueger

9.1   Introduction — 153
9.2   Structuring Decisions — 155
9.3   Automating Dependency Management — 161
9.4   Software Process — 165
9.5   Conclusion — 168
       References — 168

10. Agents for Complex Control Systems — 171
Ricardo Sanz

10.1   Introduction — 171
10.2   Examples of Complex Control Systems — 172
10.3   Control System Complexity — 175
10.4   Heterogeneity and Integration — 177
10.5   Objects, Components, and Agents — 180
10.6   Agents for Complex Control — 183
10.7   Some Examples of Agents in Control Systems — 186
10.8   Conclusions — 188
        Acknowledgments — 189
        References — 189

11. System Health Management for Complex Systems — 191
George Hadden, Peter Bergstrom, Tariq Samad, Bonnie Holte Bennett, George J. Vachtsevanos, and Joe Van Dyke

11.1   Introduction — 191
11.2   Condition-Based Maintenance for Naval Ships — 194
11.3   Details of the MPROS Software — 197
11.4   Prognostic/Diagnostic Algorithms — 204
11.5   Installation and Validation — 211
11.6   Conclusions — 212
        Acknowledgment — 213
        References — 213

PART 4   COMPLEXITY MANAGEMENT AND NETWORKS — 215

12. Current and Future Developments in Air Traffic Control — 217
Steven M. Green and Joseph Jackson

12.1   Introduction — 217
12.2   National Airspace System — 218
12.3   'Free Flight', the Future of ATM — 231
12.4   Long-Term Challenges in ATM — 234
12.5   Conclusions — 238
        References — 239

13. Complex Adaptive Systems: Concepts and Power Industry Applications   241
A. Martin Wildberger

    13.1 Introduction   241
    13.2 The Electric Enterprise: Today and Tomorrow   242
    13.3 Complex Adaptive Systems in General   249
    13.4 Simulator for Electrical Power Industry Agents   256
    13.5 Conclusion   259
    References   260

14. National Infrastructure as Complex Interactive Networks   263
Massoud Amin

    14.1 Introduction: Complex Interactive Networks   263
    14.2 Complex Interactive Networks/System Initiatives\   265
    14.3 Societal Context: Infrastructures and Population Pressure – a 'Trilemma' of Sustainability   266
    14.4 Genesis: President's Critical Infrastructure Protection Report   266
    14.5 Examples of Complex Interactive Networks   267
    14.6 Vulnerabilities   274
    14.7 R&D Objectives of CIN/SI   276
    14.8 Distributed Multilevel Control   279
    14.9 CIN/SI Program Content   280
    14.10 Conclusion   284
    Acknowledgments   285
    References   285

15. Multiscale Networking, Robustness, and Rigor   287
John Doyle

    15.1 Introduction   287
    15.2 Themes for a New Science   288
    15.3 Illustrative Examples   293
    15.4 Other Examples of Virtual Design and Uncertainty Management   299
    15.5 Conclusion   301
    Further Reading   301

16. Conclusions: Automation, Control, and Complexity   303
John Weyrauch

    16.1 Automation and People   303
    16.2 Sensing and Control   304
    16.3 Software and Complex Systems   305
    16.4 Complexity Management and Networks   305
    16.5 Future Challenges   306

Current Affiliations and Addresses of Contributors   309

Index of Names   313

Subject Index   319

Preface

Complexity has always been noted as a characteristic of our world, but today it is a defining feature. The label of 'complex' is now applied, perhaps too cavalierly, to systems and phenomena – both natural and artificial – that seem to have little in common. Yet, if the rhetoric exaggerates the reality, it does not wholly misrepresent it – and certainly not in the realm of technology.

Performance, cost, environmental impact, safety – in virtually all our technological undertakings we are attempting to optimize several of these conflicting criteria. Complexity arises in part from the challenges posed by these desires. The ongoing revolution in information technology is another contributing factor, in that it has rendered meeting these challenges technically feasible!

Although talk of complexity is rife, it is in the area of control and automation that a practice of complexity management is urgently needed. It is one matter to analyze the increasing complexity of systems or phenomena of interest; it is another matter entirely to 'close the loop' on such systems. Indeed, the complexity of control solutions can increase disproportionately to the complexity of the target system.

However, complexity is not just a problem; it is an opportunity as well. Organizations (and people) that can harness the increasingly sophisticated technologies now at our disposal and leverage them into a new generation of automation and control solutions can expect to realize substantial practical (and intellectual) benefits.

There is no shortage of comprehensive texts today that provide authoritative coverage of any specific technology. However, increasingly, developing products and solutions for automation and control systems requires the integration of expertise across several disciplines: software engineering, human factors, sensors, signal processing, control science, and many others.

In conceiving of and preparing this volume, we have been motivated by the belief that there is a need for accessible treatments of the states of the art and key trends spanning multiple disciplines. If our interactions with other technologists – designers, operators, researchers, managers, and executives concerned about complexity and automation – are any indication, this need is broadly shared.

We are hopeful that this book is a useful contribution in this context. We freely admit, however, that no ultimate grand unifying syntheses are revealed herein. Any aspirations to this end we may have harbored at the beginning of this project disappeared soon thereafter. We now stand convinced that managing the complexity of automation and control systems is a journey, not a goal. The final destination will always lie in the darkness beyond the stretch of road that books such as this one can hope to illuminate. We can but hope to light the way.

This edited volume is an outcome of a program conducted over the last three years at Honeywell Technology Center. Due to its genesis, most of the contributions are from Honeywell and have been supplemented with submissions we have solicited from selected external collaborators. We are grateful to all the authors for taking time out of their busy schedules to accommodate our deadlines.

We expect few would argue about the value – even in economic terms – of promoting interdisciplinary cooperation and seeking cross-disciplinary insight. Yet there are few organizations where such a multidisciplinary initiative could have been undertaken. We would like to acknowledge the support of Dr. Kris Burhardt, Vice President of Technology for Honeywell, and Dr. Ron Peterson, former Vice President of Technology for Honeywell Inc.

We would also like to thank Barbara Field for her extraordinary effort in helping with the editing of this volume. Finally, it has been a pleasure working with Peter Mitchell of John Wiley & Sons, Ltd. in the planning and production of this book.

TARIQ SAMAD
JOHN WEYRAUCH
Minneapolis, U.S.A.
January 1, 2000

1

Introduction: Complexity Management for Automation and Control

Tariq Samad
Honeywell Technology Center

1.1 Domain Knowledge and Solution Technologies

Whether related to passenger aircraft, oil refineries, or commercial buildings, a primary motivation for progress in science and technology has always been controlling and automating our engineered systems. Over the last few decades, we have seen substantial progress in control systems technology for these and many other applications. Improved performance, reduced design time and cost, more efficient use of human operators, ease of maintenance, reduced environmental impact, and greater human safety are some of the attendant benefits that society and industry have reaped.

For devising effective control and automation solutions, two types of expertise are required. Domain knowledge is the first of these – we can control systems only if we understand them. This knowledge can exist in various forms, explicit and implicit: domain models for conceptual design, dynamic models for control algorithms, mental models of human operators, and many others. The quality and quantity of domain knowledge that is brought to bear determines the degree and effectiveness of automation achievable. Thus, controlling a distillation column requires that we capture an understanding of its construction and the chemistry of its operation; modern flight control would be impossible without knowledge of aircraft aerodynamics and jet propulsion; environmental control for buildings is contingent on knowing building layouts and pollutant diffusion mechanisms, among other factors.

Domain expertise is not the only requirement. The availability of relevant 'solution technologies' is equally critical. The automation and control of complex engineering systems is a multidisciplinary undertaking. Sensing and actuation technologies are needed for measuring and moderating the physical world; control algorithms furnish the 'intelligence' for mapping measurements and commands to actions; effective interaction between the automation system and its human users requires knowledge of human factors and cognitive science; software engineering is essentially the new manufacturing discipline for knowledge- and information-intensive systems; system health management technology is needed to help minimize and manage down-time; and so on.

Despite the enormous impact of automation, neither refineries, buildings, nor aircraft operate autonomously today, and this is not likely to change in the foreseeable future. Many activities continue to be performed manually with inadequate assistance by control systems. Even the subsystems that are effectively automated (e.g., regulatory control, data recording, alarm annunciation) operate in isolation, and these islands of automation in many cases confound system-wide management and integration initiatives. This is true not just for the operational aspects of a control system, but for its entire life cycle – conceptual and detailed design, implementation, verification and validation, commissioning and configuration, maintenance and support, decommissioning and replacement.

There is no gainsaying the critical importance of control systems in our technological society today. Yet, compared to the full range of problems for which automation is seen as a potential solution, today's systems are quite limited. This situation is schematically illustrated in Figure 1.1.

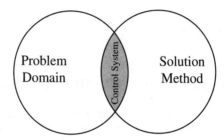

Figure 1.1 Current control systems represent a limited intersection of solution technologies and domain expertise

1.2 Complexity Management – Motivation

Much of our existing control systems technology predates a profound, ongoing revolution in computational infrastructure. Hardware, software, displays, memory, and communications are all orders of magnitude more powerful and/or capable than they were even a decade ago. One implication of these developments is that it is now technically feasible, in principle, to effect a much larger overlap than depicted in Figure 1.1. The foundation has been laid for a potentially revolutionary increase in the scale and impact of automation.

Yet the challenge is unprecedented. Realizing the possibilities for dramatic advances in control systems requires equally dramatic changes to existing approaches and methodologies for their design and operation. If there is one word that captures the multifarious, interconnected opportunities and challenges for automation systems of tomorrow, it is *complexity* – and it is the management of this complexity that is the new research need (Figure 1.2).

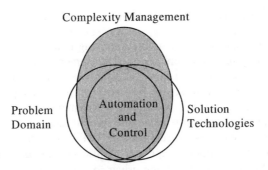

Figure 1.2 Control and automation can now play much larger roles in engineering systems. A science (and art) of complexity management is needed to exploit the opportunities and to avoid the pitfalls.

Complexity management is not so much a new discipline as a metadisciplinary perspective. The problems we need to address with automation are increasingly cross-disciplinary. For many challenging applications, we already know how to design good control algorithms, customize existing sensors, and complement automation with human operators. However, conventionally, these and other aspects of control systems have been treated in isolation, and such decoupling is no longer tenable. This assertion is not particularly novel today. Yet, although many research efforts are under way that are ostensibly tackling the management of complex systems, few (if any) are making any broad-based attempt to bridge communal and/or disciplinary boundaries.

If cross-disciplinary emphasis is one defining characteristic of complexity management, large-scale applications is another. In the process industries, we have seen the progression in control technology from regulatory to supervisory control. Plantwide optimization is the new watchword. The scale of such an undertaking is daunting: a model of a refinery might comprise more than a million nonlinear equations. Only a few years ago, research in refinery-wide optimization would have been dismissed as quixotic. It is now a leading research topic in process automation. Similarly, 'Free Flight' air traffic management, a subject at the forefront of research in aviation control, represents a substantial increase in scale and complexity from existing air traffic control techniques.

As implied by Figure 1.2, we are interested in managing the increasing complexities of *both* the problem domain for which control and automation systems are being developed and the integration of solution technologies that constitute these systems. Most of the chapters in this book address specific technologies and application domains that populate the intersectional area depicted in Figure 1.2. In some cases, the emphasis is on the problem, in others, on the solution. Ultimately, effective solutions for practical problems will require advances in our understanding on both fronts.

Before previewing the subject topics of the individual chapters, we discuss two other matters that bear on automation and complexity. First, some of the rich variety of motivations that are spurring increasing automation – automation that both causes complexity and attempts to manage it – are reviewed. Subsequently, some approaches that have been suggested in scientific and technical communities for managing complexity in engineering systems are discussed.

1.3   Objectives for Automation

The call for new, more capable automation is not being issued under a 'performance at all costs' banner. Control system suppliers and their customers are operating in multifaceted, multicriteria business and technical environments. In many domains, investment in improved control cannot be justified on the basis of improved performance (speed or accuracy, for example). Economic return on investment, as well as environmental considerations, human safety, and staff reductions, are all increasingly important drivers for control and automation. We discuss these and others in this section.

Since commercial organizations comprise a significant proportion of both the suppliers and users of new technology, an underlying driver for automation is frequently economics. However, it is useful to examine the economic motivation in some detail, decomposing it into different aspects that, to a first approximation, can be considered separable. Thus, although we do not explicitly break out profits and revenues as objectives for automation, these are often the ends served by the means discussed below.

1.3.1   Human and environmental safety

Safety is seldom the primary motivation driving the development of an engineering system, but it is a near universal concern that itself is typically addressed through technology and automation. For mature systems, safety issues can spur a large proportion of the innovation and research targeted toward the systems. We build cars primarily to transport people and their possessions, but a variety of devices are employed to ensure that this transportation does not endanger lives or the environment. Many of the recent technological developments in automobiles have been driven by the need for improving the safety of drivers and passengers – witness airbags, antilock brakes, and crumple zones. In contrast, on some performance measures – cruising speeds, cargo capacities – no improvements have been seen for decades.

Whether safety is built in (as it ought to be) or an afterthought addition (as it often is), it is typically not the reason an engineering system is developed, but an essential consideration in its development and use. But there are also systems in which safety is the primary objective and that seek to protect us and our environments against natural forces or other engineering systems. Civil structures such as dikes and bomb shelters are obvious examples. This primary versus secondary distinction may be debated, and in any case depends on what we isolate as a system. Safety is certainly the primary consideration in a bicycle helmet, but it seems natural to consider seatbelts as a safety feature within the automobile system.

1.3.2   Regulatory compliance

In the eyes of the general public, human and environmental safety is an overriding concern with new technological developments. This concern directly influences the design, development, and operation of engineered systems, as alluded to above, but it also impacts complexity in an indirect way. At all political and governmental levels – from city councils to international organizations – there is increasing oversight on technology and its products. Regulations on environmental impact and human safety have resulted in new industries being established and have affected existing ones in profound ways.

In the automotive industry, regulatory compliance has been (arguably) the single most significant factor in increasing the complexity of engine control systems. Commercial automobiles may not drive much faster today than they did three decades ago, but their tailpipe emissions of nitrous oxides and hydrocarbons are orders of magnitude lower. New sensors, electronic fuel injection systems, catalytic converters, on-board diagnostic modules, advanced algorithms that regulate fuel flow, engine speed, and ignition timing in response to instantaneous conditions – these are some of the major enhancements in engines today that have, as their primary purpose, ensuring that concentrations of regulated exhaust chemicals do not exceed mandated limits. Regulations on human safety in automobiles have also spurred technological innovation. Such laws hastened the national deployment of seatbelts decades ago, and more recently they are responsible for dual airbags in all new cars (1998 and later models) in the United States.

Although these two safety devices perform similar functions, a comparison of the two helps illustrate our theme of increasing complexity. Traditional seatbelts are purely mechanical devices. They have only a few components, none of which require high-precision manufacturing. Yet, despite their simplicity, seatbelts are enormously effective. The incremental effectiveness of airbags is relatively small and is gained at considerable additional complexity. The sensing and actuation required for effective airbag deployment have only recently become technically feasible for widespread insertion in automobiles. See Buede (1998) for a discussion of how the complexity of automobile airbags as supplemental restraints has resulted in what by some measures is a poorly designed system.

The industry/government relationship *vis-à-vis* regulation is often adversarial. Companies may believe they can ensure safety just as well with practices that depart from legislated mandates. Regulation may on occasion thus be seen as a bureaucratic nuisance, and technology directed to regulatory compliance may be considered only politically necessary.

At the same time, legislation can be a driver for new technological developments. New product and service businesses have been formed and thrive by facilitating regulatory compliance and reporting by various industries. Emissions monitoring software for power plants and other process industries is one example, and many others exist in the biomedicine, pharmaceuticals, consumer products, food, and transportation industries.

1.3.3 Time and cost to market

The design, development, and manufacturing of new technological products are complex processes, and doing them well requires time and effort. As the products themselves become more complex, these processes do so as well, and in many respects disproportionately. Doubling the number of parts in a device quadruples the number of potential (binary) interactions, and validation and testing become that much more problematic.

Yet marketplace realities are demanding that companies develop new products, of increasing complexity, in less time. Product turnovers and innovation indices are widely seen as indicators of quality in technology sectors. Good technical ideas are one ingredient, but their rapid commercialization is just as important (after all, the technical idea can be much more readily acquired than a complete product).

Fast time to market demands are not wishful thinking on the part of executive management; numerous instances can be cited where recent product development times have been considerably shortened in comparison to previous developments. Computing technologies have been central to these achievements. For example, the Boeing 777 aircraft, an exemplary complex system, was designed completely digitally, using several computer-aided design and analysis software packages. Despite being 50% larger than Boeing's previous aircraft, the 777 took no more design time or effort.

1.3.4 Increased autonomy

In many businesses, personnel costs constitute a large fraction of total expenses. Where human involvement is not considered critical, its replacement by suitably sophisticated automated substitutes can result in substantial financial savings.

The benefits of automation are of course not limited to cost reduction. Autonomous systems are being sought for higher productivity, for operation in hostile or otherwise inhospitable environments, for miniaturization reasons, and others.

Complete automation of any significant system is not feasible now and will not be feasible in the foreseeable future. Increased autonomy thus implies essentially that the role of the human is shifted from lower to higher level tasks. What used to be accomplished with a large team may now be done with a smaller team, or with one person. With improvements in navigation systems and avionics, for example, the aviation industry has been able to reduce its cockpit flight crew to two, losing the radio operator, navigator, and flight engineer. In some process industry sectors, automation solutions are being sought for some of the functions that plant operators perform. Preventive maintenance, prognostics, and diagnostics – aspects of system health management – are a particular focus of activity in this context.

Increased autonomy, and the technological infrastructure that has enabled it, also implies that larger-scale systems are now falling under the purview of automation. Sensors, actuators, processors, and displays for entire facilities can now be integrated through one distributed computing system. The control room in an oil refinery can provide access to and/or manipulate 20,000 or more 'points' or variables. In possibly the largest integrated control system implemented to date, the Munich II international airport building management system can be used to control everything from heating and cooling to baggage transport. The system controls more than 100,000 points and integrates 13 major subsystems from nine different vendors, all spread out over a site that includes more than 120 buildings (Ancevic, 1997).

1.3.5 System 'performance'

Technological systems are developed for some primary purposes: producing gasoline in the case of oil refineries, transporting people and goods in the case of automobiles and airplanes, information processing in the case of computers, organ stimulation in the case of implanted biomedical devices, and so on. The drivers discussed above have been used to illustrate how these primary purposes are not the only ones that matter, and how aspects of complex technological systems that might originally have been considered secondary from the point of view of system functionality can take on substantial importance.

Improvements in how well the system performs its primary functions are also a continuing need – if not the only one. The specific nature of these performance improvements are system and function dependent, but it is useful to consider some examples:

- *Manufacturing yield*: an obvious and ubiquitous performance metric for a manufacturing process is the quantity of within-specification product it is able to produce per unit time. Improvements in yield relate directly to increased revenues (demand permitting) and so are always driving the development of automation and control systems. In many industries, plants are distinguished firstly on the basis of their production: megawatt output in power generation units, barrels per year for oil refineries, annual units for many discrete manufacturing lines.

- *Vehicular transportation capacity*: with vehicles, measures of speed, payload (number of passengers or cargo capacity), and distance are primary. In the case of automobiles and commercial aircraft, we have seen little or no improvement on the first two counts over at least two decades. (However, the system capacity of air transportation in terms of passenger and cargo miles has increased tremendously.)

- *Information processing power*: processing speeds and memory capacity define performance in computers, and satisfying the continuing demand for growth on these counts has required increased complexity – in device physics, design methods, and manufacturing processes. At the same time, these improvements have themselves fueled the development of complex systems.

- *Energy efficiency*: this applies to all technological products, although its importance varies widely. Fuel efficiency in automobiles is of relatively less importance compared to fuel efficiency of aircraft. Even for a given system, energy efficiency can vary over time or location. Thus automobile fuel economy is much more important in most parts of the world than in the United States with its relatively low gasoline prices.

- *Miniaturization*: the utility of some functions is dependent on their encapsulation in small packages. Heart pacemakers, for example, have been around for more than 40 years, but initially they were bulky, used external electrodes, and operated on line voltage. Compact, implantable pacemakers have, obviously, been a revolutionary improvement; miniaturized sensors and actuators (electrodes), in particular, have facilitated implantation. Miniaturization is also central to the development of effective substitutes for organs such as the heart (Maslen *et al.*, 1998). For some applications, the miniaturization necessary can be dramatically greater than for these biomedical examples. Some speculative applications of 'nanotechnology' are sketched in Crandall (1996).

1.3.6 Other objectives

There are innumerable other reasons why we employ technology and develop engineering systems, and many of these can be surprising and can resist easy classification. To cite one final example, providing entertainment to passengers was an important objective in the design of the Boeing 777 – to the extent that the nominal design weight of the aircraft was deliberately exceeded by more than 5000 pounds to accommodate the in-flight entertainment system (Norris, 1993).

1.4 Managing Complexity

The call for complexity is thus being heard from all directions. However, engineering sufficiently sophisticated technological systems is no easy matter. As might be expected, there is no consensus view on how the increasing complexity of the automation solutions we need to develop, and the increasing complexity of the target physical systems for these solutions, should be managed. But while there is no consensus, there is also no shortage of ideas and theses. Some of these are outlined below.

1.4.1 Forfeiture

In an influential book, Perrow (1984) suggested that for many complex systems catastrophic accidents can never be made literally impossible, regardless of the diligence with which we design or operate the systems. Perrow considers three cases. For systems such as chemical plants and aircraft, the benefits are substantial (no feasible alternatives are known) and the freak worst-case accident will not be unthinkably catastrophic. With modest efforts we can further improve the safety of such systems, and live with the remaining risk. His second category includes large marine transports and genetically engineered systems. Here benefits are again substantial, and risks, although higher, are outweighed. His recommendation is that we invest the substantial resources necessary for minimizing the likelihood of accidents such as the Estonia ferry. The final category is reserved for systems that we must abandon, because the scale of catastrophe that can result from their failures is absolutely intolerable and, further, because less risky alternatives to these systems exist. The primary instance of this class is nuclear power.

Perrow's carefully reasoned, and selected, proscriptions can be contrasted with the more wholesale arguments for forfeiture put forth by various neo-Luddite factions (e.g., Mills, 1997). As technologists, we may dismiss these as generally uninformed by engineering or technical judgment, but society at large can sometimes give even extreme technophobic representations a sympathetic hearing.

1.4.2 Risk assessment

Perrow's approach is not a quantitative one. Indeed, in some respects, it is a reaction to a mathematically sophisticated approach to complexity management known as risk assessment. Given any system, risk assessment develops models that describe various ways that accidents can result, and it attempts to calculate the probability of each accident by considering the individual abnormal events that must happen for the accident to occur. Probabilities for these individual events are needed in this calculation. Accidents are also often quantified in terms of the monetary losses or loss of lives effected by them (and sometimes the two are explicitly related to each other). The end result is a number, often a monetary amount, which can be compared to the expense necessary in engineering a solution that will not allow this sequence of events to occur, or that will sufficiently reduce probabilities of individual events in the sequence. In this vein, Lewis (1990) considers a variety of risks to human safety and attendant statistics.

Risk assessment remains popular in some circles, but the thought of associating hard numbers with human lives or limbs, or of identifying precise probabilities for events for which statistically meaningful data is nonexistent, is inconceivable to many. Consequently, risk assessment is often a ready target, both for critics who demand the absolute operational safety of systems that, in principle, can cause catastrophic devastation (regardless of how low the probability of a catastrophic accident), and of reports decrying the quality of statistical analyses in scientific and technical studies (see Fennick, 1997).

1.4.3 Systems engineering

Whereas risk assessment and the 'forfeiture' tack focus on the issue of safety, systems engineering (e.g., Sage, 1992) also addresses other aspects of complex technological systems, such as how to design systems cost-effectively so that their reliability is maximized. Depending on the needs of specific projects, systems engineering can include systems analysis, requirements specification, system architecting, and other topics. Systems engineering (also sometimes referred to as systems management) often focuses on structured, formal process models for the design and development of complex technological systems (Figure 1.3). These models, at the highest level, specify the steps and tasks involved in the development of engineering systems, from the conceptual to the product stage, or even models for the entire life-cycle of products or technologies. Depending on the characteristics (such as complexity) of the technologies and products, different models may be appropriate. When sequential development is appropriate, 'waterfall' models can be recommended; with complex systems, however, couplings and interdependencies may be better suited for 'spiral' models (Rechtin and Maier, 1997).

1.4.4 Virtual engineering

The enormous progress over the last few decades in hardware, software, and communication technologies has resulted in the development of increasingly sophisticated computational tools for dealing with various aspects of complex engineering systems. These tools allow complex systems to be engineered 'virtually', greatly reducing the need for manufacturing physical prototypes (Figure 1.4).

The paradigmatic example is integrated circuit design – a technology that is fueling complexification requires complexity management itself, as might be expected. Putting millions of transistors on a few square centimeters is not a task that is humanly possible, and the electronics industry has been able to accomplish it only by relying extensively on libraries of more or less automated tools for placement, layout, verification, simulation, synthesis, and other tasks associated with the design and analysis of VLSI devices.

The use of conceptually similar tools is spreading to all technology-driven industries. The recent development of the Boeing 777 is a well publicized example. At one time, 238 teams were working in concert to design the 777 using 'design/build' computational tools. Full-scale physical mockups were not needed at every stage, correct part fits were ensured before manufacturing through digital preassembly, and teamwork was well coordinated across organizational and national boundaries. The design/build virtual engineering teams were also considered essential for delivering 'service ready' aircraft to airlines.

Although design is a primary application of virtual engineering, there are many others, Simulation-based training, as another example, is commonplace in many industries. Aircraft pilots, air traffic controllers, power system dispatchers, and oil refinery operators routinely hone their skills on simulators, which allow appropriate human responses to abnormal situations to be learned by trial and error, but without the potentially catastrophic consequences of trials on the real system.

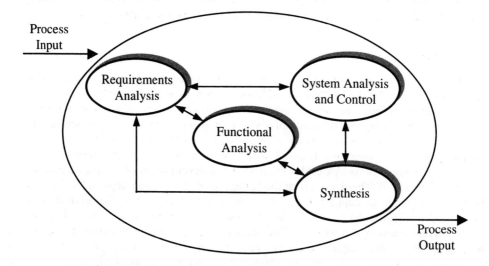

Figure 1.3 Systems engineering as a process (from Parth, 1998)

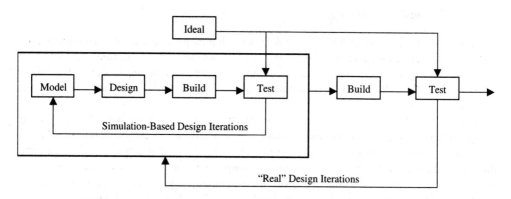

Figure 1.4 Design by virtual engineering (adapted from Doyle, 1997)

1.4.5 Biology

Every living system represents a successful solution to a complex problem – posed by the environment within which the organism must survive. The biological world thus provides innumerable examples of complexity management. These successes have been achieved through an approach unlike any of the ones noted above. No formal, structured development process was followed, no specifications written in advance, and no computer-aided design or analysis tools used!

This realization has led many researchers to propose that complex engineering systems can be developed and operated using biologically inspired methods. Computational implementations, not necessarily especially faithful, of biological evolution can 'learn' the right answers to difficult problems by a process that simulates natural selection, but since this process is simulation and not reality, without the inefficiencies of the original. The algorithms that have been developed for this field, generally referred to as 'evolutionary computing' (Fogel, 1995), are all highly flexible and customizable. They make few assumptions about characteristics of the problem compared to more traditional optimization algorithms and are thus applicable to problem formulations for which the latter cannot be used. Thus the design space need not be differentiable, convex, or even continuous; discrete and continuous variables can be simultaneously accommodated; arbitrary inequality and equality constraints can be included; and so on. Genetic algorithms are the best known instance of an evolutionary computing algorithm.

The evolutionary perspective is not limited to the development of computational analogs for natural selection. The field of sociobiology, or evolutionary psychology, views human and (other) animal behavior as largely determined by our evolutionary heritage (Wilson and Landry, 1980). Societal structures, emotional attachments, religious and political beliefs have been 'explicated' as products of nature, not nurture (and of this one natural phenomenon). This is not a universal opinion among biologists by any means (see Gould (1998) for a strong dissension), but proposals have been proffered suggesting that evolutionary thinking can form the basis for a new, fundamental integration of all scientific understanding, with all its implications for technology and society (Wilson, 1998).[1]

1.4.6 Chaos, complexity, and power laws

In most of the discussion above, complexity can be seen, at least in part, as a consequence of scale. Systems become more complex as their components become more numerous, as more couplings arise, and/or as more behaviors are encapsulated. There is, however, another sense in which complexity is used within the technical community. In many small-scale systems, certain parametric regimes can cause a transition from well behaved, 'simple' behavior to unpredictable 'chaotic' dynamics. The output of such a system may appear to be random, but the system is entirely deterministic and the apparent randomness is due to (often mild) nonlinearities.[2]

The key attribute of chaos is a sensitive dependence on initial conditions, as memorialized in the Chinese butterfly apothegm: lepidopteran flittings in Beijing can produce showers in Minneapolis (or blizzards, depending on the time of year). A variety of mathematical tools are now available to detect chaos in systems by analyzing their outputs, and chaos has been uncovered in a wide range of systems, including manufacturing

---

[1] E.O. Wilson's 'consilience' theory puts biology in the role of the central scientific discipline. Similar claims have been put forth on behalf of other disciplines too, notably physics (e.g., Weinberg, 1992) and chemistry (Silver, 1997).

[2] Perhaps the simplest example of chaos arises for the logistic difference equation, $x[t+1] = ax[t](1-x[t])$. For $a = 2.5$, the system will eventually lead to a constant value for x, given any starting value $x[0]$ between 0 and 1. For $a = 3.25$, x will oscillate forever between two values. For $a = 3.5$, a cycle of period 4 results. For $a = 4$, chaos is produced.

operations, road transportation networks, biological organisms, contagious disease epidemics, and financial markets. Systems of all sorts – small-scale and large-scale, abstract and real – can thus be analyzed through the language of nonlinear dynamical systems, the 'science of complexity' (Nicolis and Prigogine, 1989). Chaotic attractors, bifurcations, Lyapunov exponents, and other characteristics can provide insight into the complexity of a system and suggest approaches for managing the complexity.

The leading exponent of work in this area is the Santa Fe Institute (SFI), a private, nonprofit, multidisciplinary research and education center founded in 1984. SFI is described by one of its founders as 'one of very few research centers in the world devoted exclusively to the study of simplicity and complexity across a wide variety of fields. . . ' . (Gell-Mann, 1994; p. xiv). Chaos theory and nonlinear dynamical systems are among the main themes pursued, but in an interdisciplinary environment that also includes international experts in biology, quantum physics, information theory, and economic theory. Theoretical research is coupled with application explorations in financial markets, manufacturing operations, cosmology, and social sciences, among others.

Alternative and intriguing explanations have recently been proposed for some putatively chaotic phenomena – and many others (Bak, 1996). It turns out that a vast array of systems exhibit power law spectra: a quantity of interest (for example, the frequency of occurrence of an event) can be expressed as some power of another quantity (such as the magnitude or severity of the event). Thus a log-log plot of earthquake magnitude versus the number of earthquakes of at least that magnitude in some geographical region is a straight line (of negative slope).

The pervasiveness of power laws in human systems, including distributions of cities on the basis of their populations and the distribution of English words as a function of their usage rank, has been known since Zipf (1949). A wealth of new data from natural systems and observations in controlled experiments have further validated the power law model, and we now have the beginnings of a theory that may ultimately lead to a science of 'self-organized criticality', the label coined for the new field. Unlike chaos theory, which shows how low-dimensional deterministic systems can exhibit seemingly random behavior, self-organized criticality is concerned with large-scale systems. Power law dynamics arise from the statistics of the interactions between system components.

1.4.7   'Technoscience' and society

Although we have referred above to market considerations and the economics of technology development and commercialization, the focus of the discussion has been on technological issues and approaches. Another mark of increasing complexity, however, is that technology spills over into societal arenas. Managing complexity in such an environment requires an awareness of the interconnectivity between technology and society. As engineers and scientists, we are used to thinking that the primary influences are unidirectional. Technological and scientific achievements lead to societal change. Telephones, television, electric power, automobiles, aviation, synthetic fibers, gasoline, computers, etc. – our ways of life have changed in ways that were once inconceivable.

However, technology development itself does not happen in a vacuum. Government funding of scientific research is ultimately under political and societal control; grass roots movements can derail major industries; 'slick' marketing campaigns can sometimes overcome technical shortcomings; small, and not necessarily objective, advantages can snowball into industry domination. Technology certainly affects society, but it is 'How society shapes technology' (Pool, 1997) that must also be understood by technologists in today's complex world. Pool's emphasis is on nuclear power – a paradigmatic example, at least in the United States, of how society has influenced technology – but numerous other case studies are also discussed.

In a similar vein, Latour (1987) coins the word 'technoscience' to 'describe all the elements tied to the scientific contents no matter how dirty, unexpected, or foreign they seem'. Science and technology themselves are just elements of this broader, socially influenced enterprise, in which laboratory researchers, product engineers, and technology managers are parts of a network that includes consumers, heads of funding agencies, military strategists, legal professionals, and even metrologists.

1.4.8 *A pluralist perspective*

The 'attacks' on complexity outlined above are often seen as competing approaches. The philosophy behind each is distinctive, their emphases are different, and the procedures they suggest are conflicting. Yet, they also all address more or less different questions and share some important characteristics, such as the reliance on computational tools. In this sense, they can equally well be seen as complementary, and it is this perspective that this volume attempts to promote (even as it includes contributions by experts in, and devotees of, specific technologies).

1.5 Complexity and Automation – Technologies and Applications

If the breadth of this conception of technology for complex systems appears bewildering, it is! Attempts to encompass all the issues and aspects of a hypothetical future generation of automation systems cannot hope to furnish grand unifying theories. In our opinion, it would be premature to even seek a cohesive framework at this early stage of exploration, conceptual discomfort notwithstanding. We do not understand the problems well enough to suggest one comprehensive solution.

We can, however, ask some questions:

- What are the scientific and engineering disciplines that are central to automation and complexity?
- How are these disciplines addressing the increasing complexity of the technological world, and how are they themselves being transformed by this complexity?
- What insights can these disciplines furnish that are relevant for future automation systems?
- What trends toward increasing complexity are being seen in different industries and businesses?
- What solution approaches are being attempted to manage this complexity?

- What key technical and technological obstacles stand in the way of realizing the control and automation systems that are being sought in different application areas?

These and other questions lie behind the project of which this book is the outcome. This book as a whole, as well as each of its constituent chapters, suggests some answers. The reader should not assume that the answers are comprehensive or even consistent, but they illustrate the issues that all those with a serious interest in advanced automation systems must contemplate, and we are hopeful readers will find them thought-provoking.

The contributions in this book are divided into four parts which reflect what we feel are central themes for complexity management, as follows (brief summaries are also included at the beginnings of the parts):

- *Part 1 – Automation and People:* the three chapters in the first part of the book examine the impact of people, especially operators and designers, on automation systems and vice versa. Chapter 2, by Edward Cochran and Peter Bullemer, focuses on interpersonal, team-based, and cultural issues related to the operation of complex engineering systems. Next, Chris Miller contrasts actual complexity with perceived complexity and discusses the trade-offs involved in managing human perceptions of complexity. Chapter 4, by Victor Riley, presents examples to illustrate the point that perceived complexity is a strong function of system and interface design.

- *Part 2 – Sensing and Control:* developments and trends in sensors and control methods are covered in four chapters. The first, by Tariq Samad, suggests that 'active multimodels', diverse computational models that are suitable for online processing, are an important technology for realizing autonomous operation. Next, in Chapter 6, Rudolf Kulhavý reviews developments in randomized algorithms and related areas – these developments suggest intriguing new ways to deal with problems that are ill suited for the conventional analytic machinery of control science. In Chapter 7, Blaise Morton and Tariq Samad discuss the biological solution to the problem of autonomous operation in a complex, uncertain environment. Last, David Zook, Ulrich Bonne, and Tariq Samad specifically focus on trends and needs in sensing technology.

- *Part 3 – Software and Complex Systems:* Jonathan Krueger opens the third part of this volume with a review and analysis of approaches for complex software development, with an emphasis on control system software. The other two contributions, by Ricardo Sanz and George Hadden *et al.*, present examples of complex software developments. Sanz discusses agent-based technology for industrial process control in Chapter 10; Hadden et al. note the importance of system health management for tomorrow's automation systems and describe a system recently developed for naval ships in Chapter 11.

- *Part 4 – Complexity Management and Networks:* the final part of the book focuses on complex systems of the form of large-scale networks. Chapter 12, by Steven Green and Joseph Jackson, examines the air traffic management domain, where the move toward 'Free Flight' represents a dramatic increase in complexity over today's 'highways in the sky' routings. Martin Wildberger discusses deregulation in the electric power industry and suggests that complex adaptive systems may help address the challenges this business restructuring is raising. Massoud Amin discusses challenges associated with several infrastructural networks and outlines a new research initiative that is attempting to meet these challenges. In Chapter 15, John Doyle emphasizes the need for a new scientific underpinning for the analysis and design of complex real-world networks, and outlines some themes for this new science.

The final chapter offers some concluding remarks, highlighting the eclectic nature of the contributions in this volume and summarizing the contents of each of its parts. John Weyrauch recalls the systems engineering complexity associated with the development of the Space Shuttle more than two decades ago. This complexity was dramatically greater than had been faced in previous aerospace system developments, and it was finally managed through compartmentalization and exhaustive analysis. This recourse is simply inconceivable today. Complexity management for current and future engineered systems requires a revolution in multidisciplinary research.

The editors of this volume have attempted to integrate its diverse contents. All chapters have been edited for general readability, we have screened the use of jargon, and a comprehensive book-wide index is included. A shared understanding among all involved communities of the multidisciplinary aspects of complexity management is a necessary condition for attaining the objectives that now drive the development of automation systems – be they related to performance, safety, reliability, cost-efficiency, design time, or, more likely, combinations thereof. It is our hope that this book will be a small first step toward developing this understanding.

References

Ancevic, M. (1997) Intelligent building system for airport. *ASHRAE Journal*, November, 31–35.
Bak, P. (1966) *How nature works*. Copernicus Books, New York.
Buede, D.M. (1998) The air bag system: What went wrong with the systems engineering? *Systems Engineering*, **1**, 90–94.
Crandall, B. (ed.) (1996) *Nanotechnology: molecular speculations on global abundance*. MIT Press, Cambridge, MA.
Doyle, J.C. (1997) Theoretical foundations of virtual engineering for complex systems. http://www.cds.caltech.edu/vecs/.
Fennick, J.H. (1997) *Studies show: a popular guide to understanding scientific studies*. Prometheus Books, Amherst, NY.
Fogel, D.B. (1995) *Evolutionary computation: toward a new philosophy of machine intelligence*. IEEE Press, New York.
Gell-Mann, M. (1994) *The quark and the jaguar: explorations in the simple and the complex*. W.H. Freeman, New York.
Gould, S.J. (1998) Darwinian fundamentalism. *New York Review of Books,* June 12.
Latour, B. (1987) *Science in action*. Harvard University Press, Cambridge, MA.
Lewis, H.W. (1990) *Technological risk*. Norton, New York.
Maslen, E.H. *et al.* (1998) Artificial hearts. *IEEE Control Systems Magazine,* **18**(6), 26–34.
Mills, S. (ed.) (1997) *Turning away from technology: a new vision for the 21st century*. Sierra Club Books, San Francisco.
Nicolis, G. and Prigogine, I. (1989) *Exploring complexity*. W.H. Freeman and Company, New York.
Norris, G. (1993) Tailor made twinjet. *Flight International,* **144**, 32–33.
Parth, F.R. (1998) Systems engineering drivers in defense and commercial practice. *Systems Engineering*, **1**, 82–89.

Perrow, C. (1984) *Normal accidents.* Basic Books, New York.
Pool, R. (1997) *Beyond engineering: how society shapes technology.* Oxford University Press, New York.
Rechtin, E. and Maier, M. (1997) *The art of system architecting.* CRC Press, Boca Raton, FL.
Sage, A.P. (1992) *Systems engineering.* Wiley-Interscience, New York.
Silver, B.L. (1997) *The ascent of science.* Oxford University Press, Oxford, U.K.
Weinberg, S. (1992) *Dreams of a final theory.* Pantheon Books, New York.
Wilson, E.O. (1998) *Consilience: the unity of knowledge.* Alfred E. Knopf, New York.
Wilson, E.O. and Landry, S. (1980) *Sociobiology.* Harvard University Press, Cambridge, MA.
Zipf, G.K. (1949) *Human behavior and the principle of least effort.* Addison-Wesley, Cambridge, MA.

Part 1

Automation and People

Increasing automation implies that fewer people, and fewer person-hours, will be needed to accomplish the same operational tasks. This trend may appear to foretell a de-emphasis of the role of the human. In fact, however, several factors belie the prediction. Increasing automation is resulting in system behaviors that are less predictable and correspondingly harder to grasp conceptually. Personnel reductions are forcing substantial losses of expertise in many industries, and the average experience levels in large-scale complex systems are steadily decreasing. System components and behaviors are becoming increasingly coupled, thus demanding consideration of multiperson and team issues. And the cross-cultural workforce has brought with it new challenges for the designers of automation solutions.

These and other factors contradict the notion that the role of the human demands less attention today than it did yesterday. On the contrary, it has become essential to adopt a broader view of the human element in automation and control systems. System design, interface design, organizational and process design – all need to be revisited with a fresh, complexity-oriented perspective.

Part 1 of this book comprises three chapters by research and management staff at Honeywell Technology Center. Each of these examines the role of the human in complex automation systems, from different but complementary perspectives.

Chapter 2, by Edward Cochran and Peter Bullemer, notes a number of tragic outcomes of complexity gone awry, including commercial airliners, naval ships, and process plants. These failures cannot be attributed to individuals involved in the operation of these failed systems; they are 'systems failures'. The authors criticize two assumptions in traditional models of human-system interaction: that the operator has an accurate understanding of the system and that he or she has an accurate understanding of the state of the environment within which the system is operating. A revised model of what operators do is presented which emphasizes the team-based operation that is in effect in all complex systems. Ambiguities in linguistic communication, differences in understanding and degrees of knowledge, and ineffective cultures of authority are some of the hurdles that must be overcome before we can achieve any dramatic improvement in our ability to manage abnormal situations in any complex enterprise. Greater consistency in training, incident reporting, command structures, and operational procedures, and subsequently, better collaboration support are identified as primary issues that must be addressed for team-based operation of complex systems.

In Chapter 3, Chris Miller introduces the distinction between actual complexity and perceived complexity. The former refers to objective measures of the difficulty of understanding or managing a system, whereas the latter takes the human into account. The notion of perceived complexity is Miller's focus of attention – and particularly perceptions on the part of human operators. Another theme of this chapter is that an increase in the capability or 'adaptiveness' of a system can be achieved through two means: by increasing the workload of the operator and by greater automation, which implies increased unpredictability for the human. Perceived complexity can result from increases in either workload or unpredictability, but there is not necessarily a direct connection in either case with any objectively quantifiable measure of system complexity. The question then arises as to how we can reduce perceived complexity. Two strategies are discussed: first, we can free up 'operator capacity' by designing better interfaces or by otherwise improving operator efficiency; second, we can reduce the actual complexity of the system. Miller concludes by outlining two new concepts that may help manage the human impact of complexity: shared models between the human and the automation system and tasking interfaces that can accommodate different levels of abstraction in human-system interaction.

The theme of perceived complexity is continued by Victor Riley in Chapter 4. Riley's focus is on how interaction metaphors and mental models can influence the perceived complexity of a system. Three levels of perceived complexity are distinguished in this context. An interface may lack a consistent conceptual framework, it may have such a framework but this may not already be familiar to the user, or it may represent the system's functionality within a unified conceptual framework that the user readily recognizes – the ideal situation. Examples from different domains illustrate these ideas. One example, the design and use of flight management systems for commercial aircraft, is discussed in detail. Feature creep has led to flight management systems that are function-rich but difficult to use, and many pilots are calling for simpler systems. A new pilot interface that the author has developed, called the Cockpit Control Language, is described. Cockpit Control Language has arisen out of a human-centered design approach as distinct from the traditional user-interface oriented 'human-factored' methodology. The single metaphor on which it is based is the pilot's existing knowledge of air traffic control clearances.

2

Advanced Technology in Complex Systems: Automation, People, Culture

Edward L. Cochran and Peter Bullemer
Honeywell Technology Center

2.1 Introduction

Consider the following incidents, all of them failed opportunities to demonstrate excellence in the management of complex systems:

- The left engine on a British Midlands 737 fails catastrophically. The flight crew shuts down the *right* engine, and the aircraft ultimately crashes short of the runway while attempting an emergency landing.

- An air traffic controller clears a 737 to land on a runway at LAX, forgetting that she had cleared a commuter aircraft to hold for takeoff partway down the same runway. The 737 collides with the smaller aircraft with horrific results.

- The *U.S.S. Vincennes,* a guided missile cruiser engaged in combat operations, detects an aircraft leaving a commercial airport in an uninvolved country. The crew concludes the target is a threat and ultimately shoots down a commercial airliner, killing 290.

- The officers of the lead ship in a squadron of destroyers on a night sortie become disoriented. They misperceive lights on shore and, believing they have passed a dangerous shoal, turn the ship directly into the hazard. It runs aground, as do most of the ships following closely behind.

- A process operator begins to start up equipment on an oil rig following a shutdown for maintenance. Unfortunately, the maintenance crew had not yet finished reconnecting the pipes. Escaping vapors explode, leveling the platform and killing 167.

- The operator of a chemical plant disbelieves a level reading that has led to the activation of safety interlocks and prevented the initiation of a batch process. He overrides the interlock and the reactor explodes.

- A commuter airliner is returned to service after maintenance and crashes shortly afterwards, killing all aboard. Most of the bolts holding a key part of the tail had been removed but not replaced.

- An operator, sent to open a valve on one of four identical process units, opens the valve on the wrong unit, leading to a catastrophic fire.

- A cargo handler loads hazardous cargo on a commercial airliner. The cargo starts a fire, which brings down the plane, killing all aboard.

- A satellite goes into a safe mode after encountering undetected errors in software. Ground controllers compound the problem, and control of the satellite is lost for several months.

This chapter will argue that these incidents all have one thing in common: Although blame was apportioned in the resulting investigations, *individuals* weren't really at fault. The causes of the incidents were listed as human error when in fact the incidents resulted from systems failures. People were put into very complex situations, and the actions that seemed appropriate at the time had unforeseen consequences. Thus our focus will be on interpersonal, team-based, and cultural issues related to the operation of complex engineering systems.

These accidents are just a few examples from a long and tragic list of cases in which automation has led to more being expected of people than they can reasonably deliver. The complexity of the systems involved, combined with the knowledge and communications requirements in the specific circumstances, led to task demands that require a level of performance that individual humans are fundamentally incapable of sustaining – and that organizations as a whole can sustain only with difficulty.

In all these cases the result was an accident, and therefore newsworthy. However, it is certain that, although not newsworthy, nonoptimal management of complex systems is ubiquitous and has enormous financial impact in terms of waste and inefficiency. We are only beginning to understand the magnitude of the problem and the true nature of its cause.

The fundamental problem is that while we are building increasingly complex systems, at an ever more rapid pace, we are not managing the complexity of those systems very well.

2.2 What Makes a System Complex?

System complexity is a function of the number of components in a system, the number of changes that can occur within the system, the possible interactions among those changes, and the speed with which those interactions propagate. In practice, the number of components is less important than the possibilities for change in the behavior of those components, because people can learn to deal with large numbers of predictable things.

2.2.1 Strategies for managing complexity

As systems grow larger, one strategy to manage their increasing complexity is to group the components in some way so as to reduce the number of things that have to be monitored. However, if the components all behave differently, they defy grouping, and complexity increases.

Even if the behavior of the components is different, complexity can be reduced through automation as long as the behavior is predictable. This allows people to manage the automation system instead of all the components that are being controlled. When the overall system is highly reliable, this strategy works well. It does not work as well if the behavior of the components is too challenging to automate well, or if the components interact with each other in unpredictable ways. And the strategy breaks down completely when people are expected to deal with problems that the automation system cannot manage, because they end up being required to deal with both the complexity of the original system and the influence of the automation system.

It is sometimes possible to reduce the impact of systems failures enough that real-time intervention by human operators is not required. For example, the original digital computers required large numbers of operators to tend the machinery that is now superseded by the pocket calculator – when one of those breaks, it is discarded. Larger systems can be managed in the same way: modern automobiles incorporate very sophisticated computing power – many times that in the original lunar landers – but the consequence of a systems failure is rarely life-threatening.

However, refineries, airliners, utility networks, and other modern systems are not yet amenable to these measures and require human attendance. When available automation strategies are employed and the systems are still too complex for one person to manage, the typical strategy is to divide the management task among members of a team of people. Thus oil refineries have multiple processing units, each operated by a team of control room and field operators. Similarly, the air traffic control system is divided into areas, and within those areas controllers are given specific responsibilities. System designers develop interaction rules, or *procedures*, to guide the interaction of the various human operators.

The strategy of adding people as systems become more complex works for a while, but the advantages of having more people are eventually offset by the need for those people to communicate among themselves. In addition, people are often surprised by the introduction of new kinds of problems resulting from various kinds of communication lapses. Thus operators of complex systems must not only be capable of understanding the situation and developing a successful response, but they must rely on each other in developing their understanding and determining the best course of action. Operators who act on garbled information, or information that is based on incorrect assumptions of other operators, are making what they believe to be correct responses, and the problems caused under those circumstances can be particularly difficult to identify and resolve.

2.2.2 Limits to current approaches to managing complexity

There is a delicate balance to strike in increasing the number of operators to reduce individual workload, because operators must collaborate with each other to maintain a common understanding of the state of the system. As the number of people involved increases, the amount of communication required increases. If communication demands consume all the attention of the people who were added to the system to help manage it, no net benefit may be achieved. Further, the operators must all be more capable, since all the knowledge being communicated is highly specialized.

Eventually, too much is required of the individuals charged with supervising the system, and breakdowns occur. To blame these breakdowns on human error is akin to criticizing a novice juggler for her inability to juggle 15 balls at once. While standing on someone's shoulders. While on a unicycle.

Since the capabilities of individual humans are limited, and change too slowly for us to rely on evolution as a solution, we need a better option. It is unlikely that we can reduce the actual complexity of these systems. Even if we could, that strategy would only buy us some time before the next generation of systems evolves and adds complexity again. It is also unlikely that we can reduce the impact of problems in such systems. That leaves us with the need to improve the effectiveness of human operators. There appear to be only two options:

- Reduce the apparent complexity of the system to its operators, or
- Reduce the need for, or increase the efficiency of, communication among operators.

The remainder of this chapter describes several ways of achieving these goals. I begin with an overview of some of the problems experienced by individuals and teams in dealing with complex systems, which will inform our concluding discussion on promising approaches to better solutions.

2.3 The Role of Individual Humans in Complex Systems

Before we discuss strategies for helping humans manage complex systems, it is necessary to describe the problem from the operators' perspective.

2.3.1 Classical approach

Many current views of human action in complex systems use a model similar to that in Figure 2.1, which is a variation of a decision-making model adopted by the Chemical Manufacturers Association (CMA). It is notable that the trend in modeling human-automation interaction is increasingly towards an emphasis on abnormal situation management. Automation under nominal conditions is a considerably simpler undertaking – and normal conditions are, oxymoronically, a rarity in complex systems.

In this model, there is an outside world that generates events, some of which are perceived by humans, processed internally, and ultimately generative of responses – actions that initiate new cycles of activity.

This model is often used to explain and categorize the kinds of errors that can arise during different stages of the orient-evaluate-act cycle. For example, operators may fail to notice key information that is present on their displays, leading to inadequate orientation and inaccurate evaluation. Such errors can be minimized by making the information more conspicuous or reducing the workload of the operator. To identify and minimize such problems, there has been increasing emphasis on the rigorous application of human factors principles to the user interfaces to complex systems.

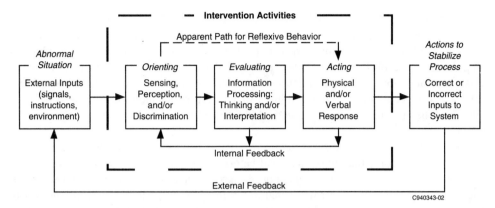

Figure 2.1 Standard model of human interaction with complex systems

2.3.2 Limitations of the classical approach

Those who apply this model in the traditional way often make two assumptions that are only approximately true. Worse, these assumptions break down precisely in the same way and for the same reasons that management of complex systems often breaks down. The result is that our traditional approaches to developing better ways to manage complex systems have led us to the brink of a large pit.

2.3.2.1 What do operators believe about the system?

The first assumption is that the operator understands the system. Thus, we tend to think that if turning on a pump lowers the level in a tank, and the operator needs to lower the level in a tank, the operator should turn on the pump. If that act is not carried out, a mistake is declared. Most of the time this way of thinking is perfectly adequate. However, the system that *really* matters is the one in the head of the observer.

If the operator does not know that turning on a pump lowers the level of a tank, or does not know that the particular tank with the problem has a pump that can be turned on, then it is pointless to expect the operator to turn on the pump. The 'mistake' is not the operator's – the operator is after all acting perfectly consistently with respect to the world as they know it. Instead, the problem is with the way the operator learned about the system. That process is not represented in the model at all.

Abnormal situations frequently arise because the way the system works in the real world is not congruent with the way it works in the world in the head of the operator (as borne out by the examples at the beginning of this chapter). In simple systems, all of the relevant knowledge can be managed by a single person, and inconsistencies are easy to spot. By definition, this is not true of complex systems. In fact, complex systems demand an entirely different approach to knowledge management, as will be seen later.

2.3.2.2 What do operators know about the state of the world (and how do they know it)?

The second assumption that users of the orient-evaluate-act model often make is related to the first. We not only expect operators to fully understand the system, we expect them to fully understand the current state of the world in which the system is operating.

We assume that there is a single set of 'facts' to be had, which should ideally serve as a reference for evaluating the performance of operators and the system itself. This assumption serves us well most of the time, especially when the facts in question concern physical things that can be observed, discussed, and agreed upon. The assumption that 'facts are facts' is clearly not true in the realms of political opinion, religion, and other cultural matters, and indeed there is a long tradition in Western philosophy to distinguish between 'facts' and 'beliefs' and to develop a method (the 'scientific method') of extending the realm of fact.

The problem is that individuals rarely distinguish between the two, and therefore it is wrong for us to assume that individuals will be orienting, evaluating, and acting solely with respect to 'the facts'. Consider the operator of a process unit faced with an unexpected and severe upset. The plant manager has stated in writing that the policy in such instances is to shut down the unit to prevent damage and to permit an orderly recovery. All operators have read and signed this policy. The operator in this instance attempts to ride out the disturbance, and, after some drastic control moves, ultimately succeeds. The 'facts' would indicate that the operator in this case made a mistake by persisting, in direct opposition to standing policy, with the attempt to recover from the upset without shutting down. However, what if the last four times this event has occurred, operators who have kept the plant running have been rewarded, and those who have followed established policy have been ostracized?

When such conflicts exist, people do the best they can, forming beliefs about what the facts are based on their own observations, knowledge, and experience. (The contributions of culture to experience can significantly affect the results, as we will see later in this chapter.) When the need to act in accordance with the 'facts' arises, the results can be unpredictable.

A different example of a problem with the use of 'facts' arises in complex systems due to the relative rarity of the kinds of problems that people have been tasked to prevent. In essence, the issue is that humans are not very good at understanding and explaining events with which they have little experience. College educated adults may believe that objects that are dropped from a moving car fall straight down, or that objects swung around on a string and released will spiral away. Humans are also fundamentally flawed statisticians and fall prey to all manner of inaccurate assessments of low-probability events, which causes problems when those assessments are acted upon as 'fact'.

Thus the problem for us as designers of complex systems is that we tend to design them as if (a) their users will understand them completely, and (b) their users will have access to 'the facts' and will be able to determine the 'true' state of the world and be able to act accordingly. As we have seen, the facts are quite malleable and the perceived state of the world may be inaccurate. Abnormal situations frequently arise for this reason.

Richard Feynman makes a similar point in a discussion of the *Challenger* accident:

> 'Let us make recommendations to ensure that NASA officials deal in a world of reality in understanding technological weaknesses and imperfections well enough to be actively trying to eliminate them. . . . NASA owes it to the citizens from whom it asks support to be frank, honest, and informative, so that these citizens can make the wisest decisions for the use of their limited resources.
>
> 'For a successful technology, reality must take precedence over public relations, for nature cannot be fooled'.

2.3.3 Revised model of what operators do

The simple orient-evaluate-act model needs to be viewed as a more complex system of interactions between perceivable events and the knowledge and beliefs of an individual, leading to the construction of an understanding of the situation upon which responses are based. This expanded model is depicted in Figure 2.2.

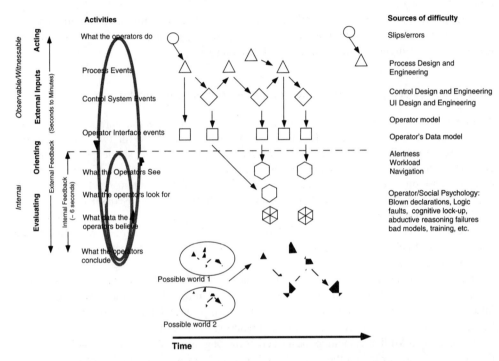

Figure 2.2 Elaborated model of human interaction with complex systems

2.4 Operations Teams and Complex Systems

Complex systems are usually managed by *teams* of individuals. As a result, all of the issues discussed above are present for each individual member of the operations team. In addition, several new issues emerge as a consequence of the need to solve problems collaboratively. Many of these issues arise from the fundamental fact that *nearly all human knowledge is socially constructed*. This is true regardless of whether the topic is rules of behavior, the operation of an oil refinery, or the landing procedure for an airport.

In general, we learn about the world from others; we draw inferences using methods that we learned from others, and we understand the current state of things based on communications with others. Even the knowledge we gain from seeing things with our own eyes is mediated by rules of interpretation that we learn from others, and by our observation of the responses of others.

The consequences are profound. For example, people may respond to a situation by disregarding the information from their own senses and instead emulate the responses of those with more seniority, experience, or authority. Designers of a system may withhold operational knowledge from operators based on perceived professional status. Alternatively, designers may provide complete information, but have it ignored by operators in response to social pressures. Operators may respond to a situation inappropriately based on misunderstandings about how the world works that are rooted in childhood. People who understand a problem may fail to speak up, even when asked, thus depriving the team of the information it needs to be successful.

The fact that knowledge is socially constructed is especially important in team settings, because the successful operation of complex systems by teams can only occur if the following requirements for information needs are met.

2.4.1 The need for a shared understanding of how the world works

As important as it is for lone system operators to have thorough and accurate knowledge of their system, it is even more so for members of operations teams, for two reasons. First, teams are responsible for larger, more complex systems, and no single person may be capable of understanding the whole system. Instead, all of the members of the operations team depend on each other's understanding of parts of the system, and gaps in that understanding are both more likely and more likely to have greater consequences.

The second reason that thorough and accurate knowledge of the system is critical for members of operations teams is related to the first: The team member with inaccurate knowledge is likely to lead the rest of the team down the wrong path *even when they presumably know better.*

There are many reasons for this. The more knowledgeable team member(s) may incorrectly yield to the authority, experience, or position of the other, may be overly influenced by irrelevant social factors such as friendship (or lack thereof), or may be shaken by stress, excitement, fear, or other emotional reaction to the situation.

In the British Midlands incident described at the beginning of this chapter, at least one of the pilots may have acted on the 'knowledge' that fresh air to the cockpit is provided by the right engine. The smell of smoke may have led to the assumption that the right engine was the one that had failed – an assumption that was not challenged by the other pilot nor shaken by the evidence of the instruments.

2.4.2 Shared understanding of the desired state of the world

It is critical that the members of the operations team develop a consistent understanding of their short- and long-term goals – the formal and informal rules and guidelines for the operation of the system they are responsible for. We have already seen that the 'facts' about the desired response to an incident may be not be the same for operators and plant managers, and indeed this is an example in which the lack of shared vision within an operations team can lead to nonoptimal results.

The same sort of problem (albeit on a much shorter time scale) happens hundreds of times a day in the United States when team members miscommunicate – when one or more members of the team fail, for whatever reason, to receive the message that another member intends to send. From that point forward, the team members are operating with respect to different views of the world. As a result, the wrong valves get opened, airplanes land on the wrong runway, ships run aground, and plants run short of raw materials.

To meet these two needs, better communication between system designers and system operators is required.

2.4.3 Shared understanding of the current state of the world

It is critical that the members of the operations team not only develop an accurate model of the state of the system they are responsible for and its desired state, but the state that the system is actually in. When members of an operations team believe different things about the state of a plant, trouble often results. (Note that post-incident review teams typically label the beliefs of the team member(s) who is determined to have been incorrect as 'beliefs' or 'assumptions' (often described as 'unfounded'). In fact, the team member(s) who are determined to have been *correct* were often working with beliefs or assumptions as well – they just happened to be accurate. Remember, all of the team members – correct as well as incorrect – act consistently with the world inside their own heads.)

When an operator believes that maintenance is complete on equipment at a plant at the same time that a maintenance worker believes that the equipment is still out of service for maintenance, an incident is likely. Whether the operator or the maintenance worker is 'correct' – that is, has a set of assumptions more consistent with the 'true' state of the plant – is irrelevant, as this changes only the nature of the incident, not the likelihood that an incident will occur. Some of the most horrific incidents in the history of technology (including the loss of the Piper Alpha oil rig with most of those on board and the $1.6 billion fatal explosion and fire at the Phillips plant in Texas City in 1989) can be traced to this precise problem.

The problem worsens when the state of the plant is unfamiliar (as it is during upsets), when the understanding has to be developed rapidly (as it does during upsets), when the team is inexperienced in working together on the particular problem involved (as it is during upsets), and when accurate understanding is important to the selection of an appropriate response, and ultimately to the outcome of the event (also as it is during upsets). Reviewers of the *Vincennes* incident described the chaotic environment on the bridge and in the ship's combat information center – the ship had been maneuvering violently and firing weapons at surface targets – and concluded that it played a significant role in the incident.

To meet this need, better communication both within operations teams and between operations teams (unit to unit and shift to shift) is required.

2.4.4 Shared understanding of the plan to reduce the discrepancy between actual and desired states of the world

Even when the operations team has a shared vision, an understanding of the how the process works, and an accurate diagnosis, it may fail to develop a shared understanding of the recovery plan. This is relatively rare and usually happens when the response is relatively decentralized and/or not amenable to detailed preplanning (e.g., firefighting). Nevertheless, since most processes are complex enough that several options are available to recover from any undesirable state, it is important that the operations team members agree on and operate consistently with respect to a single recovery plan. To meet this need, better communication within operations teams is required.

2.4.5 Opportunities for error arising from the need for communication

Given the need for every person on the operations team to act consistently and accurately with respect to the state of their process, it is not surprising that errors occur. In fact, the relatively low rate of incidents stands as testament to the effort that has been devoted to designing procedures intended to create a shared vision across process teams.

Many good analyses of the causes of individual human errors exist and need not be summarized here. In an ideal culture of team operations, most individual errors would be quickly detected and corrected, given the interlocking responsibilities and oversights of the members of the operations team. (An exception occurs when the team structure is designed to depend on the complete trust of members of the team in an experienced team leader, who must make accurate decisions in isolation. The need for this team structure is rare – formation acrobatics is one example – and when the leader of the team does make a mistake, the cost can be high: the squadron of destroyers following the admiral's ship onto the rocks is one example.)

However, most operations teams do not yet operate with respect to completely consistent perspectives, and as a result they are subject to several kinds of errors that are unique to team settings and that are extremely difficult to overcome. Awareness of these team-induced errors has been growing, and attempts are being made to address the problem (e.g., Cockpit Resource Management training initiatives in aviation), but the study of this area is still in its infancy. Several examples of these errors are described below.

2.4.5.1 Inadequately qualified assertions

Assertions are statements about our observations. Although we may believe these statements are descriptions of the way things 'really' are, in fact, all we know is how we ourselves observe them. An assertion is 'true' if we could provide a witness who would concur with our observation, and it is 'false' if a witness would dispute it. The act of making an assertion commits us to truthfulness, but assertions may be qualified and thus relieve us of the need for absolute precision. For example, the difference between a tank being 'empty', as opposed to 'nearly empty', is more critical to a pump operator than it is to a production planner.

Accurate communication requires that we be able to assess the need for such qualifications in our sharing of observations, facts, urgency, and so on, with other members of our team. This requires either significant knowledge of the other people on the team (as is the case in long standing close-knit teams) or absolute adherence to procedure (as is the case in naval carrier operations or air traffic control). Otherwise, we qualify our observations too much (leading to inefficiency and reduced trust) or not enough (leading to mistaken assumptions in others). Both of these outcomes cause problems.

Most process plants – indeed, most civilian operations of all kinds – operate as if the team has the requisite self-knowledge to communicate accurately in the absence of rigorous procedures. This is risky, as demonstrated by the large number of audio recordings of incident communications in which amplifications and clarifications of observations are required in order to get an appropriate response.

2.4.5.2 Inaccurate declaration of knowledge

Declarations are decrees – impositions of aspects of our own world, such as our beliefs, on others. They can be loosely described as determining a view of the world that those subject to the decree will abide by, 'accurate' or not. As declarations represent the views of at least one person but not others (else imposition would not be necessary), they are usefully described as valid or invalid, according to the power of the person making the declaration, as well as accurate or inaccurate.

The power we grant to some people to make valid declarations is called authority. Thus valid declarations (whether accurate or not) are made by those with authority, and invalid declarations (whether accurate or not) are made by those without authority. It is easy to understand how valid, but inaccurate, declarations can be a source of problems, as when a formation of aircraft follows the leader into the ground.

Inaccurate declarations are in fact a major cause of the progression of abnormal situations to accidents, because they are relatively frequent, and because they create a world – a set of shared beliefs – in the operations team, whether it is accurate or not. For example, in the presence of two conflicting temperature readings and no other witness, a declaration is required to determine which reading is 'correct' – and that declaration will be used to coordinate subsequent action. The declaration may be valid – the shift supervisor may have the authority to make it – but it may not be accurate. This is a critical point in any situation, because from then on the operations personnel will act in accordance with the world they have created, until it becomes so obviously inconsistent with further events that the declarations are retracted. Unfortunately, humans are very good at weaving consistency out of noisy or incomplete observations – 'obvious inconsistency' must be very obvious indeed.

2.4.5.3 Ineffective culture of authority

A more insidious kind of problem arises when the exercise of authority is ill-defined, as when incident command functions are not specified, or those with authority do not exercise it appropriately. Authority is important because it enables declarations to be valid. Declarations are one of only two ways of defining shared knowledge, and by far the most efficient in terms of time. (The other way to define shared knowledge is through a process of inquiry – questioning the sources of differences in assertions until they are explained through more comprehensive sharing of context.)

Hierarchies often exist primarily to permit declarations to be made in the presence of unresolved questions. In the absence of authority, decision making can become a social process, and anyone on the team becomes capable of making a *de facto* declaration on the basis of field knowledge, social influence, bravado, self-certainty, or other factors, whether appropriate or not. There is no time for an effective inquiry process, and a consensus 'state of the world' is thus reached in an *ad hoc* way.

Many incidents attributed to so-called 'garden path thinking' or 'cognitive lock-up' can be viewed in these terms. A human declares that a radar blip represents an attacking fighter, or that safety interlocks have inappropriately engaged, or that the landing gear is down. The declaration is not questioned, even though authority structures are supposedly in place to ensure that it is. Subsequent actions, consistent with the inaccurate declaration, lead to incidents.

Throughout the Piper Alpha oil platform fire, a neighboring production platform pumped product to Piper Alpha (literally adding fuel to the fire) because the supporting platform manager assumed that he had no authority to do otherwise in the absence of an order from

Piper Alpha to stop production. (No such order could be given: Piper Alpha's control room was destroyed in the initial explosion.)

2.4.5.4 Confusion between requests, promises, offers, and assertions

Although we should not expect operations teams to be students of discourse and rhetoric, we should understand that in the absence of clear procedures (or a team with enormous experience working together), the opportunity for confusion is rampant. Consider the opportunities for problems inherent in the confusion between requests, promises, and offers, (these distinctions can also be contrasted with assertions and declarations, discussed earlier):

- *Promises* are the linguistic acts that allow us to coordinate action with others (both implicitly and explicitly). Four fundamental elements are involved in a promise: A speaker, a listener, a condition of satisfaction, and a completion time. The act of making a promise commits the promiser in three domains: sincerity, competence, and reliability.

- *Requests* are solicitations of promises from someone else. They can be thought of as a promise of the form 'I [perhaps conditionally] promise X if you promise Y'. X can stand for tangible and intangible conditions of satisfaction, e.g., 'I will be pleased', 'I will be better able to make a diagnosis', 'I will get you safely out of this'. Y carries the form of the promise in return, e.g., '[Listener] will fetch a reading from the gauge'; '[Listener] will close valve 17 and open valve 18'. A request, when answered with a declaration of acceptance, creates a promise. If a request is declined, no promise has been made. (A request that cannot be declined is not a request; it is a declaration.)

   These distinctions are critical: In most cultures, the act of making a request implies that we are competent to assess our knowledge of the system, and it commits us to acting consistently with an implicit assertion we thereby make, that the *other* person is sincere, competent, and reliable. In everyday life, these implications are trivial. The assessment of another's expertise associated with a request such as, 'Will you pick up some milk on the way home?' is rarely a life or death issue. But when complex systems are involved, a request such as 'Can you close that valve for me?' carries social baggage. For example, it implies that I believe I am competent, and may therefore reduce the chance that the listener will question the request. In highly procedural operations such as air traffic control, communications are rigorously structured in an attempt to avoid ambiguity in such situations. Thus a declaration such as 'United 111, descend and maintain ten thousand' is explicitly converted to a request using words such as 'if able'.

- *Offers* are a request for a promise of the form 'If I promise Y, will you promise X?' where X and Y are defined as above. Thus, if it is sincere, a simple 'May I help you?' carries a social meaning such as 'I will help you [the promise] if you would be pleased for me to do so [promise as condition of satisfaction]'. The commitments, structure, and implications of making offers are very similar to those involved with making requests; only the agent of conditionality is reversed. Again, however, the implication of expertise can cause trouble: inexperienced operators offering to help out in an emergency are likely to be given tasks that exceed their expertise, unless the listener knows the offeror or procedures are in place to make the offeror's limitations obvious (which is why probationary fire fighters often wear conspicuously marked helmets, for example).

This is obviously a brief and highly simplified treatment of an extremely rigorous and extensive domain of study. Successfully transferring information from one person to another requires the successful application of myriad knowledge about ourselves and our colleagues; otherwise confusion results. We make requests of people who are not able to carry them out, raising doubts about our own competence and the validity of further requests. We unknowingly accept a request that we are unable to fulfil, calling into question our competence and/or sincerity. Ultimately, we may not know what is being asked of us, or why, or what the consequences will be for failing to meet the request.

If that were not challenging enough, the linguistic domain of *pragmatics* enables every one of the above distinctions to be rendered false; for example, an assertion ('It is cold in here!') can be a request ('Please close the window!'). An offer ('May I review your monthly report') can be a declaration ('Give me your monthly report') – especially if the speaker has authority over the listener. And so on. Pragmatics requires a further layer of social knowledge to be negotiated, one that focuses on the implicit or implied motivations, values, and characteristics of the people with whom we interact. Pragmaticians study linguistic acts from both sides – the intent of the speaker and the interpretation of the listener – and the social mediations can be complex indeed. Different cultures, of course, have different conventions for the interpretation of these acts. The fact that humans communicate so well is testimony to how finely tuned and sophisticated are our abilities to act within a social environment.

In fact, in everyday life, and certainly in the environment in which these human abilities were evolving, the fact that language is flexible and that absolute precision is rarely required is a good thing. For example, it is rarely the case that one needs to specify a color in terms of wavelength instead of a label such as 'green'. The vast majority of promises do not need several pages of explicit caveats, conditions of satisfaction, and declarations of competence. The need is usually not to describe the world precisely, but to engage the emotions and feelings of others in order to promote relationships, group efforts, social institutions, and political activity, and pragmatic constructions are indeed well suited for that purpose.

2.4.6 Summary: Teams and culture in complex systems

An operations team in a complex process, then, must understand how the process works, and its goals for the process, as well as the state of the process and the actions required to close any gap between the state of the process and its desired state. However, teams need to know more. We have argued that different kinds of linguistic acts imply social commitment, and they can only function from the background of shared social practices and a shared ability to recognize and act in accordance with those practices, which means that team members need to understand themselves, their team members, and their culture.

As a rule, those charged with designing, implementing, operating, and maintaining complex processes pay little attention to these issues, if they are aware of them at all. The result is that we handle the requirements imposed by the need to collaborate with others in operating complex systems in the same informal way that we handle the need to collaborate with others in planning a PTA fund-raiser, remodeling a kitchen, raising a family, and every other endeavor typical of life.

As a rule, people are intuitively very good at this, as the continued existence of PTAs and remodeling contractors, to say nothing of civilization itself, amply demonstrates.

Nevertheless, mistakes are made; misunderstandings occur; backtracking is necessary; failure is a possibility.

There are two primary reasons why this situation is not tolerable in complex systems:

- The complexity of the systems and the speed with which events propagate make our intuitive methods of identifying misalignment very inefficient and error-prone.

- The consequences of failure are too high.

2.5 Lessons for Managing Automation in Complex Processes

Given the previous discussion, we are left with two primary issues that need to be addressed if we are to progress further in successfully managing systems of higher complexity, with larger operations teams.

First, we need to find better ways to manage the apparent complexity – to ensure that team members have *consistent views* of all the things that we have argued they must.

Second, we must find better ways to support real-time collaboration and the exchange of information as accurately, quickly, and richly as the operations teams of complex systems will require.

2.5.1 Consistency

There are several ways to ensure consistency in the understanding of systems.

2.5.1.1 Training
Rigorous, high quality training is mandatory, especially including training for the rare events that test the operations team most severely. As technology improves, the extension of simulation-based training from domains such as aviation is becoming more economical. But good training need not be expensive: 'What if?' training, role playing, and scenario development are all very effective in building shared views and expectations.

2.5.1.2 Reporting
Incident tracking and reporting is critical to the deliberate and disciplined sharing of insights and experience gained from unusual events with everyone who might benefit from them.

2.5.1.3 Communications
A formal communications policy must be in place to ensure that consistent messages are sent from the top to the bottom and from one end of the plant to the other.

2.5.1.4 Authority
An incident command structure must describe the authority of those involved – and those involved must live up to that authority.

2.5.1.5 Procedures
Most important of all, consistency requires that the operations culture make a disciplined practice out of creating, evolving, and following *procedures*. Procedures incorporate knowledge compiled when there is time to think, so as to benefit those in situations in which

there isn't. Procedures clarify ahead of time the goals, expectations, and information requirements of unusual situations. Communications procedures ensure that the appropriate information is available to the appropriate people at the appropriate time. (Good radio procedures alone can reduce uncertainty in message content by a factor of two, and perhaps even more in an industry that often refers to equipment entirely in easily confused alphanumerics, such as 'D-159B'.)

The U.S. Navy, over the past 30 years, reduced the rate of mishaps by a factor of six, from 19.3/100,000 hours of flying time in the early 1960s to less than 3.0 in the early 1990s, through a comprehensive effort to develop best practices and instill absolute procedural discipline in using those practices.

2.5.2 Collaboration support

Once we have achieved some measure of consistency within the operations team – and only then – we can begin to develop ways for the members of the team to collaborate quickly and accurately. (The introduction of collaboration support technology before consistency is achieved is not helpful, and may in fact be harmful in that it might permit the more rapid exchange of inaccurate information.)

The goal will be to ensure that each member of the operations team has access to the right information, at the right time, for the right reasons. We are in the midst of an ambitious effort in this area as part of a multi-organization research and development program that is developing new technology for abnormal situation management in the refining, chemical, and petrochemical process industries (see http://www.iac.honeywell.com/Pub/AbSitMang/).

2.6 Conclusion

We have seen that the use of operations teams for the management of complex systems introduces several extremely challenging issues that must be solved if progress is to be made. Most of these issues result from the fact that our normal approaches to human interaction, and our skills in communicating with others, are simply not adequate to keep up with the rapid changes that occur in complex processes. Even if we could keep up with these systems as well as we do in our other daily activities, the consequences of the occasional mistakes are too great for us to tolerate.

To make further progress, effort needs to be devoted to the development of deliberate, principled approaches to the establishment of rigorous and consistent operations cultures. Only then will further advances in automation technology be likely to succeed.

Acknowledgments

The preparation of this chapter was in part supported by a grant from the National Institute of Standards and Technology's Advanced Technology Program to the Abnormal Situation Management Joint Research and Development Consortium.

The authors gratefully acknowledge the contributions made by Richard Lewis and Ian Nimmo to the ideas presented in this chapter.

3

The Human Factor in Complexity

Chris Miller
Honeywell Technology Center

3.1 Introduction

A dictionary definition of *complexity* derived from *Webster's* (1989) is 'the state of being so complicated or intricate as to be hard to understand or deal with'. This points to the fact that, in common language, we use the term *complex* to refer to virtually any problem, situation, or system with which we have difficulty. If we are to accept this definition, then *complexity management* becomes, essentially, the task of dealing with things that are hard to deal with. Although such a topic of study would certainly be useful, it might be a bit broad to tackle profitably.

This definition also points to the fact that complexity is, at least in common speech, inherently a relational concept. A situation or system is 'complex' to someone who has problems understanding or dealing with it. In fact, a system may be complex to one person who must deal with or understand it at a deep level, whereas the same system may be simple to another person who does not have to understand it so deeply.

This relational aspect of complexity brings home the importance of the human perspective. Ultimately, our interest in the complexity of a system has to do with how much difficulty it gives us. And, as our interests and understanding expand and deepen, we find more and more complexity even in previously 'simple' systems. Thus we can profitably narrow the broad investigation of 'dealing with things that are hard to deal with' to a more focused examination of what makes a system *seem* more or less complex to a human who must interact with it and the consequences of that complexity for the interaction. I will refer to this phenomenon of being difficult to understand or deal with as *perceived complexity*.

Perceived complexity is distinguished from *actual complexity* by virtue of this human element. Whereas perceived complexity focuses on those aspects of a task or situation that make it hard to deal with, actual complexity implies some more objective criteria for this difficulty. To anticipate the next section slightly, if the number of components involved in a situation is a factor in complexity, then the *actual* number of components would contribute to its actual complexity, whereas the *perceived* number of components would be relevant to its perceived complexity. The latter formulation allows for the possibility of both human misunderstanding of a complex situation and human cognitive management strategies, such as hierarchical aggregation and rule formulation, to sort and group those actual components into more manageable structures.

There is, undoubtedly, a correlation between actual and perceived complexity, but this correlation is rarely, if ever, perfect. Mismatches between perceived and actual complexity can give rise to spectacular disasters – such as the under-representation of the complex interactions between temperature, elasticity, gas expansion, and joint movements that led to the *Challenger* disaster. On the other hand, such mismatches are also behind the simplifications that make it possible for me to interact with, and profitably use, a far greater range of complex technologies than I would ever have time to understand at the level of their 'actual' complexity.

Victor Riley's distinction between functional and cognitive complexity (see Chapter 4) is similar to the perceived/actual distinction. Functional complexity refers to the complexity required to achieve a given functionality, and conceptual complexity refers to the human's ease or difficulty in understanding how to use the system to achieve that functionality.

This chapter presents a view on the nature of our *perceptions* of complexity, on why and how system complexity affects human operators, and some thoughts on how these effects might be mitigated. My primary emphasis is on the users of complex systems rather than the designers – although implications for the human-centered design of systems and their user interfaces will be drawn throughout. The first section begins by providing a definition of the components of perceived complexity – that is, what it is about systems that makes us label them 'complex'. The next section discusses the human impacts of increased perceived complexity, and the final major section presents some strategies for managing perceived complexity in the systems we build and field. Overviews of two recent related research activities are also provided.

3.2 Definitional Issues

Evolving through the course of the discussions of the Honeywell Technology Center (HTC) Complexity Working Group, we found it useful to distinguish three types or dimensions of perceived complexity: (1) *component complexity*, (2) *relational complexity,* and (3) *behavioral complexity.* I will define these terms and provide some examples below. In comparing any two problems, situations, or systems, an increase in complexity along any of these dimensions causes us to perceive that the system which is higher in that dimension is 'more complex' than the system which is lower. It is possible, though somewhat rare, to have complexity that occurs on one of these dimensions alone. It is far more common for increases in complexity to be an interaction effect resulting from complexity increases in and across all of the dimensions. When attempting to compare complexity increases across dimensions, or when attempting to 'sum' complexity effects when they occur within multiple dimensions, things get more fuzzy. I believe, however, that the dimensions are listed in more or less the order of importance for their impact on overall perceived complexity – that is, component complexity has less impact than does relational complexity, and relational complexity less than behavioral.

3.2.1 Component complexity

Component complexity refers to the simple number of components that an observer perceives a system to have and to the perceived diversity of those components; that is, component complexity decomposes into number and heterogeneity. The more components a system is perceived to have, the more complex it is. The more different types of components it is perceived to have, the more complex. Remaining purely within this component complexity dimension, we would say that a big pile of sand is 'more complex' than a small pile or a single grain – because each has more components than the others. Similarly, a battalion is more complex than a company, which is more complex than a platoon or a single soldier. Along the diversity dimension, a pile of 100 black socks is less complex than a pile of 100 socks of different colors and patterns – and a company of soldiers of mixed race, gender, class, and education level is more complex than one which is homogeneous.

With each of these dimensions of perceived complexity, as we will see below, there is an essential connection to the perceiver. What is regarded as a component (or perhaps a 'component of interest') is a function of the human doing the perceiving. To be noticed, or to be of interest, the component must be *significantly different* from its surroundings and from the background. To a crystallographer interested in the molecular composition of things, a grain of sand has immense component complexity; to a stellar cartographer interested in mapping astronomical entities, all the grains of sand on the earth are literally 'beneath notice' – below the level of the single 'component' of interest (the planet). Thus, the component complexity increases from a grain of sand to a small pile to a big one in the above example, and will only hold if the different numbers of grains are perceived and of interest to an observer. What makes a component significantly different may change in different contexts, and it is always possible that the perceiver is not noticing things that he or she should (that is, not attaching significance to components that are, in fact, important in the domain). Nevertheless, as an arbiter of *perceived* complexity, this notion of what is perceived as different or significant is critical.

I find that most people have, at best, a grudging acceptance of this dimension of complexity. The feeling is this is somehow a degenerate use of the term *complexity*. The perceived number and/or heterogeneity alone of a system's components clearly fail to capture the most important or interesting aspects of 'complex' systems. I take this as evidence for the hypothesis that component complexity alone generally has a small impact on perceived complexity, and thus we need to look at other sources and dimensions of complexity.

3.2.2 Relational complexity

Relational complexity refers to the number and variety of perceived links between components – that is, the ways in which they are related. It is similar to Perrow's (1984) notion of 'interactive complexity' – the idea that the number of ways in which components can affect each other is a contributor to complexity and, for Perrow's focus, 'inevitable' accidents. A large number of grains of sand, spread out and glued to a sheet of paper, represents a less complex system than the same number of grains free to move about on a vibrating platform. In the latter example, the ability of each grain to move about and connect with (and thereby affect the behavior of) the others represents relational complexity. Similarly, a thousand diodes in a pile represents a less complex system than each of those diodes wired individually to a power source – and this system is far less complex than if the diodes and power sources were wired together into complex circuits with multiple AND and OR links between them.

Again, what constitutes a link, or a noticeable or significantly different link, is defined in terms of the perceiver, so that one person can look at a chip with many circuits and see great complexity, whereas another may see the chip as a 'simple' unit in a much more complex whole.

3.2.3   Behavioral complexity

Behavioral complexity refers to the number and variety of states the system and/or its components (depending on the viewer's focus) can be in – that is, the number of different behaviors the system or components can exhibit. There is an obvious connection to the human interpretation of complexity here, since we are rarely interested in all of the possible states a (usually continuous) system can exhibit, but only in some subset of 'significantly' different states, where significance is a function of our goals for interaction with the system.

The concept of behavioral complexity across significantly different states grew out of, and accounts for, a curious example produced by a colleague's ancient and failing truck. Over time, the truck has become a more 'complex' system for him to drive – not because it has significantly increased its number of components or the relationships between them, but because wear and tear of the steering mechanism and other subsystems have made it possible for the truck to achieve more behavioral states (and to achieve them less predictably) than it did when new.

Note that an implication of 'significantly different' behavioral states is that it is possible to reduce the perceived complexity of a system by reducing the number of possible significantly different states. This is a partial source of the humor in Rube Goldberg designs, since these systems have high component and relational complexity for the sake of very little behavioral complexity – obviously, a poor design. More important, the notion of significant differences and behavioral complexity permits the addition of overall complexity (e.g., a complex automation subsystem – such as an automatic transmission) for the sake of reducing the *perceived behavioral complexity* that the human must deal with (that is, allowing a driver to ignore gear shifting). Note, though, that in general we are only successful in shifting complexity from one place or time to another.

Related to behavioral complexity is the concept of the unpredictability of a system. Even if we understand all the possible significantly different behavioral states a system can achieve, we may still be faced with a complex (and uncontrollable) system if we cannot predict when and why the system will achieve any given one of those states. This accounts for our tendency to label as 'complex' systems that, mathematically, are very simple but that are chaotic and therefore unpredictable. Unpredictability may be a strong enough component of our use of the term *complexity* to warrant including it as a dimension in its own right, but strictly speaking, it is not an aspect of the system but of an agent's (human designer, human user, controller agent, etc.) interaction with the system. Thus I have not included it as a separate dimension in this nascent theory of perceived complexity. We will have more to say about the human's relation to complex systems, and particularly about unpredictability, in the next section.

3.2.4 Compounding perceived complexity

As noted in the introduction, perceived complexity is a compound function of the number and variety of components, relations, and behaviors *that make a significant difference to the user* in any context or situation, but accounting for the compounding is extremely difficult. Nevertheless, a few claims are obvious given the above formulation of the relevant dimensions. It is clear that there is a logically necessary progression through the dimensions. First, a component must necessarily have a minimum of one significantly noticeable behavior (its existence), but it need not have more than one. Second, a relationship must necessarily exist between at least two components, and to be 'of interest' it must manifest itself as a change in behavior of one or the other (or both) of those components when they are in a certain relationship.

A number of corollaries follow from these proposed compounding rules. First, components may sometimes be added to a system without substantial impacts on perceived complexity – as long as these components are homogeneous, perform no significantly different behaviors, and have little interaction with other components. On the other hand, the more significantly different components in a system, and the more significantly different behaviors they exhibit, the more opportunities there are to have interactions that are significant. If all interactions produce the same behavior (as far as the perceiver is concerned), perceived complexity will be reduced. Finally, note that the above changes can be effected through *either* changing the number of components, relationships, or behaviors in the actual system, or changing the number of significantly different entities the user is aware of and has to interact with. That is, changes in perceived complexity can be effected either through changing the actual system or through changing the ways in which the user perceives and interacts with the system (though training, user interface design, etc.).

In the next section, we will examine the implications of this formulation of perceived complexity as it applies to human interactions with complex systems. Although there has been a large body of research in this area, seldom has anyone tried to explain the data on human perception of complexity in a systematic way. If the perceived complexity model described above can offer insights into these data, we may be on our way to being able to reliably predict and therefore shape perceived complexity to aid in human-machine system design.

3.3 Human Impact of System Complexity

From the perspective of a human user, any relative increase in any of the complexity dimensions described above will (if unaccompanied by the complexity-mitigation strategies described below) produce an increase in the perceived complexity of the system. Any attempt to cope with this increased complexity must, I believe, result in some blend of two impacts for humans interacting with the system: either increased workload or increased unpredictability.

3.3.1 Adaptiveness as the goal; complexity as a side effect

Why do we repeatedly strive to increase the complexity of systems we deal with? Although this may not be our objective, I claim that it is an inevitable by-product of the desire to increase the 'coverage' or 'specific applicability' of the systems we have access to. We wish to increase the number of aircraft in the national airspace – this produces an increase in component and relational complexity. We realize that we must make a discrimination among oil refinery exhausts that are more or less harmful to the environment – thus introducing a whole new class of oil refinery behaviors of interest and thereby increasing behavioral complexity. Let us call this increase in coverage of situations the *adaptiveness* of the system. We increase the adaptiveness of a human + machine system when we add new behaviors to it, make new distinctions about when certain behaviors are correct or incorrect, or increase the range of circumstances over which behaviors must be applied.

3.3.2   *Adaptiveness, workload, and unpredictability*

In human history, we have found only two ways to increase the adaptiveness of a system. The first is to put the control of system adaptation into the hands of the human operator – thereby making it 'adaptable' in Opperman's (1994) terms. Alternatively, the control can be built into system automation, thereby making the system 'adaptive'. These two strategies are related to two human phenomena: workload and unpredictability.

Conceptually, the relationship can be presented as in Figure 3.1. The implication of this view is that for any increase in adaptiveness (that is, the ability of the human-machine system to perform in an appropriate, context-dependent manner in different situations), there must be an accompanying increase in either human workload (the amount of physical, attentional, or cognitive 'energy' the human must exert to use the system) or in unpredictability for the human operator (inability to know what the system or world will do at any given time). Stretching one leg of the triangle in Figure 3.1 will inevitably result in stretching at least one of the other two legs. Since adaptiveness is generally the goal of added complexity (although systems can be complex without achieving this goal), this is equivalent to saying that any increase in complexity must affect the operator in a mixture of two ways: either (1) the added complexity must be fully controlled by the human, resulting in workload increases or (2) it must be managed by automation, resulting in unpredictability increases. These alternatives represent endpoints on a spectrum; many intermediate points and blendings are possible.

Perceived complexity can result from either increased workload or increased unpredictability (or both) precisely because these are the things that make a system 'hard to deal with'. In turn, increases in the number of components, relationships, and behaviors that must be attended to to understand or 'deal with' a system will increase one or the other of these dimensions. It is interesting to speculate, although we have no concrete data at this point, that a measure of perceived complexity might be the combined lengths of the workload and unpredictability segments in Figure 3.1 – and that a system is said to be 'complex' when the combined length of these two dimensions exceeds some threshold. Consider the following examples. Playing concert piano is an activity many would label 'complex' because it requires high workload, even though it is not at all unpredictable. Roulette is complex (as games of chance go) because the combinations of many numbers and colors make it highly unpredictable – although it doesn't produce much workload. In fact, if increased workload is spent on monitoring and understanding a system's behaviors, it might become less unpredictable – although we'd probably tend to say there is little difference in the complexity of interacting with that system due to its high workload demands (that is, the sum of workload + unpredictability has not changed).

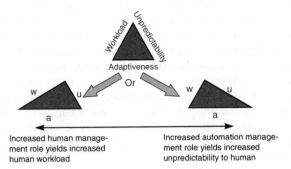

Figure 3.1 Conceptual view of the relationship between system adaptiveness, human workload, and unpredictability to the human operator

3.3.3 Complexity increases without adaptiveness payoffs

Another implication of the relationship shown in Figure 3.1 is that it is entirely possible to achieve the same sorts of human impacts that increased adaptiveness causes (increased workload and unpredictability) without any payoff in adaptiveness. One could stretch the workload and unpredictability legs of the triangle without making any change to the adaptiveness leg. Although on the face of it this might seem absurd, in fact it happens quite frequently; for example, when operators are asked to work faster (more workload), with inadequate training (more unpredictability), with degraded or faulty equipment (more unpredictability), or with inadequate communication between agents involved in the overall job (more unpredictability). It is quite likely that humans faced with such a situation will claim that their jobs have become 'more complex', even though the mechanical systems they work with may not have changed at all.

3.3.4 Mitigating complexity impacts

It is not the case that increased adaptiveness *must* inevitably and immediately result in increased workload or unpredictability to the primary user of that system. The effects of increased perceived complexity may be mitigated or shifted to other individuals or other times in the system's life cycle. Some important and useful strategies will be discussed in section 3.4 below. Nevertheless, I suspect that these strategies only buy us limited improvements in human capacity to deal with complexity. Improved interfaces and 'human-centered' automation design can only enable handling so much added complexity before they will be overwhelmed. Ultimately, our strategy must be to mitigate the complexity of the system by reducing the component, relational, or behavioral complexity it requires the human to deal with, rather than by increasing these and asking human operators to cope.

3.4 Strategies for Mitigating the Human Impacts of Complexity

First, it is important to note that the relationship between adaptiveness, workload, and unpredictability serves as a fundamental limiter on the complexity of human-machine systems that can be controlled (although, unfortunately, not on those that can be built). At some point, increases in perceived complexity produce enough added workload or unpredictability to give the user trouble in dealing with the system. Although it is extremely difficult, especially in the absence of a reliable measure of complexity, to determine where that limit lies, there are clear and reasonably well understood limits on human capabilities that must be brought to bear to interact with complex systems. These include short- and long-term memory, perceptual capabilities, attentional capacity, decision-making behaviors, fatigue, strength, and reach and grasp. In essence, much of human psychology is the attempt to identify and characterize these limits, whereas the study of human factors and ergonomics is devoted to applying that knowledge to designing systems and their interfaces so that they avoid exceeding the limits.

Although the limit may be difficult to predict, it is not hard to recognize in retrospect. When system complexity exceeds human capacity to control it (regardless of whether that 'control' is immediately input by a human controller or input years ahead of time by an automation designer), an accident can result. Since accidents are, by definition, to be avoided, the existence of an accident implies that the means available to the human and/or machine were insufficient to achieve the goal of accident avoidance. 'Managing complexity' must ultimately mean providing the means somewhere within the human + machine system to avoid accidents.

Another implication of the relationship illustrated in Figure 3.1 is that complexity-mitigation strategies must have their effect by limiting the expansion of one of the three legs of the triangle. Since, as already discussed, expanding the adaptiveness of a system is generally the goal that produces added complexity, the problem becomes how to increase adaptiveness without simultaneously overburdening human operators with excessive workload or excessive unpredictability. Not surprisingly, most mitigation strategies work by affecting one or both of these dimensions.

Before we begin discussing mitigation strategies, it is important to understand that most domains of interest contain four components that must be brought into harmony with each other: the user(s), the system (some device(s) the user controls and interacts with), the world (which affects and is affected by the system), and the interface, which presents the user with information about the world and the system and allows the user to control the system in an effort to effect desired states in the world. This relationship is illustrated in Figure 3.2. Workload and unpredictability come from these components, as described in Table 3.1; therefore, mitigation strategies must operate on one or more of these components, as described in Table 3.2.

In the next sections, we will discuss two broad classes of strategies for coping with the human impacts of complexity: (1) strategies that operate by improving the efficiency of the interface or the human operator so as to 'free up' human resources to deal with increased workload and unpredictability stemming from added complexity resident in the system and the world; and (2) strategies that operate by reducing or shifting the complexity so as to make smaller demands on the human operator.

3.4.1 'Freeing up' operator capacity through improved interface and human operator efficiency

Strategies discussed in this section share the trait that they mitigate the effects of complexity by attempting to reduce inefficiencies in either the human operator or the interface to the system and the world. When these strategies are effective, they enable an equivalent amount of work with less workload and unpredictability. Otherwise, they do not affect the complexity of the system with which the human is interacting in any way. Although these are not strategies for *reducing* system complexity *per se,* they are strategies for improving the human operator's capability to deal with workload and unpredictability more efficiently than in some baseline system – allowing the operator to tolerate greater system complexity.

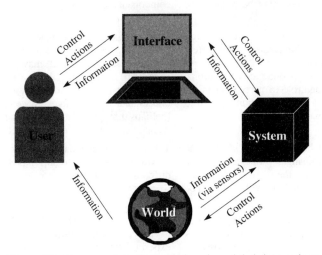

Figure 3.2 Components in the work domain and their interactions

Table 3.1 Work domain component roles in producing workload and unpredictability

Work domain component effects on workload	Work domain component effects on unpredictability
The *user(s)* may be distracted, uninterested, fatigued, disabled, etc. Thus their capacity for workload is diminished.	The *user(s)* may be ignorant of the workings of the other domain components, thereby increasing their unpredictability.
The *interface* may be inefficient (as outlined in section 3.4.1.4), causing more effort for information extraction or control actions.	The *interface* may be inconsistent, making it hard to learn or predict. Alternatively, it may be too difficult to use for the available workload, thus operator ignorance and system unpredictability will increase.
The *system* itself may require frequent monitoring or interaction, thereby increasing workload.	The *system* itself may be very complex (in any of the senses defined), making it more difficult to learn, easier to forget, and more difficult to monitor and control – all leading to increased unpredictability.
The *world* may change frequently and in unpredictable ways, increasing the need to monitor it and attempt controlling actions.	The *world* may be complex. There may be unknown (and, in chaotic systems, perhaps fundamentally unknowable) aspects of world behavior that make some unpredictability inevitable.

Table 3.2 Mitigation strategies by work domain component

Workload mitigation strategies by work domain component	Unpredictability mitigation strategies by work domain component
Users may be chosen for their workload and attentional capacities; work and environments may be structured so as to be motivating and interesting; adequate rest may be required to prevent fatigue, etc.	The *user(s)* may be trained in the workings of the interface, system, and world, thereby decreasing the unpredictability of these components for them.
Interfaces may be 'optimized' (cf. section 3.4.1.4) to convey information and effect control with minimal workload. These strategies include ergonomic considerations, migration from cognitive to perceptual tasks, addition of human or automation agents to distribute task demands, etc.	The *interface* may be made to provide more information, about the system and world (within workload limitations). It may be made more consistent, reducing both workload and unpredictability. Providing more capacity to understand the interface, system, and world, (perhaps through reduced workload or additional manpower) will also reduce unpredictability.
The *system* might be made more stable. The speed with which changes propagate may be decreased. Increasing the ease with which system behaviors are effected will also reduce workload.	The *system* itself may be made more stable, making it less likely to slip unpredictably into a different behavior.
The *world* can occasionally be changed so as to reduce workload, generally by some means of reducing the number and range of possible, intrusions on the world/system interaction.	The *world* may occasionally be controlled so as to reduce unpredictability as well. Additionally, experimentation and study of the world will improve our understanding of it and, if properly conveyed to operators via training, will reduce unpredictability for them.

3.4.1.1 Human operator selection, motivation, and fatigue

Sometimes, one of the easiest ways to increase the workload capacity of a human + machine system is by ensuring that the humans involved with the system have the right capabilities and motivation for performing the job. This is readily apparent when the capacities involved are physical: operators must not be visually impaired if the task requires visual acuity. This is to say that a person with reduced visual workload capability (e.g., a person with poor vision who must focus longer and harder on a visual scene to obtain weak signals from it) must use more of his or her 'visual workload' capability than one whose vision is unimpaired. Attentional and cognitive abilities are also important for many tasks, and, generally speaking, the more capacity workers have for these activities, the more cognitive workload they will be able to absorb and, therefore, the more complexity (of at least some sorts) they will be able to tolerate.

The effects of motivation on human capabilities are not well understood, but the fact that motivated operators can accomplish more than unmotivated ones is well documented – although whether this is the result of freeing up workload capacity or of willingness to devote more workload to a task is unclear and, perhaps for our purposes, irrelevant. Within fairly narrow bounds, motivated workers will behave as if they have more workload capacity and will therefore have some capability to absorb the additional workload imposed by added complexity (and/or of using that additional capacity to obtain information that reduces unpredictability). Care must be taken, however, not to be too quick to blame failure (even 'human error') on lack of motivation, since there is a wealth of other possible causes.

Finally, fatigue acts, in some regards, like the opposite of motivation. Fatigued operators behave in many ways as if they have less workload capacity than rested ones do, hence they will be comparatively less capable of dealing with increased workload situations.

3.4.1.2 Operator training

Operator training has the capacity to reduce unpredictability in some application domains. Insofar as human operators fail to understand important aspects of the system or the world, these aspects will be mysterious and unpredictable to them. Of course, operators can only be trained in what is understood by others in the first place; if there are poorly understood principles of the system or world, or simply unexpected aspects of either, then these cannot be trained. Training, of course, only provides benefit to the degree that it is remembered and applicable.

The type of training provided is important in reducing unpredictability as well. Broadly speaking, there are two types of training: rote learning and derivational or 'first-principles' learning. Rote learning is typically stimulus-response pattern learning: operators are trained to execute a specific type of response when a certain event or stimulus is recognized. Many aircraft procedures are of this type: engine fire responses, missile evasion, and so on. Rote learning is fast (and therefore produces some workload reduction, since it requires little cognitive processing or lengthy decision-making activities) but of limited applicability, since it is rare that all possible responses can be anticipated. Rote learning has little impact on unpredictability of the world or system to the human operator, but it can improve the predictability of the operator to other operators in a team setting. By contrast, derivational learning provides deeper understanding of the system and the world. It may require more workload during execution, but it will place operators in a better position to explain and anticipate unusual system or world behaviors (that is, it will reduce unpredictability) and therefore to understand and react to unexpected situations.

3.4.1.3 Distributed operations – use of additional operators

A final method of increasing the workload tolerance of the human(s) involved in a work domain is by adding to their number – that is, by apportioning the overall task among multiple operators. Since each operator has an individual workload capacity, the overall workload capacity of the human + machine system is increased by the addition of operators. This, however, is not a linear process. Since the work domain will generally require integrated activity and knowledge, apportioning the task among multiple operators means that each operator incurs an added burden of communication and coordination that is not present when the task is done by a single operator. These communication and coordination tasks also incur workload, and thus they eat into the gains obtained by increasing workload capacity. Worse, the distribution of knowledge about the system and world and their current states means that additional unpredictability begins to creep into the human + machine system with each added operator.

3.4.1.4 Improved interface design

Much of traditional human factors and ergonomics concerns itself with how to design interfaces that improve information flow from system to user and/or to reduce the workload users must expend in understanding or controlling the system. Improvements in either of these dimensions will also address the added human factors concern of reducing human error in interaction with a system. Although it is beyond the scope of my discussion here to provide detailed strategies for improving interface design, I will provide some thoughts on how good interface design mitigates the effects of complexity.

At its simplest, good interface design attempts to improve the information input and output (I/O) capabilities that the user has access to. This can reduce workload by making it easier to access information or controls or to identify important system states and control actions. Much good display design takes this tack by attempting to transform high-workload cognitive tasks (such as mentally computing a critical fuel level and comparing a current digital fuel value to that computed level) into lower workload visual ones (comparing a graphical fuel remaining indicator on a dial or graph to a previously computed graphical 'bug'). More simply, poorly placed, shaped, or designed controls can require excessive force to actuate, resulting in more rapid fatigue (as well as increased likelihood of errors due to either ineffective, inaccurate control actions or to a hesitancy to interact with the difficult control system). To the degree that a system interface has been 'optimized' to minimize human workload, the operators may have 'spare' workload capacity available for dealing with workload increases that stem from added complexity in the system or world.

Similarly, good interface design reduces unpredictability. On the one hand, reduced workload may provide more time or capacity to access and process additional information or control actions to understand and maintain the state of the system. Beyond this, good interface design provides the information needed to maintain awareness and control of the system. When this information is missing, obscured, or in an inefficient format, the operator may be less aware of the system's behavior – making it more unpredictable to him or her. Note that although the addition of information or control capability to an interface has the potential to reduce unpredictability, there is a limit to this effect as well. At some point, the amount of information becomes too great to monitor, and 'information overload' results – the condition of missing important conditions due to a surfeit, rather than a dearth, of information. Again, good interface design can mitigate the effects of information overload, but only to a certain point, not indefinitely. Of course, 'good' interface design implies that the information added is useful and of high payoff; simply adding information of little or no relevance provides no guarantee of unpredictability reduction.

There is a final way in which good interface design can serve to reduce both workload and unpredictability: through improving interface consistency – both within itself and with the user's 'conceptual model' of the system. Consistency reduces workload by making it easier for the operator to learn and use the interface without having to search extensively for needed information or controls. Consistency with the user's mental model reduces workload and training requirements still further and makes understanding of system behavior easier – provided that the user's mental model is accurate with regard to the system and the world. When this is not the case, the unpredictability that can result from the mismatch may be disastrous. Riley (Chapter 4) has more to say on the relationship between mental models, training, metaphors, workload, and complexity.

3.4.1.5 Intelligent interfaces and interface automation

Another way to effect workload and unpredictability reductions via the interface is by incorporating intelligence and 'automation' into the interface itself. This is approximately equivalent to distributing the task over multiple human agents (see 3.4.1.3 above). By removing the task of monitoring and controlling some aspect of the system or world from one human agent, we at least reduce workload for that agent. We generally do this only at the cost of increasing unpredictability, however, since detailed knowledge of some portion of the system has now been removed from that human operator and given to another agent. The difference here is that the agent is automated and resident in the interface itself. Alarms are a very trivial example of interface intelligence; they monitor one or more variables and report when their states reach some predefined criteria. In the absence of such alarms, the human operator would have to monitor the variables continuously to detect when they reached the stated criteria – at the cost of much more workload, but with less unpredictability. Problems with alarm floods in many current 'automated' control rooms point to problems with intelligent interfaces. First, it is extremely difficult to make them intelligent enough to truly provide for human information needs without additional human input or monitoring costs, and second, the loss of awareness of contextual information that occurs when continuous monitoring is reduced can frequently be a critical contributor to misinterpreted situations, ignored alarms, and erroneous decisions. At best, the operator must expend workload at the time of the alarm (when time and workload capacity are in shorter supply) to situate him/herself to the alarm and its context.

3.4.2 *Reducing demands through complexity reduction in system and world*

In the previous section, the strategies affected either the human or the interface to enhance human abilities to deal with existing complexity in the system and world. By contrast, each of the strategies in this section actually affect either the system or the world by reducing the perceived, significantly different components, relationships, or behaviors with which a human user has to cope. They reduce the problem of the workload and unpredictability impacts of complexity to human operators by reducing the complexity that produces those effects – at least from the perspective of the human operator. These strategies break down along the lines of the types of complexity described earlier in this chapter.

3.4.2.1 Reduced component complexity

One way to reduce complexity is by simply reducing the number or diversity of components in the system or in the world itself (for example, reducing the number of aircraft an air traffic controller must manage or the number of control loops an oil refinery board operator must monitor). Similarly, if a technology can be produced that uses fewer components to accomplish an equivalent goal (e.g., jet engines versus propeller engines), it will effect complexity reduction – at least on the component complexity front. Although, at first glance, it might appear unlikely that we could affect the number of components in the world, in fact, this is a common approach to complexity management – establishing sterile conditions in operating theaters and micromanufacturing environments is one modern example. Keep in mind that it is not the absolute number of components that is important here, so much as the number of components that are relevant to the humans in the operational environment. I will say more on this later.

3.4.2.2 Reduced relational complexity

Similarly, complexity can be reduced by eliminating some of the number or types of links that can occur between components. This is one of the ways in which air traffic control has been made a manageable task: restricting the flow of traffic into and out of given zones ensures that the number of possible intersections of aircraft vectors is comparatively small at any given time. Similarly, the use of inert gases and nonreactive materials in many chemical processes helps reduce the number of possible interactions between components by ensuring that at least some aspects of the processing environment will not participate in the reaction.

Obviously, reducing the number of components has the side benefit of reducing the number of possible relationships between those components. Again, note that the absolute number of relational links in the system or world need not be reduced; only those of interest and relevant to the operational environment need be minimized.

Two important methods of reducing relational complexity will be discussed in a bit more detail here; these correspond to physical and temporal decoupling. Physical decoupling is dealt with extensively by Perrow (1984) and essentially refers to expanding the space between components and/or to providing buffers which ensure that they will not affect each other. Increased separation of airplanes in the airspace is an example of physical decoupling, as is the use of spillways and containment barriers in industrial processing. By extension, providing additional processor capability or memory registers might be an example of 'physical' buffering for computer processing.

Temporal decoupling refers to spreading complexity over time. This is easy to see in the example of spreading a fixed volume of air traffic over time, because the result is fewer components per unit time. Other forms of temporal decoupling work similarly. The insertion of large vessels (with spare capacity) in a refining operation serves as a form of buffer that literally buys time for different operations in the process – instead of problems propagating from one unit to the next immediately, product from a problematic unit can be held in the vessel for a period of time to permit downsteam units to prepare for it. Since workload pressures increase as time to perform tasks is shortened, providing additional time means reducing the likelihood of exceeding workload capabilities. It also means providing more time for the situation to be inspected and understood, thereby reducing unpredictability.

In a different sense, many forms of automation and procedure creation may also be seen as examples of temporal decoupling. In essence, these techniques distribute complexity over time by removing some portion of the problem recognition, detection, and/or decision-making and action processes from the hands of the operator at the time of operation, and placing it instead in the hands of designers, supervisors, or maintainers who handle that complexity at some other time. When this is done successfully, the effect is to simplify the tasks of operators, generally at the expense of making the tasks of the maintainers or designers more complex. For example, choke controls and automatic transmissions make the task of driving a car substantially easier for drivers, but only at the cost of making automobiles themselves much more complex – although this complexity is only thoroughly understood and managed by skilled mechanics and not by the general population of drivers. This implies, however, that reducing perceived complexity for operators can be accomplished by hiding aspects of the actual complexity of the system from them. This hiding, of course, carries risks of misunderstanding or failing to notice complex distinctions that may prove important.

3.4.2.3 Reduced behavioral complexity

As with the other forms of complexity, behavioral complexity may be reduced either by eliminating some possible behaviors from the system or the world or by redefining the task of the operators so that some possible behaviors are no longer significantly different for them. Anything that serves to make the system more stable or to control the range of possible behaviors accessible from any current state will accomplish the former. For example, fly-by-wire controllers in inherently unstable aircraft serve to increase the stability of the aircraft orientation in response to slight changes in the world or to control inputs. 'Fail-safe' control designs in industrial plants are one way of creating a predictable failure state for many types of equipment, thereby reducing unpredictability and behavioral complexity. By reversing an example cited earlier, concerns about atmospheric flaring in oil refineries, we can illustrate how reducing the number of significantly different states operators must worry about reduces the perceived behavioral complexity of the system to them – if operators did not have to worry about the different types of gas they flare to the environment, their jobs would be substantially simpler.

Although it may be difficult to conceive how we can reduce the behavioral complexity of the world in which we operate, in fact, we do this all the time. This takes the form of reducing the likelihood of random and unwanted intrusions on the controlled process. Restricted airspace, barrier fences, controlled access to facilities, and even rodent screens used to keep animals out of sensitive equipment at oil refineries and airports are all examples of this complexity-reduction strategy. Locating a temperature-sensitive system in a temperature-invariant part of the world is a similar attempt at reducing this form of complexity and unpredictability, but the weakness of the approach is exemplified by the Space Shuttle *Challenger* disaster.

3.4.3 *General trends in complexity reduction*

This subsection contains some general conclusions about the set of complexity reduction and mitigation strategies described above.

3.4.3.1 Perceived complexity is the key to the human operator

Although absolute complexity of a system, if and when we can define and measure it, will undoubtedly have significance for overall system design, implementation, and operation, it is the perceived complexity of the system to the individual operator that increases workload and unpredictability for that operator. Hence, reducing the complexity for the operator can have payoffs in improved performance (at least locally). Furthermore, given the relational nature of complexity (that a system must be difficult for someone to deal with if it is to be called 'complex'), it is unclear whether a system that is perceived to be simple can be said to be complex in any meaningful way, except via a shift in perspective – which is tantamount to a shift in those components, relations, and behaviors that are of interest.

Of course, 'perceived simplicity' is ultimately no substitute for accurate understanding. If accomplishing desired ends with the actual components, relationships, and behaviors of the system demands a certain level of understanding of those entities, then hiding that complexity from human operators is ultimately counterproductive and dangerous. Nevertheless, it is sometimes equally unreasonable to ask operators to understand and manage that complexity if it violates their capacities. In this sense, perceived complexity should be regarded as a constraint on human-machine system design. When such a system violates human capacity, it must be reworked – using any of the human, interface, system, or world complexity-mitigation strategies discussed above.

3.4.3.2 Perils and promise of shifting complexity

Shifting complexity in time, to automation or to other human operators, has a long and honorable tradition. In fact, it can be said to be the defining characteristic of civilization – shifting labor requirements to specialist classes and/or the performing of some types of labor during some periods so that other, more time-critical types can be performed when they are most necessary. An agricultural society is almost undeniably more 'complex' than a hunter-gatherer one.

But such shifts only work in limited circumstances. First, there must be a genuine offloading of workload and unpredictability. If communication, coordination, memory loading, time to access needed information or controls, and the like, cause greater workload and/or unpredictability in the 'shifted' condition, there will be no perceived complexity reduction and no payoff in performance, safety, and so on.

Second, and particularly with regard to the use of automation and additional human agents, the added agents must be capable of performing their tasks *correctly* (which is not the same as *reliably*) under all circumstances. The inclusion of another agent in the human + machine system inevitably increases absolute overall system complexity because it necessarily adds components, relations, and behaviors that were not there before. If this new agent can successfully perform some portion of the task, then perceived complexity is shifted away from the human operator. Otherwise, as has frequently been the case with aircraft automation over the past 20 years (Billings, 1997), the operator is placed in the undesirable position of having to monitor, control, and understand both the system as it previously existed and the new agents; thus complexity increases for the operator.

Bright (1958) proposed, on the basis of a series of studies with automation available to him at the time, an inverted U-shaped curve for the effects of automation on such human operator requirements as skills, education, physical and mental effort, responsibility, decision making, and so on. A representative curve is illustrated in Figure 3.3. When confronted with 'lower levels' of automation (characteristic of early, partial, error-prone automation), requirements on many of these dimensions increase, but as the automation progresses toward higher levels on Bright's scale of competency (characterized by environmental sensitivity, action correction, anticipation of requirements, and self-correction), requirements on these dimensions decrease not just relative to initial increases, but also relative to prior requirements before the automation's introduction. Although Bright tends to have an overly rosy notion of the perfectability of automation, especially in light of history since his time, his point remains valid: when a portion of the task can be *correctly* performed by another agent or at another time, the complexity of the task is effectively reduced for the operator; otherwise, the perceived complexity increases, since new components, relations, and behaviors have been added to the overall system.

3.4.3.3 Be careful what gets done with spare capacity

Human factors, as a discipline, has recently become aware of the illusoriness of spare capacity. Not only do employers tend to reduce the workforce or set higher production goals to absorb whatever spare capacity might exist, but there is also evidence to suggest that humans demand a relatively constant level of stimulation and, if their jobs are made too safe, stable, predictable, or easy, may act to make them more 'interesting'. For example, Tenner (1997) details several reasons why increases in office automation have not resulted in uniform productivity increases. Although some of the reasons are technology-driven (the never ending need to upgrade and therefore retrain), others are human-driven (increased expectations about the 'polish' of documents, increased access to games and the Internet, etc.). Similarly, Rassmussen (1994) reports studies showing that the introduction of antilocking automobile brakes has resulted in faster and more reckless driving in several European countries, yielding a number of accidents approximately equal to what it was before the 'safer' braking systems were introduced.

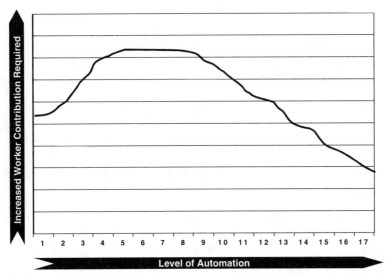

Figure 3.3 Hypothetical relationship between automation competency and operator workload (after Bright, 1958)

The notion of spare human capacity as a buffer against complexity must be carefully guarded if it is to fulfil its role. This includes not only guarding against management's cost-cutting and efficiency drives, but also against the human's own tendencies toward seeking outside stimulation and thereby losing vigilance.

3.4.4 An example of intraoperator complexity management

Sections 3.4.1 and 3.4.2 argued that perceived complexity could be reduced (or 'managed') by either (1) freeing up human resources through better interface design or adding more intelligent agents, or (2) by reducing the complexity of the system or world – which necessarily means reducing the number of entities (components, relations, or behaviors) with which a single human operator must interact. Several hypothetical examples were provided in section 3.4.2 and Table 3.2 to illustrate how this latter form of complexity reduction can be achieved.

Human strategies under complex work conditions can be seen as showing that humans regularly manage their own complexity by adjusting their attitudes about the set of components, relations, and behaviors that are important to them. In periods of low workload, they make finer distinctions, striving to optimize the job or perhaps even inventing games around their normal work variables. This has the effect of increasing complexity to some minimal level required to hold their interest (cf. section 3.4.3.3). By contrast, in periods of high workload, they will slough off some of the distinctions they might otherwise make in an effort to reduce the complexity with which they must deal.

A fine example of this type of behavior comes from Vicente (1999), reporting on work by Sperandio (1978). Vicente reports these data as evidence of alternative strategies used by air traffic controllers (ATCs) to perform their jobs. Using the model of perceived complexity outlined above and its implications for complexity-reduction approaches, however, we can go a step further and explain why these strategies are chosen. Here's the example:

'[Sperandio] found that, for a given situation and a given controller, certain operating procedures are more economic (i.e., require less of a cognitive burden) than others. Moreover, as workload increased (e.g., the number of aircraft to be controlled went up), ATCs would spontaneously relax the performance criteria they were using to perform the task. By doing so, they could adopt a qualitatively different strategy that accomplished the task goals in a less effortful fashion.

'... At one particular airport, Sperandio observed the following set of strategies:

1. S_a – When there were about 1 to 3 airplanes to control, the ATCs would 'nurse' each plane. Because the task demands were relatively low, they would calculate the optimal flight path for each individual plane, based on a number of variables (e.g., speed, course, altitude, and type of aircraft).

2. S_b – When there were about 4 to 6 airplanes to control, the ATCs would adopt a less sophisticated, and thus more economical, strategy. Instead of calculating an optimal path for each airplane, they would adopt uniform speeds and stereotypical flight paths. By relaxing their performance criteria and adopting more economic strategies, ATCs were able to control a larger number of planes in a manageable fashion.

3. S_c – When there were more than 6 airplanes to control, the ATCs would create waiting 'buffers' that consisted of streams of aircraft. When they were ready, the ATCs would then bring an aircraft off of the buffer and towards the runway at a generally uniform speed and descent path. Because the load is so great with this many aircraft, the performance criteria are once again relaxed so that the only primary concern is safety rather than efficiency' (Vicente, 1999, pp. 217–218).

It seems clear that the controllers in this study were systematically reducing the complexity they faced by altering the set of entities they had to deal with. Since they had no control over the number of components of interest (in fact, the number of airplanes was increasing), they coped with the problem by reducing the number of relationships and behaviors they concerned themselves with. In strategy S_b, controllers reduced the number of possible behaviors for each aircraft by neither permitting nor offering them optimal flight paths, but instead assigning them to prespecified and common behaviors.

Strategy S_c represents a further reduction, this time in relational complexity. By requiring that each aircraft maintain a position in a 'waiting buffer', the controller greatly reduces the number of relationships that the aircraft can be in with regard to each other. In strategy S_b, aircraft were assigned fixed speeds and paths, but due to variability in where and at what speed they entered these paths, the controller still had to worry about maintaining separations and safe altitudes, and so on. In strategy S_c, most of these relational concerns are done away with by assigning the aircraft a position in a holding pattern that is known to be safe and then admitting one aircraft at a time into the landing pattern.

This example provides a simple demonstration of the explanatory power of the model of perceived complexity laid out previously. Moreover, it suggests that perhaps interfaces and training programs should be designed to permit a great degree of flexibility in allowing operators to make and collapse distinctions between components, relations, and behaviors, and so as to encourage safe 'complexity-collapsing' strategies in high-workload periods and productive 'complexity-enhancing' strategies in slow periods.

3.5 New Approaches to Managing Perceived Complexity

This section briefly describes two recent research thrusts we have undertaken to manage the human impacts of complexity.

3.5.1 Shared models as a means of managing complexity

A major source of unpredictability in current interfaces and automated systems (as well as, arguably, in human team coordination) stems from the mismatches between the intentions and understanding of the agents participating in the endeavor. The Airbus A320 stick controls are a potential example. Here both the pilot and co-pilot may input stick commands, but when these are done simultaneously, the aircraft control system averages them. This makes the following scenario possible. In an emergency situation, close to an unexpected obstacle, both pilot and co-pilot grab their sticks. The pilot inputs commands for maneuvering right around the obstacle, while the co-pilot commands maneuvering to the left. The aircraft control system, unable to know why either command was input, simply takes the average of the two inputs and crashes into the obstacle. Although this design has been criticized extensively, it is worth noting that no such accident is known to have occurred. Similar, but more realistic, anecdotal information is emerging about conflicts between the suite of automation and decision aids on modern commercial aircraft. For example, in one reported incident, the Ground Proximity Warning System (GPWS) told pilots to 'climb, climb' to avoid approaching terrain, while the Traffic Collision Avoidance System (TCAS) told them to 'descend, descend' to avoid oncoming traffic. In the absence of an understanding of the global situation and the pilot's goals and intents, each system simply does what it has been built to do: present corrective actions to avoid the types of problems it can detect. So-called 'mode errors' are an even more common and well documented example of pilot/automation mismatch (Billings, 1997). Here complex interactions of automation modes and world states produce unexpected and sometimes dangerous 'automation surprises' – automation behaviors that are unexpected by the pilot.

In a series of projects at the Honeywell Technology Center, we have been striving to provide a solution to this mismatch between understanding and intentionality for the system and the human operator(s). Our approach is to provide all agents (human and automation) with a shared knowledge of each other's intentions and a common view of the world. I refer to this 'shared view' as a *knowledge backplane* because it serves to coordinate and distribute knowledge (as opposed to simple data or information) across the agents in the endeavor. The knowledge backplane is used to achieve increased coordination (and hence less unpredictability) among users and automation.

The integration desired here is no longer integration in the physical, spatial, or even strictly functional sense that characterized the physical cockpit integration efforts of the 1960s and 1970s. Rather, it is integration at a 'knowledge' layer. The goal is to take what are currently independent agents and have them behave in a coordinated fashion. Since the agents being integrated are increasingly 'smart' and goal-oriented, it's not surprising that integrating them means getting them to share their knowledge and coordinate their goals. From the crew's perspective, it must be as if the GPWS system 'knows' that it is inappropriate to climb since there is another aircraft just above, and therefore it refrains from sounding its standard alarm and instead (combining input with TCAS), tells the pilot to veer right. There is a step beyond 'mere' intrasystem integration required here too. For systems to truly behave in an appropriate fashion, they must be knowledge-integrated not merely with each other, but also with the pilots and the pilots' knowledge of the world and their intentions or goals. A knowledge backplane enables this form of integration of world knowledge and intentionality because it enables individual systems to 'plug in' to it, and both get the knowledge they need for intelligent, coordinated functioning, and deposit the knowledge they have that others need.

What would a knowledge backplane look like? Task models (e.g., Plan-Goal Graphs, Task Networks, Goals/Operators/Methods/Selectors (GOMS), Operator Function Models) are one example; they are task-centered but describe the world states under which tasks are necessary or viable and the world states that are expected to result from executing a task. A recent HTC project, conducted in an oil refinery domain, used a system-based model called the Plant Reference Model to perform a similar function. We suspect that any instantiation of Rassmussen et al.'s (1994) Abstraction Hierarchy implemented as an active description of states combined with causal and functional knowledge about component interactions could serve the same purpose. Below are some proposed defining characteristics for a knowledge backplane:

1. It is a model whose representation *integrates* world states, system states, plans, and goals.

2. It has both static and dynamic variations:
- Its static variation describes all possible states that will be detectable and discriminable and therefore sharable.
  — It has both breadth and depth.
  — It has to be at least as broad as the set of 'contexts' over which you want to integrate (a 'context' is any unique, valid combination of world + system + intent + goal states).
  — Increased depth will increase precision of integration (making it increasingly appropriate to context) but will be harder to build and probably harder to use to discriminate individual states.
- Knowing a state at any level enables links to 'appropriate' states at other levels – although perhaps not unique links. That is, knowing where you are in the model allows you to envision where you can and ought to or might want to go next. Thus the model provides a sort of predictive capability.
- A certain set of states may be flagged and combined (again, statically) as the expected states for some portion of the future (e.g., the mission plan).
- The backplane's dynamic variation maintains a 'current' context (and possible near-term transition contexts) on an ongoing basis throughout use. This is accomplished through some mixture of human input and interpretation of sensed data. At the high end, this is 'intent inferencing'; at the low end, it requires the operator to report his/her intentions.

3. Any specific context has to be interpretable by all systems (akin to a football team's playbook – each player knows what he is supposed to do when a given play is activated, and coordination is thus an emergent function stemming from all agents, *including the human* sharing the same playbook).

4. Individual automation systems can maintain their own notion of how to behave given a 'context' on the backplane (information requirements, display configurations, radio modes, planning responses, etc.).
5. Automation systems can be tasked (either by humans or by higher order automation systems) on the basis of known contexts.
6. Operators can view (portions of or summaries of) the backplane for an understanding of the whole system's understanding of current context.
7. It has to have an explicit existence – in software, in a shared databus, in the system context (e.g., on board the aircraft).

For more information on the utility of a knowledge backplane, see Miller and Levi (1994). For information on how it has been used to support information management in rotorcraft, see Miller (1999).

3.5.2 Tasking interfaces as a way around the triangle

A second concept we have recently begun exploring for mitigating the effects of workload and unpredictability is *tasking interfaces* or *adaptable autonomy*. Tasking interfaces use a task model (as described above) and an intelligent planner to allow human operators to specify the tasks to be accomplished by automation at a variety of levels. When tasks are specified at a very low level of detail, the planning system stays out of the loop (or perhaps reviews the user-created plan for possible conflicts or errors). Alternatively, when only high-level goals or partial plans are provided by the operator, the planner takes over and fleshes out the plan to an operational level in keeping with the instructions provided by the user. Although this project is only in its initial stages, more information on the concept and initial implementation can be found in Miller and Goldman (1997).

A tasking interface of this sort does not eliminate the dilemma presented in Figure 3.1, but it mitigates it by allowing operators to flexibly choose various combinations of workload and unpredictability for interaction with their automation, depending on their current situation and available resources. This situation is illustrated in Figure 3.4. Presumably, when the situation is quiet and there is sufficient time available, and/or when automation is not deemed trustworthy, the operator can use the tasking interface to develop detailed plans for the system to execute. This constitutes accepting higher workload in exchange for reduced unpredictability. Alternatively, when workload is high and/or when the automation is trustworthy, the operator might be willing to give it more authority to create its own plans – thereby trading increased unpredictability for decreased workload.

3.6 Conclusion

This chapter has provided the beginnings of a theory of *perceived complexity*. We began by examining, and attempting to account for, how the label 'complex' gets applied to systems and situations. We saw that 'complexity', at least as it is used in normal language, is inherently a perceived quality; a thing may be complex to one person yet simple to another. Very generally speaking, we tend to label as 'complex' those things we have difficulty understanding or dealing with – and dictionary definitions tend to support this.

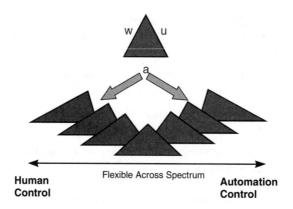

Figure 3.4 Flexible interactions provided by a tasking interface

We found it useful to decompose perceived complexity into three subcomponents:

- *Component complexity:* the simple number of components 'of interest' a thing has and the diversity of those components;

- *Relational complexity:* the number and diversity of links between components, where a link is the ability of one component to connect with another in some manner 'of interest';

- *Behavioral complexity:* the number and diversity of distinct behaviors 'of interest' that the system can exhibit.

Increases in any of these types of complexity will result in an increase in complexity to a human perceiver, although they are listed in rough order of increasing magnitude.

The inclusion of the 'of interest' phrase in each of the above definitions points back to the perceived quality of complexity. What counts as a component, link, or behavior involves a segmenting of the world – something that can be done in infinite ways, depending on the observer's interests. This qualification accounts for the variable perception of complexity among observers with different interests or viewpoints.

Complexity increases are generally a by-product of the attempt to increase the 'adaptiveness' or coverage and sensitivity of accurate system behaviors, yet increases in adaptiveness ultimately produce human impacts in some mixture of two ways: increased workload and unpredictability. It is also possible to increase unpredictability and workload for a human interacting with a system without producing any increase in adaptiveness of the human-machine system. This situation might also be referred to as a 'complexity increase' by those experiencing it, even though the system itself has not changed at all.

Perceived complexity increases to the immediate users of the system can be mitigated through a variety of strategies, which can be composed into two broad types. First, some strategies improve the efficiency of the human operator or the interface to 'free up' resources for dealing with the effects of complexity. Although these have no direct effect on the complexity in the system, they make it easier for the operator to tolerate higher levels of complexity. Second, some strategies reduce the complexity of the system or the world or distribute that complexity away from the human operator. A combination of strategies can be used to explain how air traffic controllers manage their own perceived complexity – by effectively shifting the set of components, relations, and behaviors in their perceived work domain to maintain a roughly constant level of perceived complexity.

References

Billings, C. (1997) *Aviation automation: the search for a human-centered approach.* Lawrence Erlbaum Associates, Mahwah, NJ.
Bright, J.R. (1958) Does automation raise skill requirements? *Harvard Business Review,* **36**, July–August, 85–98.
Miller, C. (1999) Bridging the information transfer gap: measuring goodness of information 'fit'. *Journal of Visual Languages and Computing,* **10**(5), 523–558.
Miller, C. and Goldman, R. (1997) 'Tasking' interfaces: associates that know who's the boss. *Proceedings of the 4th USAF/RAF/GAF Conference on Human/Electronic Crewmembers,* Kreuth, Germany.
Miller, C. and Levi, K. (1994) Linked learning for knowledge acquisition: a pilot's associate case study. *Knowledge Acquisition,* **6**, 93–114.
Opperman, R. (1994) *Adaptive user support.* Lawrence Erlbaum Associates, Hillsdale.
Perrow, C. (1984) *Normal accidents: living with high-risk technologies.* Basic Books, New York.
Rassmussen, J. (1994) Risk management: adaptation and design for safety, in *Future risks and risk management* (eds. B. Brehmer and N. Sahlin). Kluwer, Boston.
Rassmusen, J., Pejtersen, A. and Goodstein, L. (1994) *Cognitive systems engineering.* Wiley, New York.
Riley, V., DeMers, B., Misiak, C. and Schmalz, B. (1998) A pilot-centered autoflight system concept. *Proceedings of the 17th Digital Avionics Systems Conference,* Bellevue, WA.
Sperandio, J. (1978) The regulation of working methods as a function of work-load among air traffic controllers. *Ergonomics,* **21**, 195–202.
Webster's encyclopedic unabridged dictionary of the English language (1989) Gramercy Books, New York.
Tenner, E. (1997) *Why things bite back.* Vintage Books, New York.
Vicente, K. (1999) *Cognitive work analysis: toward safe, productive and healthy computer-based work.* Lawrence Erlbaum, New York.

4

Perceived Complexity and Mental Models in Human-Computer Interaction

Victor Riley
Honeywell Technology Center

4.1 Introduction

Many in the aviation and industrial control communities are calling for systems to be less complex and more usable. Unfortunately, the operational environments of these systems necessitate a certain minimal level of functional complexity. The systems themselves must be complex because the environment is irreducibly complex. Fortunately, though, there are strategies that can allow designers to reduce the perceived complexity of systems while maintaining the required level of functional complexity. The personal computer (PC) industry, having grappled with these problems over the past two decades, offers some very useful lessons in this area. In this chapter, we look at the gap between how the designer and the user think about systems, the use of metaphors to reduce the perceived complexity of the system, and the need to address usability problems at the level of system functionality rather than only at the interface. An extended discussion of flight management systems in commercial aircraft is also included

4.2 Mental Models and the Basis of Consistency

Although the PCs of today are far more complex than those of the early 1980s, users perceived the earlier PCs to be more complex. To use a PC at that time, people had to learn some rudiments of computer science and the obscure syntax of the operating system, usually DOS. For novice users, this presented significant obstacles to learning, and errors were common (such as reversing the source and destination drives in the 'copy' command, thereby deleting the file intended to be saved and saving the one intended to be deleted).

How is it, then, that thousands of people who know nothing about computer science can use a computer today that is much more complex, underneath, than those of two decades ago? The common answer to this question is the graphical user interface. Unfortunately, though, that answer is somewhat misleading, as demonstrated by the hundreds of hard-to-use applications that have graphical user interfaces.

The real answer to this question has to do with mental models (Norman, 1988). The user's mental model is his or her concept of how the system operates. Basically, the mental model can dictate perceived complexity on three levels:

1. If the functionality of the system cannot be represented within a single, unified, and internally consistent conceptual framework, the user will have to learn the unique attributes of each of the system's functions, and he or she will perceive the system as being very complex. In Miller's terms (see Chapter 3), the user's mental model must match the component complexity of the system.

2. If the system's functionality can be represented within a single, unified, and internally consistent conceptual framework, the user need only learn the high-level organizing principles of the framework. The user can then infer how individual functions operate based on these higher level principles. This reduces, but doesn't minimize, perceived complexity.

3. If the system's functionality can be represented within a single, unified conceptual framework that is *externally consistent with a domain the user already knows,* the user can apply knowledge he or she already has to infer the operation of system functions. If that representation is completely consistent with the user's existing body of knowledge, he or she can use the system with virtually no training and will perceive it as being minimally complex.

Notice that these three levels (shown in Figure 4.1) of perceived complexity and their corresponding principles of conceptual consistency do not imply differences in actual functional complexity. Theoretically, three systems can be equally complex at the functional level yet have different levels of perceived complexity. Indeed, functional complexity can be inversely related to perceived complexity (and, as a practical matter, often is). The real key to reducing perceived complexity, then, is in how the system's functionality is defined and represented to the user, rather than whether the interface is graphic or not. The graphic user interface merely improves the bandwidth of user-system communication (by enabling more information to be transferred between the user and the system in a given amount of time), enabling the use of a broader range of conceptual frameworks.

4.2.1 Some examples

Consider the design of a blender that I used to own. Its control panel consisted of a row of buttons, each with a label like 'blend', 'mix', 'liquefy', 'pulverize', and so on. If you want to use the blender to make something that isn't explicitly labeled, such as a milkshake, what setting do you use? Lacking an obviously appropriate setting, you might try to identify a setting that is at least close in blade speed to the one you have in mind. But the selection of labels doesn't establish a clear mental model of the relationship between the settings and the blade speed. Is the blade speed for 'mix' higher or lower than for 'grate?' Assuming that it is higher, and assuming that 'liquefy' is higher than 'pulverize', are the relationships across these settings linear? Is each higher selection proportionally higher than the one before, or is there a large jump in blade speed between two particular adjacent functions? Lacking this information, the user must learn the blade speeds associated with each individual setting on a setting-by-setting basis. Obviously, this would take much more time than if a higher level organizing principle governed the relationships between the settings so the user could infer which one would best suit his or her needs. An internally consistent mental model, then, would be supported by a blender that indicated revolutions per minute for the buttons, with recommended uses for each setting, and with the buttons arranged in a consistently ascending order with proportional jumps between each setting. Although the actual rpm measure may not be meaningful to most users, it would at least allow the user to infer that the step sizes are constant, and the presence of known applications as milestones might enable the user to estimate the desired setting better ('I want something between chopped and pureed, closer to pureed'). It would also eliminate one of the fundamental problems of the current design, which is that different settings have qualitatively different applications ('grate' is used for dry foods while 'whip' is for wet).

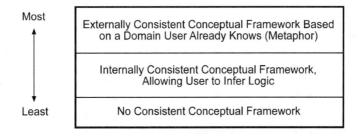

Figure 4.1 Three levels of mental model application. The lack of a consistent conceptual framework compels the user to learn and remember the unique operation of all the system's functions and components, so the user's mental model must be as complex as the component complexity of the system. Application of a consistent conceptual framework can reduce complexity by allowing the user to learn general principles and infer specific operations, but this still requires the user to learn the general principles. Application of a metaphor that builds on the user's existing knowledge eliminates the need to learn system-specific information, thereby minimizing or even eliminating training time.

Let's look at some additional examples to illustrate the notions of mental models, conceptual frameworks, and internal and external consistency.

The lack of such a consistent conceptual framework leads to a high level of perceived complexity. The definitions of the functions and their representations are essentially arbitrary from the user's perspective, and the user must learn and remember how the system works on a function-by-function basis. This is analogous to many users' experience with DOS; they had to memorize the available commands and their syntax on a command-by-command basis. Although DOS did, in fact, have a consistent syntax, many users actually used so few of the commands that the higher level syntax wasn't readily apparent to them. Consequently, their ability to use the system relied on their ability to memorize apparently arbitrary commands. As the functionality they needed grew, the number of new commands they had to learn simply added to the complexity of their mental model. Lacking a higher level conceptual framework, the user had no opportunity to make this process more efficient by learning the framework instead of the commands themselves. In the terms of Miller (see

Chapter 3), the user's mental model of the system was driven by the component complexity of the system. The user's knowledge of the system had to grow proportionately with the number of system components or, in this case, functions.

Perceived complexity can be reduced somewhat by providing an internally consistent higher level conceptual framework. The user still has to learn the framework, and the organization of the framework may still be arbitrary, but at least the user can transfer knowledge from one part of the system to another, reducing his or her learning and memory load. An example of this is the standardization of controls in cars. Facing the problem of people changing cars and finding the unfamiliar controls difficult to use, the automotive industry adopted a set of standards governing the placement and design of controls and indicators (Morcom, 1978). Although not all of the controls and indicators abide by the standards all the time, there is enough commonality among the most often used functions that most people can find their way around a new car with relative ease. Note, however, that even though the standard set provides consistency, it is still essentially arbitrary. There is no high-level organizing principle that dictates that the headlight brightness selector be a lever on the steering column. The type and placement of the control and the assignment of its function are not inherently natural, so the driver has to learn the standard. Once learned, however, the standard becomes a conceptual framework that the driver can apply to other cars. Most other cultural control and display conventions (such as up for on, clockwise for increase, and so forth) work this way. A design that does not abide by these standards will be perceptually more complex than one that does, even though the standards are only internally consistent, because the user has to learn and remember unique information about that design. Furthermore, violation of cultural conventions and user expectations can promote errors because users often fall back on their existing knowledge and habits.

The most powerful means of reducing perceived complexity is by making as much use as possible of external consistency. This is based on the principle that perceived complexity is largely a function of what people know; a subject or domain that one knows well will seem less complex than a subject that one doesn't. For example, a mathematician knows enough of his or her domain to see the high level organizing principles that establish meaningful relationships across the domain and therefore can draw useful inferences from higher level principles to lower level operations. Math may therefore seem less complex to that person than, say, chemistry, even though the two may be of roughly equal complexity. Reducing perceived complexity by relying on the user's existing body of knowledge takes advantage of this principle. Finding a body of knowledge that is common among all the likely users of the system enables the designer to apply a metaphor to the definition of system functions and their operational logic. Once again, we can look to the PC industry for an example of this.

4.3 Metaphors as a Means of Reducing Perceived Complexity

Contrary to popular belief, the graphical user interface does not make a computer easy to use because it is graphic. Many application developers have released software that they assumed would be easy to use because they used a graphical user interface, only to find that users had tremendous difficulty with it. The inability of a graphical user interface to make a system easy to use without the application of an appropriate metaphor was well demonstrated by

Abbott (1995), who tested one designed for a conventional flight management system. Although the underlying flight management functions were unchanged, the graphical user interface provided pop-up windows, menus, and other tools to make the system's underlying functions more accessible. After finding that the new interface made no significant difference in either training time or errors, Abbott concluded that simply changing the interface was not enough. We believe that the underlying functions themselves have to be recast into a conceptual world the user already understands. That's what the metaphor does.

The classic example of an appropriate metaphor is the desktop metaphor used in the PC user interface. Again, note that it is the metaphor, rather than the graphic nature of the interface, that makes the modern PC usable by so many novices. Virtually every potential computer user is already familiar with the world of paper documents, so using file folders as a metaphor for directories, documents as a metaphor for files, and garbage cans or recycling bins as a metaphor for the act of deleting a file merely takes advantage of the users' existing body of knowledge. To the extent that the on-screen world is consistent with the real world the user is already familiar with, someone can use the system with little or no training and can infer the underlying operational logic of the system based on the implications of the metaphor. This is perhaps the most powerful means of reducing perceived complexity. And again, note that perceived complexity is decoupled from the true underlying complexity of the system, as the Windows operating system is far more complex than DOS.

That said, there are two important points to make about the desktop metaphor and the graphical user interface. The graphical user interface is not an end in itself, but rather a means: it enables the application of the metaphor by increasing the communication bandwidth between the system and the user. Interestingly, the graphical user interface and the desktop metaphor were originally developed to overcome one of the most egregious by-products of the early DOS applications, the use of modes (Cringely, 1992). Modes are necessary when the functionality of the system exceeds the bandwidth of the interface (the information transfer needs between the user and the system exceed the information transfer rate of the interface). In the case of DOS word processing applications, for example, users had to learn special keystrokes to select between 'insert' and 'overstrike' modes. If the user made an error, another unexpected mode might be invoked leading to unpredictable results. Or, the user might type away thinking the system was in 'insert' mode but instead losing text because it was really in 'overstrike' mode. The graphical user interface enabled the required functionality to be expressed in another way, by distinguishing between text to be saved and text to be operated on (formatted, deleted, or whatever) through the use of highlighting. This is an example of increasing the bandwidth of communication.

We will further examine this topic later on; however, it brings up the other important point about the desktop metaphor, which is that the metaphor is far from complete. Highlighting to indicate text to be operated on is not a natural consequence of the desktop metaphor because there is no direct real-world analog (other than, perhaps, the highlighter pen, which doesn't change the text directly but merely indicates to an editor what portion is to be changed). As the desktop metaphor was applied to the operating system after the fact, the operating system functionality was more or less force-fit into the metaphor. Functions that the metaphor does not cover continue to pose usability challenges.

To get around this problem, PC designers have adopted a hierarchy of design strategies. The most powerful is to try to fit as much functionality as possible into a single, all-encompassing metaphor. For those functions that do not fit gracefully within the metaphor, a secondary metaphor might be adopted. For example, to distinguish between mutually exclusive selections and selections the user adds together, radio buttons and check boxes are used; the metaphor is that everyone knows you can only select one radio station at a time, so the radio button format communicates exclusivity, whereas there are no such natural restrictions on check marks, so they can be used for inclusive selections. Lacking any externally consistent metaphors, the next most powerful strategy is to adopt common user interface conventions that the user must learn but that apply consistently to all relevant applications and functions. The Windows standards of the three buttons that minimize and expand the window and close the application (the lower bar, full screen icon, and X box) are examples of these, as are pull-down menus. Note, however, that these conventions must comply with any relevant existing cultural conventions. Note also that this three-tiered approach to design strategies conforms to the hierarchy of externally consistent, internally consistent, and arbitrary conceptual frameworks illustrated in Figure 4.1.

4.4 Airplanes à la Mode

The PC industry managed to get around the problem of modes with the help of the graphical user interface. The graphic nature of the interface increased the bandwidth of the communication channel between user and system, enabling the application of metaphors and more powerful design conventions than would be available solely through text. However, modes remain a significant problem in another important domain, the commercial aircraft flight deck. Usability is an important concern here both because of the safety-critical nature of flight and because pilot training time represents a major expense to air carriers. Consequently, there has been much research on human factors issues in this environment.

Two aspects of the autoflight system that are the most difficult for pilots to learn and use are the flight management system (FMS) and the mode logic. The primary pilot interfaces to the autoflight system are through the autopilot controls, usually located on a panel just below the windscreen, and the Control Display Unit (CDU), a combination text display and alphanumeric keypad through which the pilot programs the FMS. A typical glareshield controller is shown in Figure 4.2 and a typical CDU in Figure 4.3. The FMS allows the pilot to enter a flight plan for a trip, consisting of geographic positions the aircraft is to fly to and instructions at each position regarding the altitude and speed the aircraft should fly at to get there. During the flight, the pilot enters strategic flight plan changes into the FMS CDU by editing the stored flight plan. But if an immediate or short-term trajectory change is required, the pilot accomplishes this by manipulating the autopilot controls on the glareshield controller to adjust heading, altitude, speed, and so forth. In this case, the modes tell the aircraft whether to fly to FMS targets or autopilot targets, and they can be applied independently to each axis of flight (thrust, altitude, and latitude). The pilot selects the modes from the glareshield controller and reads the active and armed modes on the Flight Mode Annunciator, a text display located at the top of the Primary Flight Display.

These systems, the FMS and the mode logic, illustrate the potential gaps between how the designer thinks about the system and how the user thinks about it, and demonstrate the need to attack usability problems at the level of functionality rather than at the interface. Sarter and Woods (1991) have found that it takes around 1200 hours of line operations experience before a pilot feels comfortable with the autoflight system, and even then there are significant gaps in their knowledge of how it operates. This represents about a year and a half of experience for a typical airline pilot.

Airplanes à la Mode 65

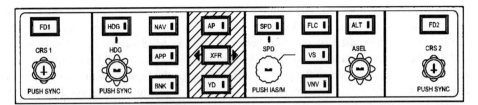

Figure 4.2 Typical glareshield controller showing autopilot controls for short-term flight path interventions. FD stands for Flight Director, a guidance cue for the pilot. HDG stands for Heading, and this region contains lateral modes, including NAV (following a complex lateral path) and APP (Approach). The middle region selects which autopilot is controlling the aircraft (AP). The Speed (SP) region controls speed modes, including some vertical modes (Flight Level Change and Vertical Speed), and the ASEL (Altitude Select) region allows the pilot to set the target altitude. This configuration is for a small business aircraft; large transport aircraft have many more selectors and options, but the essential functionality of selecting lateral, vertical, and speed targets and modes is common.

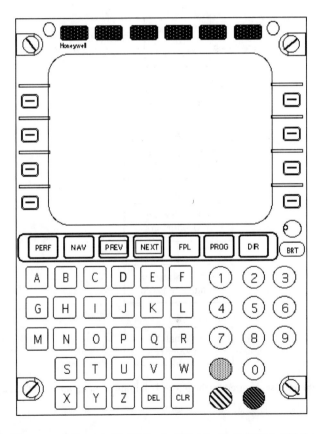

Figure 4.3 Flight Management System (FMS) Control Display Unit (CDU). This layout is also for a small business jet, but the overall configuration of screen, line select buttons around the side bezels, function keys below the display, and alphabetic and numeric keypads is common to all contemporary CDU designs.

A look at how the pilot interacts with the FMS itself explains part of this difficulty. (This process is shown in Figure 4.4.) When the pilot receives a clearance from the air traffic controller, he or she must first decide whether to use the FMS or the autopilot controls, based on whether the clearance requires a delayed response or an immediate response. If it requires a delayed response and the pilot decides to change the flight plan in the FMS, he or she must then decide which function the FMS provides that would best support the needs of the clearance. Because there is typically not a one-to-one match between clearance requirements and autoflight functions, this can require some effort and is often the topic of debate between pilots. The difficulty here is that the FMS represents the flight path in terms of waypoints (positions) and leg types (connections between positions), and the pilot must translate the clearance requirements into those terms. This means that a relatively straightforward operation such as intercepting an airway (a predefined route) may require figuring out the inbound course, entering the 'from' waypoint, defining the positions at which the aircraft will intercept and leave the airway, and deleting any now irrelevant waypoints.

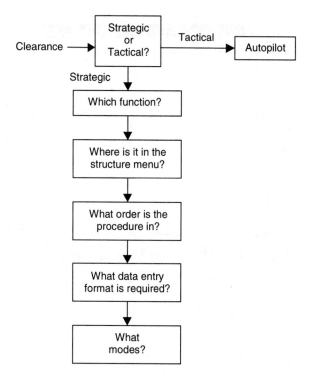

Figure 4.4 Thought process required by the pilot when implementing an air traffic control clearance

Once the pilot determines which functions are best suited to the clearance requirements, he or she must then find them in the page architecture. This requires selecting a menu page from a set of function keys on the faceplate, then drilling down through the menus using line select keys at the side of the display (see Figure 4.3). Because the interface doesn't allow presentation of all the options at once, many options must be hidden within menus, and the pilot must remember what path to follow to access the desired functions.

Once the pilot has located the appropriate functions, he or she must then invoke them in the proper order. This can also require some effort, because again there is no natural mapping between the contents of the clearance and the logical order of the procedure. For example, if the clearance is to 'hold on the 250-degree radial at twenty miles from Farmington', the pilot must first define a new waypoint name, place it in the proper position in the flight plan, define its position in relation to one or more known positions, and finally start the Hold procedure. The FMS's requirement to express every instruction in terms of known waypoints often causes the pilot to have to mentally turn around the clearance he or she heard and perform the steps contained in the clearance in reverse or some other order.

After successfully accessing the appropriate functions in the appropriate order, the pilot must then enter any data or parameters required by any of the functions. Again, this can be essentially arbitrary. For example, latitude and longitude positions may require different formats, with some systems requiring the pilot to enter the cardinal direction before the degrees, and others, the reverse. And the separator between the latitude and longitude parameters might be a slash, a period, or another symbol that the pilot must remember by rote.

Finally, the pilot must select the appropriate mode so the aircraft flies the instructions as intended. If the pilot makes a change to the flight plan and fails to select the appropriate mode, or selects that mode but it is unavailable because its preconditions haven't been satisfied, the aircraft won't follow the new instructions. Mode-selection errors have been implicated in a number of fatal accidents. For example, an Airbus A320 crashed in Strasbourg, France, in 1992 when the flight crew apparently attempted to select a 3.3-degree flight path angle approach to the airfield but erroneously entered a 3300-foot-per-minute vertical speed descent (Sarter and Woods, 1995).

Given the tremendous amount of functionality available to the pilot through the FMS, one can see that there is a correspondingly tremendous amount to learn about it. The pilot must know what all the functions are, how they work, how to access them, what the appropriate logical orders are, and what the data requirements are, and must remember this information for everything the system can do. Although this requires a significant amount of learning, it is only half the battle; the pilot must also learn how the mode logic works.

The most complex aspect of the mode logic has to do with vertical flight path control and mode transitions, in particular, the conditions under which transitions can legally occur. Vertical navigation is not only the most complex, it is also the most critical axis of control. Although aircraft are typically separated by miles laterally, they may be separated by only one or two thousand feet vertically, and the penalties for violating altitude assignments are more severe than for lateral constraints.

Sherry and Polson (1998) estimate that the mode logic of a current generation transport aircraft can be represented by between 1500 and 3000 situation-action rules, based on the aircraft's position in relation to the flight plan, speed envelope, optimum trajectories, and other factors. They also estimate that vertical guidance for a particular aircraft contains 289 such rules, and that many of these are not documented in any of the training materials or evident on the mode displays. Although many of these rules may be truly insignificant to the pilot, many of them are not. Let's look at a few examples that demonstrate the functional complexity of the mode logic in modern aircraft.

In the McDonnell Douglas MD-80, pilots discovered that if they entered in a new altitude and then adjusted the vertical speed to make the climb or descent transition smoother, the altitude target would be cleared (Wiener, 1988). This is a good example of the difference between how the pilot and the designer think about the task of changing altitude. The pilot sees no reason why adjusting the rate of climb or descent should cause the altitude target to disappear, but to the designer of the system, however, adjusting the vertical speed implies that the aircraft should change to vertical speed mode. In altitude change mode, the aircraft is climbing or descending to reach a specific altitude, but vertical speed mode simply instructs it to climb or descend at a particular rate until told otherwise. The absence of an altitude target in vertical speed mode was responsible for the deletion of the altitude target when the pilot adjusted the rate of climb or descent, a consequence that was responsible for many altitude busts when the aircraft was new to service, before pilots understood that they couldn't perform the task in that particular way.

In the Boeing 737-300, if the pilot attempts to abort the takeoff below 64 knots with the autothrottle engaged, the autothrottle will continue to try to provide takeoff thrust; the pilot must manually disengage the autothrottles. At and above 64 knots, the pilot can override the autothrottles (Sarter and Woods, 1991). Also, the rules governing how the pilot can disengage the approach mode (which allows the aircraft to land on autopilot) after localizer and glide slope capture (that is, capture of the lateral and vertical guidance signals down to the runway threshold) can be difficult to remember: the pilot must push the Take Off/Go Around buttons, turn off both flight directors and the autopilot, or retune the VHF radio. Reselecting the Approach mode button or attempting to select another vertical mode will not disengage the approach mode. Sarter and Woods found that only 15% of the pilots in their study could remember one of these methods, and that all the pilots thought they could disengage it in ways that actually would not work.

In all aircraft and in most modes, the pilot can disengage the autopilot by moving the control column or side stick; once the column or stick has been deflected a certain amount, the aircraft assumes that the pilot wants to take over manual control and disengages the autopilot. In the A300, however, this rule does not apply below 1500 feet above ground level, a factor that may have led to an accident in Nagoya, Japan. The crew in this accident apparently inadvertently selected the go-around mode, then attempted to continue the approach by manual control. Although the pilot was pushing the control column forward to reduce pitch and stay on the glideslope, the autopilot was attempting to increase pitch to go-around by adjusting trim. Eventually, the aircraft was so far out of trim that it crashed short of the runway (Sekigawa and Mecham, 1996). This is an excellent example of violating cultural conventions: the unique violation of a standard may have created a false expectation for the pilot, who may have expected the autopilot to disconnect when the control column was moved rather than having to manually disconnect it on the autopilot controls.

In the Airbus A320, the autothrottle will control thrust during a go-around maneuver if the maneuver is initiated above 100 feet above ground level. If it is initiated below that altitude, the autothrottle will disengage even if the pilot puts the throttle handle in the Take Off/Go Around position. If the pilot doesn't realize that the autothrottle is not engaged, the aircraft will speed up without bound, potentially overspeeding the flaps, because the throttle levers are in manual mode and set to full power. In a study by Sarter and Woods (1995), only one pilot out of 18 was able to accomplish a go-around below 100 feet without difficulty; all of the others found that the automation behaved differently than they expected. The pilot who successfully completed the maneuver simply turned off all the automation, hand flew the maneuver, and turned on the automated systems one by one after verifying that each one worked as he expected.

These examples all illustrate why FMS operations and mode logic take so long to learn: there are many rules that govern their behaviors, and the underlying functions and operational logic can appear arbitrary to the pilot. Without a consistent conceptual framework governing the behaviors and operations, the pilot must learn and remember each one individually. This is similar to the problem facing the novice computer user in the age of DOS. Obviously, the autoflight system is a good candidate for complexity reduction.

Indeed, many pilots are now calling for such systems to be simpler. Different carriers require different features, however, so the systems used by any individual carrier are likely to provide capabilities and functions that it will never use. And the complex nature of the environment requires that the system provide complex functionality. Indeed, much of that functionality was provided specifically to improve usability. For example, early FMSs did not enable pilots to easily intercept radials going outbound from navigation beacons; recognizing that pilots often needed to do that, manufacturers included it in later systems. The same is true of sidestepping to the parallel runway on final approach, a very cumbersome operation in early systems that was made much easier in later ones. But the list of system capabilities provided to meet real operational needs has been expanding by adding new functions, and over the years, the number of functions that pilots have had to learn has grown correspondingly. Hence we find ourselves between the proverbial rock and hard place: the number of functions provided to improve usability has accrued to an extent that systems have become hard to use.

4.4.1   Interface bandwidth

When we reach this impasse, we can only reduce complexity by reducing the number of functions available (and thereby compromising the functionality of the system), by increasing the bandwidth of the interface so more information can be transferred between user and system in a given amount of time, or by redefining the underlying functionality and functional logic of the system to match users' expectations and build on their existing body of knowledge. As stated earlier, the last approach is the most powerful, but the second option is often a prerequisite. For this reason, we'll start our pursuit of a conceptually simpler system by increasing the interface bandwidth so we have as many options for redefining the functionality as possible.

Consistent with its use in information theory, we use the term *bandwidth* to represent the data transfer rate between the user and the system. Obviously, a text-based system such as DOS or the CDU is very limited in both directions. We could calculate the bit rates of text-only versus graphical interfaces, but merely considering the number of actions required to select functions in each environment should demonstrate that the graphical interface provides a much higher bandwidth.

Note, however, that the bandwidth available to the system is much greater than that available to the user. The system can convey a tremendous amount of information to the user through a graphic display, but the user's data transfer rate to the system is still relatively limited. The graphical user interface improves it by providing more means of coding information (as in the highlighting example discussed earlier), but the user's input means are still constrained.

This is even more true on the flight deck. In the past two decades, many new technologies have been introduced on the flight deck: traffic collision avoidance systems, enhanced terrain avoidance systems, the CDU itself, head-up displays, the integrated flight instrument format of the Primary Flight Display, the Navigation Display (which shows a moving map of the flight), synthesized and digitized speech alerts and warnings, synoptic systems displays, and a cursor control device. Note that only one of these technologies, the cursor control device, is an advanced means for the pilot to transfer data to the aircraft, and that, at the time of this writing, it is only being used for a very limited number of functions in a single aircraft type (the Boeing 777). All of the other technologies are intended to enhance the ability of the aircraft to transfer information to the pilot. The pilot's relative inability to effectively communicate his or her intent to the aircraft remains one of the greatest sources of problems on the flight deck, yet it is perhaps the most overlooked area of research and development. Most work is devoted to improving display technologies. In fact, a review of the last 10 years of publications in the two largest conferences devoted to human factors and aviation human factors shows that display-related papers outpace control-related papers by 540 to 123 in general human factors (including desktop, medical, industrial, aviation, and other types of systems) and 159 to 6 in aviation human factors.

4.5 Human-Factored Systems vs. Human-Centered Systems

Lacking an effective means for the pilot to communicate intent to the aircraft in his or her own terms, the problem must be addressed at both the interface level and the functional level. These two levels represent, for me, the difference between 'human-factored' solutions and 'human-centered' solutions. Human-factored solutions typically are pursued at the level of the interface and involve display formatting, menu design, control device configuration, and so forth, whereas in human-centered solutions, the underlying functionality and functional logic are defined from the user's perspective. The application of human-centered design has significant implications for the system development process. Typically, functions and their operational logic are defined based on the designer's view of the operational requirements, and human factors considerations are brought in at the end in order to design displays and controls so the operator can interact with the already defined functionality. The reason this traditional approach has led to so many usability problems is that a large gap may remain between how the designer thinks and how the user thinks, and the interface must bridge this gap. Usually, the gap is large enough and the interface tools so limited that training is critical for building the bridge. Consequently, complex systems with limited interfaces require extensive training. The operating budgets of system operators in many industries could be reduced substantially if training could be reduced or eliminated. Since interface technology remains relatively limited, we are compelled to attack the problem at the level of functionality, and as stated earlier, the most powerful means of doing this is with the application of an appropriate metaphor.

Recall that in the PC domain, the metaphor was applied after the operating system functions had already been defined. This is why the desktop metaphor doesn't make all of the computer's functions easy to use. Ideally, the metaphor would be selected first and the system's functions and operational logic would be defined to fit within it. This way, the metaphor would cover all the functionality provided by the system, and the user could draw on his or her existing body of knowledge to infer the operation of every system function.

This is the approach we have taken at Honeywell in designing a new pilot interface, called the Cockpit Control Language, for the autoflight system.

Since the desktop metaphor is not appropriate for the required functionality of the flight deck, we selected the air traffic control clearance as the basis for our metaphor. This has several advantages: every potential user of our system already knows how to interpret clearances; the full range of clearances provides the full range of functionality required to define the flight path; and there is a direct mapping from the pilot's input (the heard clearance) to the pilot's output (the programming of the system). The Cockpit Control Language, then, allows pilots to enter the meaningful elements of clearances in the order they are heard, without having to translate the intent of the clearance into an arbitrary set of functions dictated by the system logic. Consequently, any pilot can learn to use our system within 10 to 15 minutes, instead of the many weeks and months required to learn to use a traditional flight management system (Riley *et al.*, 1998). The new system maintains the full operational complexity of the original system, but users perceive it as being much less complex. Furthermore, the same functional logic applies to all the devices on the flight deck, reducing component complexity, and there is only a single source of guidance for the autoflight system. This has the significant side benefit of eliminating mode management.

The physical interaction with our system is much like it is today. For example, when given a simple altitude clearance, the pilot can turn the altitude selector knob and push or pull the knob to enable the altitude change, just as is done today. However, rather than thinking in terms of autopilot controls being on the glareshield controller and FMS controls on the CDU, and of modes to select between them, the pilot can think of everything in terms of 'targets' and 'actions'. 'Targets' are the physical parameters that define the flight path, including altitudes, speeds, positions (of all kinds), procedures, and even the entire flight plan. 'Actions' define what the airplane is to do with a target: fly 'to' it, fly 'away from' it, do something 'at' it, stay 'below' or 'above' it, and so on. The actions and targets are defined so the pilot can flexibly combine them into any order that makes both syntactic and semantic sense. This allows the pilot to enter clearance parameters in the order they are contained in the clearance. In other words, the pilot doesn't have to translate from the logic of the clearance to the logic of the system.

You may recall our earlier example of constructing a holding pattern around an unpublished waypoint, a difficult procedure for pilots to remember and execute. We can use this to demonstrate the difference between current FMS logic (designer logic) and Cockpit Control Language logic (pilot logic). This example also illustrates how the Cockpit Control Language can be used to handle nonstandard air traffic control instructions.

The clearance we will use is, 'Hold at twenty miles southwest of Farmington'. Usually, a hold clearance will be given around a published waypoint, and the pilot simply needs to select the waypoint from the navigation database and insert a holding pattern there. But because there is no waypoint in the desired position, the pilot must define an artificial one. Furthermore, a standard clearance of this type would specify a radial from the known waypoint on which the pilot is to hold (such as, 'Hold at twenty miles on the Farmington 220-degree radial'). As our example clearance contains no specific direction, however, the assumption underlying the interpretation of the clearance is that the pilot is to hold along the planned route in the southwest quadrant of Farmington.

The combination of nonstandard phraseology and instructions with the need to build an artificial waypoint before defining the holding pattern makes this a very difficult clearance to implement with current generation equipment. Furthermore, holding clearances are typically given when sector traffic is high and traffic flow is limited, often by weather. So when the pilot receives such a clearance, he or she does not have much time to figure out how to comply with it.

The procedure for implementing this clearance with current generation equipment, as described earlier, is to first enter a made-up waypoint name, then define the actual position of the new waypoint in terms of known waypoints in the navigation database. This can be done in terms of place/bearing/place/bearing or place/bearing/distance from the known positions. Once the waypoint has been created, named, and its position defined, the pilot must place it in the proper sequence in the flight plan. Only then can the pilot place the holding pattern at the new waypoint. As the order of the procedure doesn't correspond to the order of the clearance, it can take the pilot a long time to figure out what the proper procedure is.

In contrast, the pilot using the Cockpit Control Language need only listen for the actions and targets contained in the clearance and enter them in the order he or she hears them. In the case of 'Hold at twenty miles southwest of Farmington', the pilot need only enter a command string consisting of 'HOLD TWENTY SOUTHWEST FARMINGTON'. The system will parse the string syntactically and add any redundant connecting words, transforming it into 'HOLD AT TWENTY MILES SOUTHWEST OF FARMINGTON', then parse it semantically. At this stage, the information required to interpret the clearance correctly will be applied so the waypoint can be properly positioned. Note that the thought process this requires cuts out most of the steps shown in Figure 4.4.

The general principle of stringing together actions and targets in the order they are contained in the clearance will apply to all Cockpit Control Language procedures, and any appropriate interface device can be used. When the altitude knob is turned on the glareshield controller, an altitude target is entered into the command string. When the pilot selects graphic objects on the navigation display, the corresponding identifiers are entered into the command string. The pilot could conceivably enter the command string using speech recognition, or by typing it in with a keyboard. In this way, the same underlying operational logic applies to all the devices through which the pilot interacts with the autoflight system.

The Cockpit Control Language, then, reduces perceived complexity in several ways. It takes advantage of the pilot's existing knowledge about air traffic control clearances and uses it as a metaphor for the autoflight operating system. This is why any pilot can learn to use the system with just a few minutes' introduction to it. It also provides a consistent conceptual framework, based on the single metaphor, to all autoflight system devices and interfaces. Finally, because the command string is a single source of guidance, it eliminates the need for the pilot to manage modes.

4.6 Conclusions

By establishing a unified, conceptual framework for a system's functions based on the user's existing body of knowledge, the metaphor provides a powerful means of reducing perceived complexity. The overriding principle that enables the reduction in perceived complexity is one of consistency; the metaphor provides external consistency with something the user already knows. When an appropriate metaphor is not available, internal consistency can still establish the basis for a unified conceptual framework; this enables the user's mental model to be much less complex than the full component complexity of the system, but it still requires the user to learn some arbitrary principles of system operation. Without either of these types of consistency, the user's mental model must be as fully complex as the component complexity of the system. The bottom line, then, is that very complex systems can be made to seem very simple, and perceived complexity can be decoupled from actual complexity, if the system's functions can be represented to the user within a consistent, coherent conceptual framework, ideally one that is consistent with what the user already knows.

Usable systems in the future will require that human-centered design principles be applied at the level of system functions and operational logic to fully exploit the power of metaphors. This means that designers should attempt to understand how the user thinks about the environment and tasks he or she has to perform, and they should attempt to identify a body of knowledge that all potential users of the system share and that can form the basis for an appropriate metaphor. The ultimate result of such an effort will be a system that does not maintain the traditionally large gap between how the designer thinks about the system and how the user thinks about it, that does not demand that the user learn to think like the designer, and that does not require extensive training to bridge that part of the gap which the interface cannot span on its own.

References

Abbott, T.S. (1995) The evaluation of two CDU concepts and their effects on FMS training. *Proceedings of the Eighth International Symposium on Aviation Psychology,* 233–238.

Cringely, R. (1992) *Accidental empires.* Harper-Collins, New York.

Morcom, R. (1978) Safety and economy in standardization. *Papers of the Society of Automotive Engineers,* Australasia National Convention, Hobart, Australia.

Norman, D. (1988) *The psychology of everyday things.* Basic Books, New York.

Riley, V., DeMers, B., Misiak, C. and Schmalz, B. (1998) The Cockpit Control Language: a pilot-centered avionics interface. *Proceedings of HCI-Aero 1998,* Montreal, Quebec.

Sarter, N. and Woods, D. (1991) Pilot interaction with cockpit automation: operational experiences with the Flight Management System (FMS). Cognitive Systems Engineering Laboratory, The Ohio State University.

Sarter, N. and Woods, D. (1995) Strong, silent, and 'out-of-the-loop': properties of advanced (cockpit) automation and their impact on human-automation interaction. Technical Report 95-TR-01, Cognitive Systems Engineering Laboratory, The Ohio State University.

Sekigawa, E. and Mecham, M. (1996) Pilots, A300 systems cited in Nagoya crash. *Aviation Week and Space Technology,* **145**(5), July 29.

Sherry, L. and Polson, P. (1998) Implications of situation-action rule descriptions of avionics behavior. *Proceedings of the International Conference on Human-Computer Interaction in Aeronautics,* Montreal.

Wiener, E. (1988) Cockpit automation, in *Human factors in aviation* (eds. E.L. Wiener and D.C. Nagel). Academic, San Diego, CA.

Part 2

Sensing and Control

Sensing and control have always been central to the automation of engineering systems – in this respect they differ from several of the other disciplines discussed in this volume. Regardless of the complexity of the system, the effectiveness with which we can regulate its operation stands in direct proportion to how well relevant variables can be measured and to the sophistication of the processing behind the actions taken on the system. The fact that sensing and control predate computing does not render them anachronisms. Their fundamental importance persists, even if their current manifestations bear little resemblance to those of decades past.

The four chapters in this part of the book address new trends and directions for sensing and control. The many intriguing new ideas that are discussed bear evidence to the continuing vitality of these established fields.

Chapter 5 emphasizes the importance of computational representations, or models, of the behavior of a system to be controlled. It notes that autonomy – often the principal objective for automation and control technology developments – implies an ability to react appropriately to unforeseen situations. Thus, precompiling control actions is not a workable strategy, and hence models must be available that are suitable for online manipulation. In addition, unitary models cannot be expected; knowledge about a complex system will be distributed among many heterogeneous models. Some roughly independent dimensions can help characterize relevant models: the source of the knowledge represented (first-principles understanding, empirical data, or heuristics), the complexity of the model (temporal and algebraic complexity are distinguished), and the domain of competence of the model (a model can be localized for specific operating conditions and/or for specific system components). Important aspects of multimodel research include real-time execution issues, multimodeling architectures, uncertainty management, and model adaptation. The chapter also discusses an ongoing research project concerned with the autonomous control of uninhabited aerial vehicles. A multimodeling approach is being pursued incorporating multiresolution route representations; multiple vehicle flyability models; and terrain, target, and threat models.

The topic of Chapter 6 by Rudolf Kulhavý is randomized algorithms, a radical alternative to modern control and optimization science. The analytical, calculus-based tools that are the prized products of modern control and optimization are of little help for many of the problems we now face. These problems may include combinations of continuous, discrete, and symbolic variables; they may have pervasive uncertainties; and they may require search over huge spaces. In such cases, any hope for exact solutions is forlorn and simplifying assumptions (e.g., linearizations) are unacceptable. For problems that are irreducibly complex, we must perforce settle for approximations. *Approximate,* however, does not mean *ad hoc.* Ideas from statistical simulation, Monte Carlo estimation, evolutionary computing, and statistical learning theory provide a firm theoretical foundation for randomized algorithms. The resulting solutions may not come with deterministic guarantees, but they are optimally efficient in a meaningful sense and quantitative (albeit probabilistic) performance characteristics can be proven. Randomized algorithms are now being applied to many fields, including robust control design, predictive control, nonlinear filtering, Bayesian estimation, network computing models, and time series prediction.

Chapter 7, by Blaise Morton and Tariq Samad, is concerned with the biological solution to complexity management. Although our conventional control systems technology bears little resemblance to the design of biological control systems, the success of the latter in dealing with complexity is unquestionable. This fact highlights the importance of a deeper understanding of the architecture and function of, particularly higher level, biocontrol. The central nervous system (CNS) of vertebrates is an ideal candidate for study in this context, and considerable literature is now available that illuminates such aspects of vertebrate CNS as its modular structure, the functions performed by different modules, the information processing pathways, and its integration with sensory organs and muscular tissue. It is, of course, important to highlight that biological systems have drawbacks as well. We would like to find approaches for automation that realize the benefits of biological systems but avoid the drawbacks. Although many previous attempts at biologically inspired control technology have focused at either neural or cognitive phenomena, architectural principles of central nervous systems have rarely been exploited. The chapter illustrates CNS architectures for both vertebrates and invertebrates and outlines the sensory-motor processing involved. The controversy regarding consciousness in machines is briefly reviewed.

Chapter 8, by David Zook, Ulrich Bonne, and Tariq Samad, is about sensors, a control system's window on the world. The authors define a sensor as a device that converts a physical stimulus into a readable output, illustrating the definition with several examples of engineered and biological sensors. The design of sensors is driven by desired improvements on one or more of surprisingly many performance features and attributes: signal-to-noise ratio, reliability, safety and intrinsic safety, accuracy, response time, dynamic range, cost, power consumption, size, electromagnetic interference immunity, and so on. Recent trends and developments in sensor technology include the increasing use of signal processing for compensation, typically for reducing cross-sensitivity to secondary variables; multivariable inferential sensing, which allows sensing solutions to be developed for parameters that are infeasible to directly measure online; and self-checking and self-compensating sensors that enhance reliability and reduce maintenance costs. The chapter also discusses biological sensors with specific reference to olfaction and chemical sensing; in these modalities, our best man-made sensors are often no match for nature. In conclusion, visions for improved safety and efficiency through sensor-enabled automation in automobiles, commercial aircraft, health care, asset management, and other areas are outlined.

5

Active Multimodeling for Autonomous Systems

Tariq Samad

Honeywell Technology Center

5.1 Introduction

Domain knowledge is a central requirement for control and automation. Current control solutions fulfil this requirement in various specific ways, but generally they incorporate their designers', implementers', and users' understanding about the application domain. Automation is a multifarious proposition, and its different aspects – feedback and supervisory control, control system design, prognostic and diagnostic procedures, operator display, etc. – demand different types of knowledge. Even within one area, the knowledge required often needs to be differentiated; thus system dynamics often vary drastically with the operating regime of a physical system. The various, and varied, knowledge requirements are currently satisfied by different people, with different backgrounds, working independently on tasks that are considered separable. Since automation remains a patchwork affair when seen in the context of the overall system – be it a process plant, aerospace vehicle, or a large building – the absence of coordination or integration, although a suboptimal condition, is not a showstopper.

As we attempt to unfurl the automation umbrella over ever-larger parts of the design and operation of target systems, the creation, capture, and exploitation of domain knowledge become increasingly critical. Domain knowledge can adopt a variety of forms, both implicit and explicit. When considered in appropriate generality, such knowledge includes schematics and layouts detailing system structure and design, mental models of human operators, manuals for troubleshooting and maintenance, dynamic input-output models used in control algorithms, and many other representations.

All such representations can be considered models in a general sense. Our interest in this chapter is more specifically in models for real-time, autonomous control of complex engineering systems. Implications of this perspective are discussed in the next section, which introduces the notion of active multimodels and also lists several examples of models of interest. The following section outlines the numerous ways models are used in real-time control systems. Next, various characteristics and dimensions of models are reviewed, followed by a discussion of selected multimodeling topics. Before concluding, we present an example of an active multimodel-based control system currently in development for potential application to autonomous aerial vehicles.

5.2 Active Multimodels

Progress in control can be viewed as progress toward autonomy. Our engineered systems are multilevel, complex entities. From the lowest level – single-loop control – to the highest – enterprise-wide optimization – concepts of feedback, dynamics, and adaptation are relevant. The degree of autonomy can be related to the levels through which operations are largely automated. Requirements for knowledge of the system's dynamics, regulation through feedback control, and adaptation to changing conditions can, in principle, be fulfilled through manual or automatic means. In all industries and application domains, we can see steady advances toward automation, and hence autonomy.

Commercial aviation provides an illuminating example. In the early days of flight, control automation as we know it today was nonexistent. The pilot received sensory inputs directly (through visual and vestibular channels) and operated the actuators. The first step toward automatic control was the development of what we now refer to as inner-loop flight controllers. These allowed pilots to specify higher level commands to the aircraft, such as the desired pitch, roll, and yaw, with the controller responsible for the subsecond-scale feedback loop for sensing and actuation. The next step was the development of controllers that could take as input a heading command and automatically produce a sequence of desired states based on predefined 'handling qualities'. Today, all commercial aircraft have flight management systems (FMSs) on board that remove even this level of responsibility from the pilot (under normal conditions). A sequence of waypoints can be entered through the console of the FMS and the aircraft will automatically fly the route (Figure 5.1).

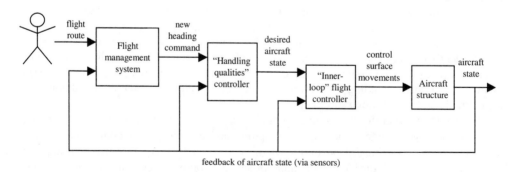

Figure 5.1 Control loops in commercial aviation today

Increasing automation requires increasing articulation and representation of domain knowledge within the automation system. Today, when aircraft can fly from point A to point B with no human interaction, it is in large part because we now have available explicit representations (models) of the dynamics of the aircraft under different conditions: the rate of fuel burn at different altitudes, speeds, and payloads; terrain topography; locations of navigation aids; special use airspace regions; etc.

Thus control and automation of complex systems cannot be based on any single, unitary concept of model. A treatment of models in the full generality of the term is an important topic for complexity management, where the breadth of activities that must be considered stands in sharp contrast to the discipline-centric focus of classical methods.

Related to the increasing use of models, another trend can be discerned, and illustrated in an aerospace context. High-fidelity models are commonly used today for designing flight control laws. These models are developed through a combination of first-principles understanding, wind tunnel experiments, and experience with similar vehicles, and are the basis for the design and validation of flight control algorithms. These algorithms themselves generally do not explicitly include high-fidelity models – computational limitations preclude their on-board use and a substantial degree of human expertise is still needed in the design process. Developments in information technologies and in algorithmic disciplines (notably control science and computer science) are likely to change this situation. Some signs of change are evident already. Advanced military combat aircraft now often have dynamic models embedded within the on-board control algorithm. In the process industries, model-predictive control technology, which integrates a predictive model of the system to be controlled (e.g., a distillation column in an oil refinery), is now well established. These models are still limited in various ways – they are generally simplified versions of high-fidelity models and are valid only over a small operating regime – but the trend is clear: increasing amounts of knowledge, in the form of models, will be incorporated within control systems. This should not evince much surprise given the objectives of increasing autonomy. Autonomy requires an ability to react appropriately to *unforeseen* situations. If detailed and accurate models are not accessible online, such an ability is not possible.

Thus, a fundamental research need in modeling for autonomous systems is *active multimodeling*: the online use of a variety of model types. As noted earlier, our interest here is specifically in models for online autonomous control. This excludes many other important uses of models, such as for system design and operator training. However, the control relevant restriction still covers enough territory to appear overwhelming. Examples of control relevant models include the following:

- Nonlinear first-principles models of systems or subsystems that are based on the physics and chemistry of their underlying phenomena
- Linear regression models derived from analysis of steady-state system input/output data
- Fuzzy rules expressing a human operator's predictions of system behavior
- Linear state-space models developed by linearizing first-principles models
- Neural network models for inferring properties of a system for which sensors are not available

5.3   Uses of Models

The importance of models for automation solutions can be gleaned from the variety of functions and features that are enabled by accurate, efficient computational models. The principal uses of models in control systems are reviewed below. The discussion refers to types of models that are explained in the next section. It should be noted that no one model can be expected to serve all these purposes, so it is not surprising that in current practice models are developed and used independently for the various wholly or partly automated functions.

5.3.1 Feedforward and feedback control

All advanced control today is model based. In many cases, the model is part of the controller and is used by the control algorithm in calculating the control output. The model can be used for both feedback control and feedforward control. In the former, the system output is measured, and the controller uses the model to compute a control move that minimizes the error between the current output and the desired or reference output (subject to other criteria and constraints). In feedforward control, feedback measurements are not used; instead a control output trajectory is computed that, under nominal conditions, is expected to improve the performance of the closed-loop system.

Most models used as part of control algorithms today are linear, and generally a number of different linear models, indexed by some operational conditions (e.g., the electricity load for a power plant boiler), are needed to capture system behavior over the necessary range.

5.3.2 State estimation

In many applications, including aerospace systems and chemical manufacturing, sensors are not available to measure all the 'states' of the system. Some important dynamic information remains 'hidden'. Dynamic models are often used to estimate these hidden states – the models define relationships between these states and the measured variables, permitting the former to be estimated based on the latter. The best known example of a state estimator is the Kalman filter.

An example of state estimation that has become popular in the process industries in particular is inferential modeling (also sometimes referred to as inferential sensing). In many processes, we are interested in predicting values for parameters that are infeasible to directly measure in real time. Examples include octane number for petroleum refining and particulate emissions for coal-fired power generation plants. Sensors for these quantities are unavailable, too expensive, or unreliable for long-term use. An alternative is to infer, or estimate, these parameters based on readily measurable variables (e.g., temperatures, pressures, and flows). Empirical models (and in some cases first-principles ones) are now often used for this purpose.

5.3.3 Operational optimization

The core control concepts of dynamics, feedback, and optimization are now being applied to higher levels of systems. In the process industries, control loops have been successfully closed around process units such as distillation columns and boilers. The next step is the optimization of the enterprise. Whereas multivariable control of a distillation column today is based on a single linear model, it is inconceivable that the same approach can be extended for enterprise-wide or plant-wide optimization. Multiple models, and more complex ones, will be necessary. Similar considerations apply in many other industries. In commercial air transportation, further improvements in flight control are likely to be evolutionary, whereas revolutionary improvements are foreseen through model-based optimization at the level of airline operations and air traffic management.

5.3.4 System health management

For many complex systems, equipment failures and other abnormal situations are a greater threat to safety and performance than suboptimal control performance. Predicting potential equipment failures in advance of their occurrence, detecting such failures when they occur, identifying the natures of faults, adjusting operating procedures to minimize the adverse effects of faults – all of these elements are important for system health management.

Models are central to accurate system health management. For example, a potential fault can be detected based on an observed divergence between the measured value of some parameter and the predicted value based on a model that (in the absence of system faults) is known to be accurate.

5.3.5 Performance monitoring

Based on some mathematical knowledge about a system, we can in some cases determine what maximum level of performance can be achieved through closed-loop control (e.g., in terms of the error variance for a continuous process). Then, by monitoring the performance of the system, we can identify when the controller needs to be readjusted. Performance monitoring has been a topic of research in the controls community for some time, and practical applications are now starting to appear.

5.4 Types of Models

As discussed above, models serve a variety of functions in control systems. Not surprisingly, then, the subject of modeling has attracted considerable attention on the part of scientists and engineers interested in automation. As a result of their efforts, many different types of models have been developed, specialized for different purposes and situations. In this section, we discuss some ways that models can be characterized, under three broad headings:

- The source of the knowledge that the model represents (first-principles, data-driven, and heuristic models)
- The complexity of the model (temporal and algebraic)
- The domain of competence for the model (localized to operating conditions and/or to system components)

These considerations suggest several characterizing dimensions that are largely orthogonal. So one can have, for example, a first-principles dynamic linear model of a particular subsystem that is designed for a particular operating point. For a technical tutorial on many of the model types reviewed here, see Misra (2000). We do not consider stochastic models in detail here, but see Chapter 6 for an extensive discussion.

5.4.1 Knowledge sources

All models are representations of knowledge about a system. This knowledge can come from different sources. We distinguish here between first-principles, data-driven, and heuristic models.

5.4.1.1 First-principles models

For many systems or parts thereof, we can articulate mathematically the underlying physics and chemistry of their operation. First-principles models are often expressed as nonlinear ordinary differential equations (ODEs) based on known laws of conservation (of energy, momentum, mass, etc.), difference equations (for discrete systems), or, when modeling of temporal evolution is not required, as algebraic equations. First-principles models can often be quite complex. In the process industries, models of individual units (such as a distillation column or a catalytic cracker) can comprise 100,000 equations – with the dynamics excluded or greatly simplified.

Whereas ODEs express the temporal evolution of 'lumped' parameters – such as the position or temperature of an object – partial differential equations (PDEs) are needed to capture spatiotemporal variations of systems for which the lumped-parameter approximation is inappropriate. Solving PDEs is substantially more computationally intensive than solving ODEs, but for limited-scale problems it is now feasible, and PDEs are increasingly being explored for control. Precision control for space structures, flexible materials (deformable mirrors, for example), and sheet-forming processes (paper machines, for example) are potential applications.

Different domains differ greatly in the feasibility and importance of first-principles modeling. High-fidelity dynamic models of aircraft are commonly used for flight control law design, but are still rare in the process industries where our understanding of many phenomena is still incomplete. In such cases, models are often based on analysis of data, and in fact 'first-principles' aircraft models also have a significant empirical component.

5.4.1.2 Data-driven models

Data-driven or empirical models are derived from data instead of mechanistic understanding. The data can be generated by designed experiments and by the normal operation of the system. Data-driven models are usually abstractions of data; the model results from a statistical analysis of the data, which itself is typically then discarded.

With advances in information technologies and statistical algorithms, the notion of 'data as model' has been gaining currency. It is now feasible to have large volumes of historical data collected and available online for 'data-centric' modeling, control, and optimization: the online database is queried and recent samples most relevant to the current situation form the basis of the model calculation. (For example, a localized model can be created in real time, on-the-fly, customized for and accurately representing the current state of the system.)

Data-centric control is not intended to supplant model-based control. Mathematical models are still required; for example, the data itself, especially when it is sparse, will not always be reliable, and the recorded operation of the system may not have included the region of the state space currently being visited. Data-centric control implies a multimodeling framework – and constitutes an important component of it.

5.4.1.3 Heuristic models

The distinction between first-principles and data-driven models is often made, but there is another source of knowledge for model development that is less often recognized. Heuristic models are formal representations of knowledge articulated by a human operator or engineer. They are usually expressed in the form of rules. Fuzzy logic has become a common approach since it provides a bridge between symbolic, linguistic expressions and the continuous domain within which the systems being modeled are operating. Fuzzy logic has had significant success for small-scale systems – typically single-input, single-output. Extension to larger-scale problems requires further research.

5.4.2 Complexity

Models also differ in their complexity – the level of detail that they capture. It is helpful to discuss model complexity under two separate headings: temporal and algebraic complexity.

5.4.2.1 Temporal complexity

All real systems are dynamical in that their outputs at any instant of time are not uniquely and precisely determined by their inputs at that instant (even for systems that are completely deterministic). The behavior of a dynamical system depends also on its 'state', a function of its history. Dynamic models are essential if details of the behavior (e.g., transients) of the system are to be predicted.

Models of systems often ignore dynamics, however, and represent only the steady-state behavior. Steady-state representations are sufficient for many applications, and using detailed dynamic models in such cases would not only be a needless expense, it would drastically affect the quality of solutions. An autonomous aerial vehicle that needs to plan a transcontinental route, for example, need not and should not have to consider in advance such short-range dynamical factors as the precise subsecond-resolution trajectory it will fly in making a turn at a waypoint along the route.

The line between algebraic and dynamic models can sometimes seem gray, but in fact there is a wide gap between them. Dealing with a model of the form $y = f(x)$ is a qualitatively different (and considerably simpler) proposition than dealing with a model of the form $y[t] = f(y[t-1], x[t-1])$, about the simplest nontrivial type of dynamical model.

5.4.2.2 Algebraic complexity

Models also vary on another dimension of complexity: linearity versus nonlinearity. This is an orthogonal dimension to temporal complexity: both steady-state and dynamic models can be linear or nonlinear. The assumption of linearity has been the foundation on which control science, statistics, and signal processing, among other disciplines, have developed into analytical, rigorous fields. In the case of feedback control, for example, we can synthesize controllers that can literally guarantee that performance and stability criteria will be met given a linear model for the system to be controlled (the 'plant') – even under a range of environmental disturbances and variations in the plant. In the general nonlinear case no such certainties exist, and in fact no methods exist for automated synthesis of feedback controllers for arbitrarily nonlinear plants.

So, although no real system is truly linear,[1] there are considerable benefits to be gained by employing linear models wherever this is reasonable. The successes of modern control theory are testament to the fact that mild nonlinearities are amenable to theoretical treatment based on linear system concepts.

As we deal with larger scale systems and less constrained problems, the existing body of theory is of less help. Even as simple a phenomenon as switching between two control modes is unavoidably nonlinear. Furthermore, even when linear approximations can be made to work, they often extract a significant performance penalty.

Whether we rely on a linear (-ized) or nonlinear model thus depends on several factors:

- How nonlinear is the system?
- What level of performance is required?
- What degree of certainty regarding stability/performance is required?

For an autonomous system, these questions do not have fixed answers. Performance requirements, safety assurance, even an appropriate mathematical characterization, depend on the operating conditions, the system's goals, and the system's current configuration, among other factors. Both linear and nonlinear models are necessary.

5.4.3 Domain of competence

Finally, we also note that an individual model for all but the simplest problems will not accurately describe the whole system for all time. Instead, a model is typically appropriate for some part of an overall system and/or for some regions of operation. The domain of competence for a model can be limited in two ways, which we refer to as operational and componential localization.

5.4.3.1 Operational localization

Models are typically applicable to specific operating regions of a system. The primary example of operational localization is linear models for control systems that are explicitly intended for particular steady-state conditions. For flight control design, for example, it is common to identify a number of models, each customized for a particular altitude and air speed. Different controllers are then designed for each condition based on the condition-specific model.

Models can be specialized in a number of other ways as well. In many systems, there is redundancy in sensors and/or actuators, so that some degree of control can be exercised through different sensor/actuator combinations. Different models can be developed for different feasible combinations; which is used is then determined by the operating mode. Failures in specific sensors or actuators can thus be accommodated, at least in principle. Similarly, different models for different damage modes of a system are desirable for fault-tolerant operation.

---

[1] I know of one, fundamental, exception: the wave function in quantum mechanics is linear, apparently by necessity and not simplification (Weinberg, 1992, pp. 85–89).

5.4.3.2 Componential localization

Models may cover one or more components of an overall system. The larger the scale of a system, the greater the necessity of developing separate models for different components of it. For large-scale systems, the components themselves will be composite entities, and their models may well be composite as well.

5.5 Multimodels

The discussion above implies that automation and control of complex systems must rely on numerous, heterogeneous models. In this multimodel environment, the primary research issue that arises is how to integrate the diverse variety of models so that we can extract the full value from the storehouse of knowledge comprised of the multiple models – even in the face of its inevitable inconsistencies, deficiencies, and inaccuracies. In this section we comment on some aspects of multimodel research.

5.5.1 Facets of models

In the context of control and automation systems, the term *model* generally refers to a representation of the behavior of the physical system (or some part or aspect of the physical system) that we are interested in controlling. Complexity management is a multidisciplinary field, and this concept of model is sufficiently complex so that the usage of the term across the associated disciplines is quite varied:

- Control science and engineering: a model is typically a set of mathematical formulae that capture some important characteristics of the behavior of a physical system. This chapter has primarily emphasized this notion of model.
- Real-time execution environments: as computational entities, models have processing, memory, and communication requirements. For online execution of multiple models, these requirements must be satisfied – generally a nontrivial undertaking given resource constraints of control system platforms.
- Software architecture: models can be viewed as objects, with inputs, outputs, and functional capabilities abstractly specified.

These viewpoints are complementary, but different facets of models are being foregrounded (Figure 5.2). Yet the intradisciplinary research needs are substantial enough that the connections are rarely investigated. The cross-disciplinary synthesis is, in our view, an absolute necessity for progress in active multimodel technology. Model composition for online control, for example, requires the integration of all these facets.

5.5.2 Hierarchical decompositions

A popular approach in engineering design is hierarchical decomposition. The real system is conceptually decomposed into subsystems that are in turn similarly decomposed, and so on, until elementary system components are reached for which individual models are available or can relatively easily be developed. This approach has met with much success, but it has some drawbacks that are likely to become increasingly serious as we target more complex systems:

- 'Natural' decompositions of many systems are not *strictly* hierarchical. Biological organisms, for example, can usefully be thought of as composed of structures at various levels of aggregation, but we often find pathways between modules that span multiple levels.

- Appropriate structurings for one aspect of a system may be inappropriate for another. A logical breakdown (which components affect which others) can be different from a spatial one (which components are physically near which others). Analogously, an electrical connectivity diagram may need to represent the same elements as a process flow diagram, but the best structurings of these components may be different in the two cases.

- Ultimately, many of the systems we deal with are continuous. Decompositions ignore this fact. This neglect is benign for many purposes, but as we attempt to develop more detailed models in the interests of higher performance, problems can arise. For example, a compositional model of aircraft structure may not be well suited for modeling flutter and other macroscopic structural phenomena.

Figure 5.2 Facets of a model

5.5.3 Uncertainty management

We have already noted that models, as encapsulators of domain knowledge, are critical for automation systems. As these latter attempt to bring larger scale and more complex problems under their purview, model accuracy becomes both more critical and more difficult to ensure. Developing models from noisy data, characterizing model uncertainty, and developing control solutions that are robust to modeling error are some of the issues involved.

Control science has made substantial progress in uncertainty management for linear systems. For reasonable (but parametric) representations of model dynamic uncertainty we can now design control laws that guarantee stability and performance. In statistics, robust linear regression methods are now available that analogously are relatively insensitive to noise, outliers, and bias in data. However, technology for uncertainty management in nonlinear systems is undeveloped. Some recent theoretical work has illuminated directions for algebraic models (Barron, 1993; Vapnik, 1995), but no similar results are available for nonlinear dynamic models.

For multimodeling, uncertainty management poses additional complications. Uncertainty representations for different types of models will be different and for the most part incommensurable. This is still *terra incognita*.

5.5.4 Model update and adaptation

Model updating and adaptation has been a central research area for conventional control models for decades. In adaptive control, for example, considerable theoretical progress has been made. Yet little of this research has been turned to practice. Exceptions exist, but only an infinitesimally small fraction of control algorithms operating today have any adaptation capabilities. If adaptation was a 'solution looking for a problem', then this lack of industrial success could be attributed to its relative practical irrelevance. On the contrary, however, practical adaptation techniques have been a continuing demand voiced by virtually all industries employing control systems.

Full-scale, real-time adaptive control is likely to remain little more than an academic research specialty. A more pragmatic outlook would view model update and adaptation as presenting a spectrum of possibilities, from within which solutions must be customized for individual applications. Whether (or under what situations) updates are done online or offline, autonomously or under human supervision, gradually or abruptly, are application-specific matters, and the technology must be developed to support the various possibilities.

Research in adaptation in this pragmatic vein needs to address several problems that are external to the adaptation algorithm (narrowly construed). These include:

- How can we effectively assess the performance of current models without making unrealistic assumptions about model fidelity in the first place?
- How can we practically detect or ensure sufficient excitation (or, more generally, sufficient signal-to-noise ratio) for model update calculations?
- How can smooth transitions be effected from outdated to updated models, whenever such are desired?

A conventional take on the complementary natures of robust design and adaptation is that the former is generally a conservative and hence safe approach, whereas adaptation, although capable of realizing performance improvements, is inherently dangerous. There is some truth to this characterization, but one qualification to be noted is that adaptation can improve safety by accommodating faults that would be impossible to cover under a robust design formulation. In any case, managing uncertainty through the sole device of highly robust control and optimization will be unacceptably conservative for many applications.

5.5.5 Multimodel mechanisms

Approaches for integrating multiple models have become an increasingly popular topic in control research. We briefly review below a few of the suggested mechanisms.

- First-principles and empirical models are complementary in that they use different sources of knowledge for a common end (e.g., prediction of a system's output variables). Where the mathematical laws that govern the operation of a system are known and resources are available, scientists and engineers often develop first-principles models as a first step in the design of control algorithms. These models can then be evaluated against data taken from the system. If the model fit is less accurate than desired, one possibility for improving the fit is to develop an empirical model that attempts to predict, not the system output itself (which might be too complex to enable an empirical model to be conveniently developed), but the error of the first-principles model. By superimposing the outputs of the first-principles and empirical model, a model of greater accuracy may be obtained than by adopting a unitary approach. Neural networks are a popular approach for empirical modeling in this context (Figure 5.3a).

  Another integration of neural networks and first-principles modeling can be noted here. Many first-principles models include tables of properties (for example, air density at different altitudes, temperatures, and pressures; or the melting point of a particular chemical compound at different temperatures and pressures). Instead of looking up values in tables and calculating interpolations, neural networks have been used to develop smooth nonlinear functions that can replace the tables entirely, saving processing time and memory.

- In modeling the operation of a complex system, different models are often developed for different operating conditions. In flight control design, as noted earlier, these different conditions can relate to altitudes and airspeeds of the vehicle. The individual models are sometimes linearizations of a global nonlinear model, but they can also be simpler nonlinear models. In using such models in real-time control systems, discrete switching between different models as operating conditions change (e.g., as an aircraft changes velocity) is undesirable – the vehicle's dynamics do not change instantaneously due to gradual changes in operating conditions. By blending the outputs of the different individual models through a supervisory function (Murray-Smith and Johansen, 1997), an overall smoothly varying model can be developed that is still computationally tractable and is easier to develop than a full-scale nonlinear model (Figure 5.3b).

- There are cases where the behavior of a system changes abruptly. We often refer to these changes as mode transitions, and they can arise as a result of different higher level commands to the system, traversal of critical points within the operational range, or faults in the system. The field of hybrid dynamical systems – an interdisciplinary area that draws upon control, computer science, operations research, and artificial intelligence, among others – focuses specifically on understanding mode-dependent temporal behaviors.

  A simple hybrid dynamical system model is shown in Figure 5.3c. The ellipses represent different operational modes with customized models. The arcs represent mode transitions triggered by conditions. A diagram such as this one could represent, perhaps, the operation of a simple batch chemical reactor, distinguishing between reactor charging, heating, and product draw, with potential for reversion to recharging or reheating during product draw.

- Multiresolution and multiscale models provide another interesting example of multimodeling mechanisms. Our demands for accuracy and precision are not uniformly high in all cases. Consider predictions of system behavior. Especially for complex systems, attempts at long-term accurate predictions are likely to be wasted effort.

Predicting or computing the route of an aircraft to within seconds and meters hours into the future is impossible. Unpredictable wind patterns, to choose one obvious uncontrollable disturbance, will vitiate the prediction. On the other hand, greater accuracy is possible, and desirable, for the very immediate future. Similarly, models of vehicle dynamics at different resolutions are needed for different time scales, multiresolution terrain models are desirable, and so on.

One generic representational scheme for multiresolution modeling is the wavelet transform, which provides an efficient way to represent a signal, such as a trajectory over time or an image, at multiple temporal and frequency scales. The next section presents an example based on our recent research that discusses this topic in greater detail.

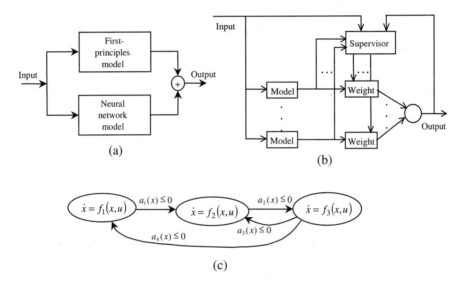

Figure 5.3 Selected multimodeling mechanisms

5.6 An Example: Active Multimodel Control of UAVs

We now discuss some ongoing work at Honeywell Technology Center that is attempting to demonstrate the benefits of active multimodeling for a challenging domain: the autonomous control of uninhabited air vehicles (UAVs), specifically UAVs that are intended for high-performance military applications. Some further details on this work are available in Koenig et al. (1999) and Krause et al. (1999). To motivate this discussion, we first present a hypothetical scenario.

5.6.1 Autonomous UAV scenario

Figure 5.4 shows a route plan for a wing of three aircraft (one piloted and two uninhabited), generated as part of the advance planning for a military mission. The aircraft leave a base and head for a target along a cruise segment of flight where winds-aloft data can influence the route plan. The wind model provides data needed to compute and update the optimal trajectory. All vehicles fly in close formation, each relying on sensor data that feeds models of its peers and the collective dynamics of the formation. At point A during the ingress, all three aircraft drop into terrain-following mode, where winds are generally not a factor but weather can be.

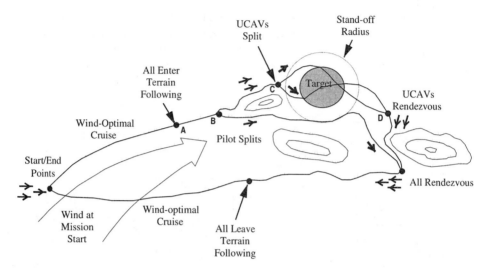

Figure 5.4 Normal ground mission plan

At point B the piloted aircraft splits from the UAVs, as it only needs to stand off from the target and identify and/or designate the target for the UAVs guided weapons. The UAVs themselves then may split at point C to conduct independent bombing runs across the target. Mission coordination would put the two UAVs across the target at a predetermined time spacing, and the piloted aircraft at its closest point at about the same time as the UAVs are making their bomb runs. Each vehicle uses models for its peers to estimate their flight paths and coordinate the maneuvers. Later there is a rendezvous (in space and time) at point D, followed by a return to the start point along another cruise segment.

A demonstration of the active multimodel capability might also include some small weather system for the pilot to avoid after the split of the UAVs (Figure 5.5) yet still make her time mark for target identification. Additionally, a pop-up target-of-opportunity may present itself after rendezvous. At point E the pilot dispatches a UAV to destroy the target, yet still requires it to catch up and arrive with the wing. Finally, the winds may have changed, making a route replan advantageous. Other sorts of unforeseen situations can require maneuver and route reoptimization at much shorter time scales: an incoming missile can require extreme evasive maneuvers, damage to the plane may limit its dynamic capabilities for high-performance flight, and so on.

5.6.2 Multiresolution active modeling

In our research, we have been addressing the important issue of what sorts of models and algorithms are needed to ensure that trajectories can be optimized in real time during the

An Example: Active Multimodel Control of UAVs

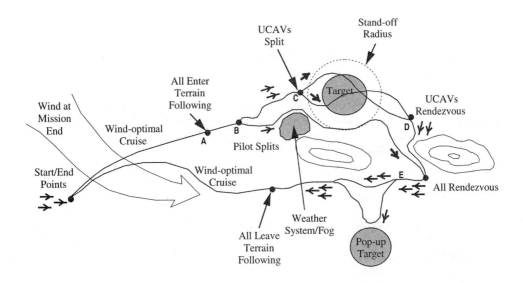

Figure 5.5 Resulting mission using dynamic replan with multiple active models

execution of a mission. A fundamental conclusion is that multiresolution models and route representations are required.

Near-term trajectories need to be computed with high accuracy with detailed models. However, we cannot just address the near term; the overall objectives of the mission – including return to base – must be considered. Further out segments of the route can be 'optimized' at progressively less accuracy. Since the situation can change in ways that cannot be anticipated between now and a future point, we should not rely on fine resolution details for future route projections. In a sense, the route generated should be more 'robust' to perturbations for more further out segments than for the nearer term ones. Computing resources should be expended accordingly, with relatively more processing dedicated to the near-term trajectory segment. One unified formulation and approach is needed so that nearer and longer term considerations can mesh with each other.

Our approach for addressing these requirements is based on a novel, wavelet-based multimodel scheme for vehicle trajectories. Wavelets are an inherently multiresolution signal composition and decomposition tool. Higher and lower frequency components of a signal are represented separately, not just in frequency space, but also in the time domain. Thus high-frequency components at different times can be distinguished. More important for our purposes, the time/frequency components can be selectively suppressed. Thus both low- and high-frequency factors can be permitted for the short term, and for progressively further out segments of the signal (the route trajectory), more and more higher frequency coefficients can be excluded. The result is a unified representation that automatically structures a route so that it is more complex in its initial stages and increasingly smoother in its later ones.

The level of detail in the trajectory optimization over the course of the route can be adjusted for current and projected future conditions and updated dynamically as the situation changes. Satellite surveillance could result in new threats being identified that would affect the UAV at some future point, necessitating low-altitude terrain following in the threat region. The vehicle could dynamically update the relevant part of the trajectory, further optimizing it at higher resolution for terrain following.

The multiresolution nature of a wavelet-based trajectory representation is shown in Figure 5.6 (top). Each 'box' in the figure represents a particular time-frequency localization of the overall route. For example, the top left-hand box captures the detailed dynamics of the first eighth of the route duration, and the wide, shallow box at the bottom represents the low-frequency content (or coarse approximation) of the whole trajectory. Note that, unlike the Fourier transform, the frequency steps are exponential, not linear, and that temporal localization is also obtained.

5.6.3 Route optimization

With a wavelet-based route representation, the trajectory optimization problem becomes one of identifying the wavelet coefficients. Essentially, we must compute a weighting factor for each of the boxes in Figure 5.6 (top) so that when the multiple wavelets are combined, the generated trajectory satisfies mission requirements (hitting a target, avoiding threats, not flying into terrain, avoiding severe weather), while ensuring that the maneuvers are flyable and, perhaps, that some mission-specific criterion (e.g., minimum-altitude flight or minimum fuel expenditure) is optimized.

We are exploring two algorithms for trajectory optimization, both of which exploit the wavelet-based multimodel representation. One is an evolutionary computing algorithm that performs a directed random search over the space of wavelet coefficients. The other is an interior point method. In both cases, we constrain the optimization to some subset of the full wavelet representation. Typically, details of the trajectory are only considered for the immediate route duration. For example, the coefficients associated with the shaded boxes in Figure 5.6 would be subject to the optimization; the others would be fixed at zero.

The labeled boxes in the lower part of Figure 5.6 show that multiple models, at different degrees of resolution and/or fidelity, are also needed for the vehicle (to determine what maneuvers are feasible for the vehicle to execute) and for the terrain. Similarly, models for threats, targets, weather, and so forth, will also need to be incorporated.

Figure 5.7 shows an initial result from this project, a route optimized by the evolutionary computing algorithm for a UAV (the diamond symbols represent the sampled route). The large spoked ellipse is a threat region to be avoided, the smaller spoked area (approximate coordinates: 30,50) is the target. The mission criterion in this case was low-elevation flight while reaching the target and avoiding the threat. Flyability constraints were represented as turning restrictions. Elevation is represented as contours in the figure. The terrain data is from north of Albuquerque, New Mexico.

The evolutionary computing algorithm is an anytime algorithm – the more computational resources available, the better the route produced. It is also inherently parallel, since many potential perturbed routes can be evaluated simultaneously, and it is sufficiently flexible to accommodate general evaluation criteria and models.

5.7 Conclusions

As we attempt to develop larger scale and more complex automation and control applications, models for control systems must be considered in their full variety. The key

Conclusions

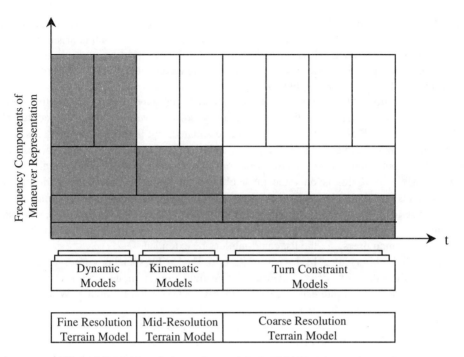

Figure 5.6 Multiresolution active models for UCAV maneuver generation

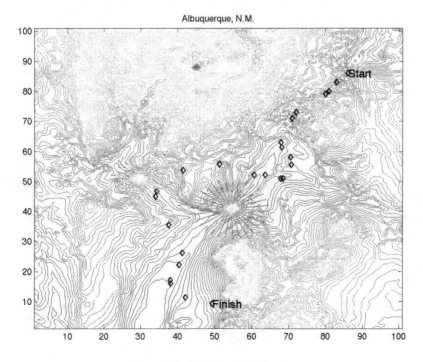

Figure 5.7 Optimized UAV route

research challenges for managing the complexity of future control system applications are not in the development or use of any one type of model. Instead, research is urgently needed in the multimodeling area. All control systems and solutions rely on multiple models for their design and operation. Successful complexity management, however, requires that all modeling activities be viewed within a unifying perspective.

Some of the important research issues that stand in the way of practical multimodeling for complex systems have not been satisfactorily solved even for unitary models. The challenge posed by these issues cannot be underestimated, but there are hopeful signs, most notably the dramatic advances in information technology. For example, CPU speeds have increased to a point that online execution of complex, computationally intensive algorithms is now feasible. And this is only the most obvious of impacts. Object-oriented software methodologies are naturally suited to multimodeling, the international communication infrastructure (e.g., the Web) facilitates integration of remote knowledge sources, graphical user interfaces allow complex and even disparate information to be comprehensibly presented to human users, ... the list can doubtless be extended.

We conclude this chapter by revisiting its title. Fueled by societal, corporate, and even intellectual dictates, the drive toward increasingly autonomous engineered systems continues inexorably. An interesting question in this context is the following: What are some of the key enabling technologies necessary for realizing autonomy?

The thesis of this chapter is that one such enabler is representations of knowledge – representations that can be algorithmically processed to determine appropriate responses to the variety of situations, the majority of them unforeseeable, that any autonomous system will have to face. These responses will require knowledge of the capabilities of the system itself, the natures of various disturbances to which it may be subject, the goals and objectives it is attempting to satisfy, and so on. We have used the term a*ctive multimodels* to capture the concept of diverse and numerous representations that must form the basis of control, optimization, and decision making online, in real time.

In fact, the history of progress in technology is also the history of progress in active multimodels. The earliest model-based control systems incorporated unitary models; today's systems are able to control aircraft, refineries, paper mills, power plants, commercial buildings, and innumerable other engineering systems by employing several models. In a sense, the next step in autonomy may seem to imply an evolutionary step in active multimodels. In another sense, however, the development of a system that is close to 'truly' autonomous – such as an uninhabited aircraft – is a revolutionary prospect given the technology available now, where the human element is generally always considered available to handle situations that the automation system designers did not predict. For the autonomous systems that are on drawing boards today, a cohesive, integrated framework for a much richer and more numerous model set remains an outstanding research need.

Acknowledgments

This chapter describes research supported in part by the U.S. Defense Advanced Research Projects Agency (DARPA) under contract number F33615-98-C-1340. The multivehicle UAV scenario was developed by Dan Bugajski.

References

Barron, A.R. (1993) Universal approximation bounds for superpositions of a sigmoid function. *IEEE Transactions on Information Theory*, **39**(3), 930–945.

Koenig, W., Cofer, D., Godbole, D. and Samad, T. (1999) Active multi-models and software enabled control for unmanned aerial vehicles. *Proceedings of the Association of Unmanned Vehicle Systems International*, Baltimore, MD, July.

Krause, J., Samad, T. and Musliner, D. (1999) Multiple vehicle mission management: coordination and optimization. *Proceedings of the NATO Symposium on Advanced Mission Management and System Integration Technologies for Improved Tactical Operations*, Florence, Italy, September.

Misra, P. (in press) System modeling, in *Perspectives in control: technologies, applications, new directions* (ed. T. Samad). IEEE Press, Piscataway, NJ.

Murray-Smith, R. and Johansen, T.A. (eds.) (1997) *Multiple model approaches to modelling and control*. Taylor & Francis Ltd., London.

Vapnik, V. (1995) *The nature of statistical learning theory*. Springer-Verlag, New York.

Weinberg, S. (1992) *Dreams of a final theory*. Pantheon Books, New York.

6

Randomized Algorithms for Control and Optimization

Rudolf Kulhavý

Honeywell Technology Center and Institute of Information Theory and Automation, Czech Republic

6.1 Introduction

A number of problems that industry and businesses are facing today are *hard* to deal with using analytical, calculus-based tools. Such problems include

- Estimation of parameters in nonlinear or otherwise complex models,
- Design of feedback control laws in nonlinear control problems,
- Engineering design of complex and not completely understood manufacturing processes,
- Design of scheduling rules in ATM switches,
- Buffer allocation in production lines,
- Estimation of rare event probabilities in high-reliability systems,
- Routing policy in communication networks,
- Logistic system and policy design in global transportation networks.

The common characteristics of such problems are lack of structure (resulting from a combination of combinatorial, discrete, and symbolic variables), inherent presence of uncertainties (requiring time-consuming averaging), and huge search spaces (not easy to parametrize and prone to combinatorial explosion).

Hard problems can be solved only approximately. The purpose of this chapter is to show that randomized algorithms yield a consistent and systematic framework for the design and analysis of *approximate* solutions to such problems.

6.2 Statistical Simulation

If we cannot solve the problem *exactly*, we must settle for solving it *approximately*. If we cannot solve *all* instances of a problem, we must settle for solving *almost all* of them. This is a recurring idea in most recent approaches to solving hard problems.

6.2.1 Sampling from a population

One does not need to canvass the whole electorate to learn the percentage of votes for presidential candidates. A sufficiently precise answer can be obtained by addressing a much smaller sample of voters. The trick is that the *sample* mimics the structure of the whole *population* in all its essential characteristics (age, sex, education, religion, ethnic origin, etc.).

The fact that we can learn all essential information about the population by analyzing its sample is now used routinely in opinion polls, retail surveys, environmental monitoring, and statistical process control, to name but a few applications. Substitution of a collection of samples for the whole population effectively reduces the size of the original problem.

6.2.2 Monte Carlo simulation

Repetitive computer simulation of a system model – starting from various initial conditions and running with different sets of parameters – has become one of the engineer's major tools. Systematic exploration of the parameter space, however, involves prohibitive computations, even for problems of moderate complexity. We can move much further if we replace the 'population' of all possible parameter values with a 'sample' of their representative values.

The prerequisite is to know the *probability distribution* of key quantities (such as parameters or states) of the system under study. Once the distribution is known, the simulation proceeds by randomly sampling from it. The desired result is taken as an average of the simulation outcomes over the number of simulation runs. Furthermore, we can predict the statistical error of this average result, and hence estimate the number of trials needed to achieve a prescribed error limit.

The above forms the essence of the Monte Carlo method, whose origin dates back to the 1940s. Monte Carlo simulation has been used since then in applications as diverse as nuclear reactor design, quantum chromodynamics, radiation cancer therapy, traffic flow, stellar evolution, econometrics, Dow-Jones forecasting, oil well exploration, and VLSI design.

6.2.3 Stochastic models

The models entering the Monte Carlo simulation must be *stochastic* – to define the probability distribution of the variables of interest. This may seem a restrictive and artificial assumption, especially as most of the first-principles models of physical and chemical processes are represented by a set of *deterministic* differential equations. A closer look shows, however, that data always exhibit some kind of random behavior. This may be due to a physical noise affecting the measurements, but more typically, it is a consequence of the incomplete knowledge of all the relevant variables. The 'hidden' variables affect the system behavior in a way that cannot be fully predicted.

Consider, for instance, a continuous stirred tank reactor where the temperature and concentration inside the reactor are measured but the feed temperature and concentration are not monitored, being assumed to be constant. Any variation in the feed parameters produces a deviation from the ideal model. As a result, we have two sources of randomness – the imprecision of sensors and the imprecision of the model. Both can be modeled as external noises corrupting the process states and the sensor measurements.

6.2.4   MC computation engine

Monte Carlo simulation can be applied even in cases that have no apparent stochastic content, such as the evaluation of a definite integral or the inversion of a system of linear equations. One can still pose the desired solution in terms of probability distributions. Even though the transformation may seem artificial, this step allows the system to be treated as a stochastic process for the purpose of simulation, and hence Monte Carlo methods can be applied to simulate the system.

This is crucially important for implementation of statistical methods that involve operations with probabilities, such as fitting a probability distribution to given data and computation of marginal and conditional distributions of selected variables from a given joint distribution. The probabilistic operations entail computation with functions (infinitely dimensional objects) and multivariate integration (extremely difficult in high dimensions). All these operations can be approximated using sample-based computations; for instance, the expectation of a given function can be replaced with its sample average.

6.3   Managing Complex Models

Consideration of increasingly complex models stresses the importance of consistent handling of uncertainty in estimation. Although the use of complex models generally decreases the approximation error between the actual and model behaviors, the estimation uncertainty typically increases – we are left with more degrees of freedom but the same amount of data.

Choosing a compromise between the approximation error and estimation inaccuracy (known as the bias-variance trade-off in mean squared error estimation) is an inevitable step in any statistical modeling. In the past decades, engineers usually preferred minimizing the estimation error, even at the cost of adopting simplifying model assumptions such as linear dynamics and Gaussian stochastics.

This approach can hardly be followed when trying to solve some of the hard problems mentioned in the introduction. Rather, we must get used to the fact that estimation will leave us with significant residual uncertainty about the resulting model. This situation underscores the need to have a systematic framework for describing and propagating the uncertainty (as the new data arrive). It is increasingly accepted that 'probability' and 'probability calculus laws' are ideally suited for the job.

6.3.1 Bayesian paradigm

It sounds very natural to describe the random behavior of data through its probability distribution. One can describe, however, in terms of probability even the uncertainty of assigning such a distribution. Probability thus appears in a twofold role – as a relative frequency of possible outcomes of a certain experiment in the long run, and as a measure of uncertainty of unknown (not necessarily stochastic) quantities.

Jacob Bernoulli, Thomas Bayes, and Pierre Simon Laplace came up with the idea of 'inverse' probability as early as the 18th century. Much later, in this century, it was shown that the rules of consistent reasoning under uncertainty coincide essentially with the laws of probability calculus.

The symmetric view of probability, usually ascribed to Bayes, has turned out to be useful in many areas. It has brought some unique features into statistics, creating its self-contained branch. It has put artificial intelligence on a firm basis and set up a reference solution for the design of expert systems. It has been applied successfully in physics, engineering, and statistics to ill-posed (underdetermined) problems, which commonly appear in system identification, fault detection, econometrics, control, medical diagnosis, geophysical exploration, image processing, and synthesis of electrical filters or optical systems.

The pros and cons of the Bayesian paradigm can be well illustrated in the case of stochastic adaptive control. Suppose we are to control a dynamic system that depends on some unknown parameter. We have two options then. Either we settle for a point estimate of the parameter, or we take its uncertainty into account. In the latter case, the expectation in a cost function applies to both the stochastic behavior of the system and the uncertainty of the parameter. This converts the original problem into a hyperproblem that has no more unknowns. Its information state is formed by the probability density function of the original state *and* the parameter, conditional on the observed data.

The beauty of the result is that it looks as if all uncertainty has vanished. As soon as the prior density is set, the state evolves in a definite way, governed by the laws of probability theory. The appeal of the solution is offset, however, by the immense dimension of the information state. Unless the problem has a finite-dimensional statistic, there is no feasible way of updating the full information state, and an approximate posterior density has to be evaluated numerically.

Herein lies the opportunity for randomized algorithms. Their bottom line is simple: the posterior distribution is approximated with a long enough sequence of samples, and probabilistic calculations are replaced with the corresponding sample averages. Their implementation can, however, become a nightmare in high dimensions.

6.3.2 Why Monte Carlo?

We have already indicated that the Monte Carlo method is a fairly old invention. The recent wave of renewed interest may therefore be a bit surprising, though there are good reasons for such a comeback.

First, the steady increase in computing power accompanied by the relative decrease in computer prices has made the routine implementation of Monte Carlo simulation affordable on fairly standard PCs.

Secondly, the methods based on linear models have approached their performance limits. This does not mean that these methods should disappear from our toolboxes, just that we cannot expect any more surprises and significant improvements in the state of the art. This has created a serious incentive for trying new paradigms.

Thirdly, powerful new sampling algorithms have appeared. The Monte Carlo efficiency is determined by the complexity of generating sample values from a given distribution. Although in the one-dimensional case the problem represents a textbook exercise with numerous possible solutions, in higher dimensions it was for a long time practically infeasible. The breakthrough in this area has been the advent of Markov chain Monte Carlo (MCMC) methods.

6.3.3   Classical Monte Carlo

Before we proceed to MCMC algorithms, we recall two traditional techniques of sampling from complex probability distributions.

6.3.3.1   Importance sampling

The trick behind importance sampling (Tierney, 1994) is the drawing of samples from a *proposal* distribution that is similar to the target one but considerably simpler to sample from. After a collection of samples is drawn from the proposal distribution, the sample set is processed so as to become a set of samples from the target distribution. The idea behind such post-processing is extremely simple – the target samples are obtained by resampling the intermediate samples with probabilities proportional to the ratio of the target and proposal probabilities.

The technique itself is a simple consequence of elementary probability calculus laws. It appeared explicitly as *sampling/importance resampling* in connection with drawing missing data patterns (Rubin, 1988). Soon afterwards it was applied – under the name *weighted bootstrap* – to the Bayesian estimation of the posterior density (Smith and Gelfand, 1992; cf. Efron, 1982).

Importance sampling works efficiently if the auxiliary distribution is close enough to the target distribution. In the opposite case, the result of importance sampling can be a degenerate set of samples – repeating just a couple of representative points – whereas the majority of the intermediate samples never get a chance. This behavior is only stressed in high dimensions.

Importance sampling is ideally suited to recursive Bayesian parameter estimation where the prior distribution is a natural candidate for the proposal distribution. It can even be extended to recursive state estimation, where the samples are additionally propagated through the state equations.

6.3.3.2   Rejection sampling

Rejection sampling (Ripley, 1987) also starts by drawing samples from a simpler proposal distribution. However, in contrast to importance sampling, the samples are immediately accepted or rejected – using a random mechanism based on comparison of the proposal and target probabilities.

Once again, rejection sampling works well if the proposal distribution is similar to the target distribution. The acceptance rate decreases to zero exponentially fast as the dimension of the underlying space increases. Hence, rejection sampling cannot be recommended for generating samples from high-dimensional distributions, but it is a very useful method for one-dimensional (sub)problems.

6.3.4 Markov chain Monte Carlo

The first algorithm of the MCMC type appeared in statistical physics as early as the 1950s (Metropolis *et al.*, 1953). It was three decades, however, before the MCMC algorithms were reinvented in global optimization and in Bayesian image restoration. In the early 1990s (Gelfand and Smith, 1990), the MCMC algorithms penetrated Bayesian statistics. Since then, the MCMC methods have become a *de facto* standard tool for approximation of probabilistic computations.

All the MCMC methods are based on a very simple idea. Suppose you need to draw a sample from a complex multivariate probability distribution. Instead of direct sampling from the distribution, you

1. design a Markov chain (stochastic process that is fully described by probabilities of all possible state-to-state transitions) whose stationary distribution coincides with the target distribution,
2. simulate the Markov chain in your computer, and
3. take the simulated values as samples from the target distribution (neglecting perhaps samples coming from the initial 'burn-in' period).

As there are many possible ways of building such a Markov chain, various MCMC algorithms have been proposed over time. From the multitude of methods and algorithms published in the last decade, some major classes of MCMC algorithms can be distinguished.

6.3.4.1 Metropolis sampler

We have noted that the techniques of importance and rejection sampling are efficient only if the proposal distribution is similar to the target distribution. In high dimensions, the design of an appropriate proposal distribution becomes a truly challenging task. The Metropolis algorithm, instead of looking for a simpler approximate distribution, makes use of a proposal distribution that depends on just the current state.

More specifically, the algorithm simulates an auxiliary stochastic process that suggests possible candidates for the samples. At every step the candidate samples are either *accepted* or *rejected* with probabilities determined by the stochastic process and the target distribution. The two most frequently seen clones of the Metropolis sampler (Tierney, 1994) are *random walk chains*, which use a random walk as a sample-proposing stochastic process, and *independence chains*, which take the candidate samples from a fixed probability distribution.

The Metropolis algorithm was originally introduced by Metropolis *et al.* (1953) for computing properties of substances composed of interacting individual molecules. Hastings (1970) relaxed the symmetric Markov chain assumption, which made it possible to consider the option of sampling from a fixed distribution. Since then, numerous modifications to the basic algorithm have been proposed, which make use of proposal distributions that give faster movement through the state space.

6.3.4.2 Example 1

A simple method of signal validation is to model the distribution of signal differences through a convex combination of three normal distributions with close-to-zero, moderate, and large variances corresponding to sensor failure, normal operation, and gross error, respectively. Assuming for simplicity that the variances can be fixed beforehand, the model has two unknowns – the probabilities of sensor failure and gross error. Estimates of these two probabilities can serve as an indicator of sensor health.

Figure 6.1 shows the posterior probability density function of the unknown probabilities, evaluated over a very dense grid of the parameter values. Note that many points in the parameter space are assigned zero probability. Evaluation at these points is avoided if only a sample from the posterior density is simulated using a Metropolis sampler. Given the sample, the posterior density can be approximated, as illustrated in Figure 6.2.

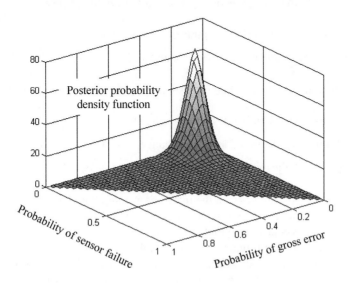

Figure 6.1 An example of the posterior probability density function for the unknown parameter being composed of the probabilities of sensor failure and gross error

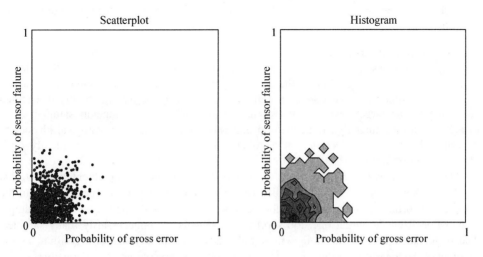

Figure 6.2 Sample probabilities of sensor failure and gross error drawn from the posterior probability density function using a Metropolis sampler

6.3.4.3 Gibbs sampler

The algorithm samples component-by-component from the full *conditional* distributions rather than directly from the *joint* distribution. The conditional distributions are typically one-dimensional and thus much easier to sample from. The Gibbs algorithm can also be extended to the case of a parameter vector composed of several blocks; sampling then goes block-by-block.

The Gibbs sampler is a natural option in the estimation of complex hierarchical or structured models. Consider, for example, the problem of joint estimation of state and parameter values. The Gibbs sampler suggests basically switching between

1. sampling of states given the previous parameter values, and

2. sampling of parameters given the previous state values.

Similarly, estimation of change time for a signal proceeds by taking samples of

1. the initial level given the terminal level and change time,

2. the terminal level given the initial level and change time, and

3. the change time given the initial and terminal levels.

The Gibbs sampler appeared first in the image processing literature (Geman and Geman, 1984). In the early 1990s, it was presented to the Bayesian community (Gelfand and Smith, 1990), where it has quickly become a tool of choice.

Note that to sample from the one-dimensional full conditionals, one can use the Metropolis algorithm where the candidate sample is taken from a kernel-smoothed estimate of the target density based on the previously accepted samples. This is an example of so-called *hybrid* algorithms, which try to combine the advantages of different sampling algorithms.

An idea similar to the Gibbs sampler is used in the *hit-and-run algorithm* (Schmeiser and Chen, 1991), where sampling is made in randomly chosen directions rather than dimension-by-dimension.

6.3.4.4 Langevin sampler

The algorithm is based on simulation of a Langevin stochastic differential equation, driven by the gradient of the logarithm of the target density and perturbed by a standard Brownian motion. The solution to the Langevin equation is known to be asymptotically distributed according to the target distribution (Rossky *et al.*, 1978). The Langevin sampler is an example of a molecular dynamics method that was used originally in statistical physics to compute phase space trajectories of a collection of molecules which individually obey classical laws of motion (Heermann, 1990).

The Langevin equation needs to be discretized before it can be implemented in a digital computer (Amit *et al.* 1991). The choice of the discretization period crucially affects the sampler performance; too long a period results in a significant deviation of the sample distribution from the target one, too short a period calls for unnecessarily many samples. The major advantage of the Langevin sampler is the simplicity of the algorithm, as the gradient of the logarithm of the target density can often be precomputed analytically.

6.3.4.5 User's choices
The application of MCMC algorithms requires solving a number of practical issues such as:

- What kind of MCMC algorithm best fits the problem under study?
- How long a simulation is 'long enough'?
- How long should the initial 'burn-in' period be?
- Is it better to run one long simulation or a number of shorter ones?
- How should the starting point (if required) be chosen?
- Can another parametrization be of help?

There are no simple answers to these questions. The truth is that the MCMC technology provides a set of tools that need to be used thoughtfully and with care. There is no free lunch solution to the general sampling problem. On the other hand, no other technology can compete currently with MCMC in the complexity of problems resolved.

6.4 Managing Complex Designs

Many human-made system problems, such as manufacturing automation, communication networks, computer performances, and resource allocation problems, involve combinatorics rather than real analysis. The objective of optimization is proposing an ideal configuration rather than tuning the design parameters. The search for optima of functions of discrete variables is known as *combinatorial optimization*.

Even today, many large-scale combinatorial optimization problems can only be solved approximately, which is closely related to the fact that these problems have been proven NP-hard. Such problems are not solvable by an amount of computation effort that is bounded by a *polynomial* function of the size of the problem. When the optimization algorithm yields a globally optimal solution in a prohibitive amount of computation time, an option is to use an approximation algorithm, which yields an approximate solution only, but in an acceptable amount of computation time.

6.4.1 Simulated annealing

Simulated annealing is a modification of the Metropolis algorithm where samples are drawn from a probability distribution parametrized by a scalar parameter, called the temperature or control parameter. The temperature is decreased between subsequent Markov chains so that it approaches zero asymptotically. The sampling distribution is related to the optimized function in such a way that it concentrates for decreasing values of the temperature over the function's local optima. If the cooling scenario is sufficiently slow, simulated annealing is able to locate even the global optimum.

Metropolis *et al.* (1953) originally proposed their sampling algorithm for an efficient simulation of the evolution of a solid to thermal equilibrium. It took about 30 years before Kirkpatrick *et al.* (1983) and, independently, Cerny (1985) realized that there exists an analogy between 'minimizing the cost function of a combinatorial optimization problem' and 'slow cooling of a solid until it reaches its low energy ground state', and that the optimization process can be realized by applying the Metropolis algorithm.

By substituting cost for energy and by executing the Metropolis algorithm at a sequence of slowly decreasing temperature values, Kirkpatrick *et al.* (1983) obtained a combinatorial optimization algorithm, which they called simulated annealing. The algorithm is now also known as Monte Carlo annealing, statistical cooling, probabilistic hill climbing, stochastic relaxation, or probabilistic exchange.

Solutions, obtained by simulated annealing, do not depend on the initial configuration and have a cost usually close to the minimum cost. Furthermore, it is possible to give a polynomial upper bound for the computation time for some implementations of the algorithm. The general applicability of simulated annealing is sometimes undone by the computation effort. A number of modifications and improvements in the MCMC vein have been suggested since the pioneering 1983 paper.

6.4.1.1 User's choices

The practical implementation of simulated annealing requires defining an annealing schedule, which involves the choice of initial temperature, deciding how many iterations are performed at each temperature (in other words, how long the Markov chain is to be), and deciding how much the temperature is decremented at each step as cooling proceeds. The cooling scenario required for the algorithm to end up (with probability one) in a global optimum is overly slow for most optimization problems. Much faster scenarios are thus used in practice, such as a geometric one proposed first by Kirkpatrick *et al.* (1983).

The choice of a particular cooling schedule affects the final cost achieved with simulated annealing. Randelman and Grest (1986) showed empirically that the expected final cost for a traveling salesman problem depends logarithmically on the *cooling rate*, defined by the ratio of a temperature decrease and Markov chain length, both of which are assumed to be fixed for simplicity. This implies that simulated annealing leaves no room for cheating; short Markov chains call for slow cooling, and fast cooling requires longer Markov chains (i.e., more samples are to be generated by the Metropolis algorithm). The cooling rate defines then how close one can get on average to the global optimum.

Simulated annealing has been used in various combinatorial optimization problems in such diverse areas as computer-aided design of integrated circuits, analytical chemistry, molecular modeling, image processing, code design, neural network theory, scheduling problems, etc. (Laarhoven and Aarts, 1987).

6.4.1.2 Example 2

Figure 6.3 illustrates the application of simulated annealing to batch recipe optimization. The batch process was pyrrole synthesis subject to hard constraints on minimum product amount, maximum by-product concentration, and maximum feedstock concentration. The objective of optimization was to minimize the makespan indicated in Figure 6.3 through a contour plot. The actual makespan was not known until the recipe was applied. The area outside the apparent 'valley' corresponds to infeasible batch recipes. The globally optimal recipe is marked with a diamond. The simulated annealing used a Metropolis sampler with independence chains drawn from a kernel-smoothened density estimate based on the previous batch runs. The global behavior of the optimizer depends on the bandwidth used for kernel smoothing. The left-hand plot shows the optimizer behavior for a bandwidth that yields a good trade-off between the speed of search and the number of off-specification batches. The right-hand plot illustrates the same optimizer behavior for a very small bandwidth, resulting in 'cautious' search with no lost production.

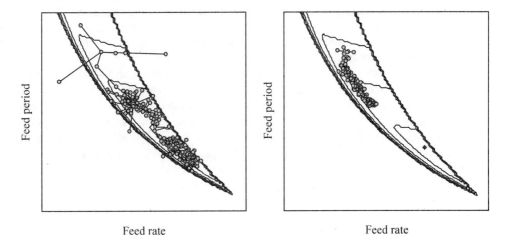

Figure 6.3 Application of simulated annealing to batch recipe optimization. Circles denote recipes proposed, lines connect past recipes with their descendants obtained through random perturbation resulting from kernel smoothing.

6.4.2   Evolutionary algorithms

Evolutionary algorithm is an umbrella term for any computer-based solution using evolution as a key element in its design and implementation. In the optimization context, evolutionary algorithms maintain a population of candidate solutions that evolve according to rules of selection and reproduction. Each candidate solution in the population is assigned a measure of its fitness. The algorithm proceeds by choosing a subpopulation of high-fitness candidates for offspring production achieved typically through recombination and mutation of selected candidates followed by the selection of survivors. The final selection gives rise to a new population (generation) of candidate solutions. The procedure repeats until a stopping condition (time, fitness) is satisfied.

Two typical representatives of evolutionary algorithms are *evolutionary programming* (Fogel *et al.,* 1966) and *genetic algorithms* (Koza, 1992). The typical genetic algorithm approach involves encoding the problem solutions as a string of representative tokens, the genome. In evolutionary programming, the representation follows from the problem. As a result, evolutionary programming is well suited for optimization of continuous functions, whereas genetic algorithms are better adapted for combinatorial optimization. In evolutionary programming, the only source of diversity is mutation. In genetic algorithms, diversity is guaranteed primarily through recombination (i.e., by exchanging genes between two 'parents' to produce two 'children').

6.4.2.1   Evolution and randomness
Selection, recombination, and mutation always involve a certain element of randomness. This is present in the random selection of individual solutions from the population, the

random choice of crossover points, the random perturbation (i.e., mutation) of the population, and the random selection of survivors to form a new generation. Evolutionary algorithms often use a stochastic tournament concept based on randomly sampling several solutions from the population and then selecting one with the highest fitness. The use of a random mechanism gives a chance (though not an equal one) to each solution in the population. As a result, evolutionary algorithms do not tend to become stuck in a locally optimal solution and can often find a globally optimal solution.

Viewed as randomized algorithms, evolutionary algorithms and simulated annealing have much in common. Simulated annealing using a Metropolis sampler with independence chains or with *parallel* random walk chains can be regarded as an example of evolutionary programming with very specific definitions of the fitness function and the selection and mutation operators. Note that the statistical theory behind MCMC algorithms provides significantly more guidance in defining the key parameters of stochastic optimization. On the other hand, the very loose framework of evolutionary algorithms provides the user with more freedom to exploit application-specific features of the optimization problem.

6.4.3 Ordinal optimization

In many real-life problems, performance evaluations are corrupted with large 'noises', which reflect the inherent random nature of the system being optimized. The performance measure is then defined as the *expectation* of the underlying utility or cost function. For such problems, even statistical simulation can appear a prohibitively expensive and time-consuming solution.

In *stochastic* optimization, one faces – in addition to the combinatorial explosion or NP-hard limitation – the problem of efficient calculation of the expected performance. In complex cases, the expectation can be evaluated only numerically, typically by sampling. Uncertainty (confidence interval) of such an estimate decreases with the square root of the number of samples. In other words, for each order of magnitude increase in the accuracy, two orders of magnitude increase in computing cost must be paid. This may become a prohibitive burden.

The concept of ordinal optimization is based on the observation that if we take enough samples of the performance measure, order the performances, and take the best N designs under this ordering, there is a high probability that among them we will find designs that are *good enough* (i.e., belonging to the top designs for the original problem). The above holds even if the performance values are quite simple (noisy) estimates of the actual performance. The designs found this way can serve as initial points for subsequent, more precise analysis.

6.4.3.1 Why it works
Technically speaking, ordinal optimization is based on two facts. First, 'order' converges exponentially fast while 'value' converges much slower (with the square root of the number of sampled designs). That is, being interested primarily in ordering, we do not need to spend so much time on the performance evaluation. The fact that it is much easier to compare values than to determine the actual difference is what distinguishes the 'ordinal' optimization from the traditional, 'cardinal' one.

Secondly, the probability of *not* finding a good enough design by randomly searching the set of possible designs decreases exponentially fast with the size of the good-enough set. The probability that *none* of 1000 uniformly sampled designs belong to the top 5% of the design space is of an order 10^{-23}. The probability that *less than five* of 1000 uniformly sampled designs belong to the top 5% of the design space is still of an order 10^{-17}. These figures, which apply regardless of the size of the design set (Ho and Lau, 1997), confirm our common sense: the more we soften the goal, the more probable (easier) it is to find a solution.

In a nutshell, ordinal optimization replaces the *best for sure* with the *good enough with high probability*. It works even with a blind search (i.e., uniform sampling of the search space). Naturally, by taking advantage of the specific structure of an optimization problem, the search behavior can be significantly improved.

Ordinal optimization was introduced by Ho *et al.* (1992) for the analysis of discrete event dynamic systems. The actual benefit of ordinal optimization depends on the shape of the performance value distribution, which may be skewed in a particular problem toward bad designs. As a result, getting within 1% of the best in order does not necessarily imply getting within 1% of the optimal value. This is critical if the performance measure quantifies the actual financial performance. The absolute size of the design set needs to be taken into account as well. Getting within 1% of a search space of 10^{10} points can still be 10^8 away from the best.

Apparently, ordinal optimization can serve as a fast route to 'good enough' starting points for subsequent analysis. Only this property is likely to give it a lot of use.

6.5 More Applications

Statistical simulation has been successfully applied in many areas in the last 10–15 years. Often the authors of the pioneering contributions were unaware of similar efforts going on in the other fields. Only recently, a bigger picture has started to appear, indicating that we are witnessing a real shift of paradigm that is likely to dramatically reshape our current view of modeling, control, and optimization.

6.5.1 Bayesian estimation

The seminal paper by Gelfand and Smith (1990) has brought the Gibbs sampler to the attention of Bayesian statisticians. The impact of Gibbs and Metropolis samplers has been enormous, and has truly revolutionized the entire field. The Bayesian methodology, interpreting everything unknown and uncertain as random, was notorious for extremely difficult computations. Very few results could be computed for models other than linear and Gaussian ones. Since 1990, hundreds of long-standing problems that were considered infeasible before have been successfully resolved. The MCMC methods have allowed statisticians to focus their attention on careful modeling – considering complex hierarchical, nonlinear, and non-Gaussian models.

6.5.2 Nonlinear filtering

State estimation for processes with highly nonlinear and possibly non-Gaussian dynamics has been a challenge to control theorists since the late 1960s. Randomized algorithms have been applied to the problem since the early 1990s, providing a conceptually simple solution, even though their practical implementation may call for additional tricks, in addition to lots of computation power. Gordon *et al.* (1993) applied *importance sampling* to a multiple-target-tracking problem, basically turning the estimation problem into parallel simulation of a sufficient number of state trajectories. O'Sullivan *et al.* (1993) proposed to solve the target-tracking problem using the *Langevin sampler,* considering the option of massively parallel computation of the gradient of the logarithm of the state density.

6.5.3 Bayesian image restoration

Bayesian estimation of the whole image is an extremely high-dimensional problem where the size of the unknown parameter vector, which is determined by the number of pixels to be restored, is typically on the order of 10^5–10^6! Geman and Geman (1984) resolved the problem by introducing a two-dimensional version of the Gibbs sampler, taking advantage of local correlation of data over a small enough neighborhood of the current pixel. The paper has profoundly influenced the Bayesian statistics, image processing, and neural networks communities.

6.5.4 Robust control design

In robust control design, given a family of plants and a family of controllers, the objective is to find a single fixed controller that performs reasonably well for almost all plants. The randomized algorithm proposed by Vidyasagar (1997) to find a '*probably approximate near minimum*' of an objective function proceeds by taking a sufficient number of samples of plants and controllers, evaluating the sample average of the objective function for each controller, and finding the controller minimizing the sample average.

There have been many other attempts to use randomized algorithms in control and optimization, including an MCMC implementation of predictive control (Chen *et al.,* 1993) and a randomized implementation of dynamic programming (Rust, 1993).

6.5.5 Network computing models

Recently, Microsoft Research has established a new fundamental research group that focuses on a statistical physics approach to discrete probability theory, combinatorics, and theoretical computer science. As computer systems and networks become increasingly large and complex, statistical physics appears to be the appropriate language to describe their operation. Probabilistic methods can be employed to predict the average behavior of these systems and to identify potential phase transitions. Combinatorial methods can be used to 'count' configurations contributing to a given type of behavior.

6.6 Practical Aspects

The unprecedented power of stochastic simulation does not come for free. The theoretically optimum procedures can rarely be implemented in practice without making further compromises.

6.6.1 Probabilistic behavior

Compared with closed-form analytic solutions or iterative calculus-based optimization schemes, which currently prevail in industry, stochastic simulation has a probabilistic behavior, which requires from the end user some caution in interpreting its results.

The time available for managing complex models or designs is rarely sufficient for approaching the global optimum. The search is often stopped well before that.

Due to its probabilistic nature, stochastic simulation may return to the same question a (more or less) different answer, depending on the particular random sequence drawn. Stochastic simulation may even fail occasionally, being increasingly vulnerable to the quality of the (pseudo)random sequence generated by the computer.

The human check of results produced by stochastic simulation is welcome or even required. Otherwise, the algorithm needs to be complemented with additional measures to make it more robust, as suggested below.

The fact that the current manufacturing conditions and business environment are changing very quickly makes things only worse. Attempts to solve modeling and optimization problems once and for all make little sense here. Rather, we need a way of steadily moving in the right direction – improving system performance and adapting to changes at every step.

6.6.2 Robust simulation

As so often happens in engineering, process knowledge and problem understanding make life a bit easier than it may seem from a general theoretical perspective.

Any prior information or structural assumptions on the problem under study can significantly limit the search space. Time spent on gathering and incorporating such information is almost always rewarded by faster and more reliable simulation.

Instead of trying to explain the behavior of a complex process using a single, global model, it is often sufficient to fit its response to the current operating conditions and current control scenario. The use of multiple or local models radically reduces the complexity of the original problem (see Figure 6.4).

Similarly, instead of trying to construct a global decision rule, it is often sufficient to search for a local decision rule that works well for the current operating conditions. 'Local design' appears here as a natural counterpart to 'local modeling'.

Data representing past experience need no longer be ignored. The current performance of computer, database, and communication technology allows us to retrieve in real time any part of relevant process history, including controls applied in the past.

With process history available, optimization can start from the best design applied in the past under similar conditions. This option turns standard iterative optimization into optimization with iterations 'spread over time'. Starting from best practices eliminates repeating old mistakes, whereas changes to the design are made only when they are likely (in a probabilistic sense) to improve process performance.

Last but not least, models and designs that might result in lost or off-spec production, or equipment failures and breakdowns, have to be eliminated explicitly by force. Such a rule-based protection layer – always met in practical applications – can also be interpreted as a special kind of prior information about the problem.

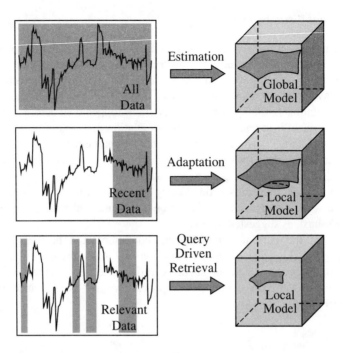

Figure 6.4 Comparison of global, local-in-time, and local-in-data modeling approaches. The plots show time series of data; the data cubes indicate the model-based data fits.

6.6.3 Scope of applications

Stochastic simulation has already become a powerful tool for offline multidimensional data analysis, offline estimation of complex, hierarchic, or otherwise structured models, and advanced decision support. In these cases, human experience and intuition can easily be involved if the simulation results leave the end user with too much uncertainty.

Penetration of stochastic simulation into *automatic* control and decision making is likely to be slower. Apart from purely technical issues, such as computer performance sufficient for real-time applications (and the cost of it), there are liability issues and psychological barriers as well.

'No bridge is built to survive everything'. Actually, no technical work is built this way; engineering design has always been a compromise between functionality and price. Software, however, has been perceived for years as a domain where one is in 'full control'. This no longer seems to be true, even if we put aside the vital issue of verification of increasingly complex computer programs. The problems encountered currently in industry and business are of such complexity that we cannot hope for their analytical solution. The end user thus must face the fact that software can fail occasionally, simply because feasible solutions have to be approximate, incomplete, and cost compromising.

6.7 Conclusion

Randomized algorithms are no panacea, but they offer a systematic and consistent approach to hard problems. The Monte Carlo computation engine is likely to turn our good old *error-free* computers, where the problem has been solved exhaustively or not solved at all, into *error-controlled* probabilistic machines, where the problem is always solved (with high probability), only with greater or smaller error.

While the key algorithms are now ready to use, a number of practical issues remain to be resolved. For instance, in safety-critical applications, such as flight control or control of complex chemical reactions, the occasional drop in algorithm performance (due to their random nature) needs to be detected and eliminated using a conventional stand-by controller. The convergence of the sample distribution to the target distribution has to be monitored. Supervisory logic needs to be developed so as to decide automatically about the initial conditions, lengths of simulation runs, algorithm termination, resetting, and so on.

References

Amit, Y., Grenander, U. and Piccioni, M. (1991) Structural image restoration through deformable templates. *Journal of the American Statistical Association*, **86**, 376–387.

Cerny, V. (1985) Thermodynamical approach to the traveling salesman problem: an efficient simulation algorithm. *Journal of Optimization Theory and Applications*, **45**, 41–51.

Chen, L.-S., Geisser, S. and Geyer, C.J. (1993) Monte Carlo Minimization for Sequential Control. Technical Report No. 591, School of Statistics, University of Minnesota.

Efron, B. (1982) *The Bootstrap, Jacknife and other resampling plans*. SIAM, Philadelphia.

Fogel, L.J., Owens, A.J. and Walsh, M.J. (1966) *Artificial intelligence through simulated evolution*. Wiley, New York.

Gelfand, A.E. and Smith, A.F.M. (1990) Sampling based approaches to calculating marginal densities. *Journal of the American Statistical Association*, **85**, 398–409.

Geman, S. and Geman, D. (1984) Stochastic relaxation, Gibbs distributions, and the Bayesian restoration of images. *IEEE Transactions on Pattern Analysis and Machine Intelligence*, **6**, 721–741.

Gordon, N.J., Salmond, D.J. and Smith, A.F.M. (1993) A novel approach to nonlinear/non-Gaussian Bayesian state estimation. *Proceedings of IEE-F*, **140**, 107–113.

Hastings, W.K. (1970) Monte Carlo sampling methods using Markov chains and their applications. *Biometrika*, **57**, 97–109.

Heermann, D.W. (1990) *Computer simulation methods in theoretical physics* (2nd ed.). Springer-Verlag, Berlin.

Ho, Y.C. and Lau, T.W.E. (1997) Universal alignment probabilities and subset selection for universal probabilities. *Journal of Optimization Theory and Applications*, **93**, 455–490.

Ho, Y.C., Sreenivas, R.S. and Vakili, P. (1992) Ordinal optimization in DEDS. *Journal of Discrete Event Dynamic Systems*, **2**, 61–68.

Kirkpatrick, S., Gelatt, Jr., C.D. and Vecchi, M.P. (1983). Optimization by simulated annealing. *Science*, **220**, 671–680.

Koza, J.R. (1992) *Genetic programming: on the programming of computers by means of natural selection*. MIT Press, Cambridge, MA.

Laarhoven, P.J.M. van and Aarts, E.H.L. (1987) *Simulated annealing: theory and applications.* Kluwer, Dordrecht, The Netherlands.

Metropolis, N., Rosenbluth, A.W., Rosenbluth, M.N., Teller, A.H. and Teller, E. (1953) Equations of state calculations by fast computing machines. *Journal of Chemical Physics,* **21**, 1087–1091.

O'Sullivan, J.A., Miller, M.I., Srivastava, A. and Snyder D.L. (1993) Tracking using a random sampling algorithm. *Proceedings of the 12th IFAC World Congress*, Sydney, **5**, 435–438.

Randelman, R.E. and Grest, G.S. (1986) N-city traveling salesman problem: optimization by simulated annealing. *Journal of Statistical Physics*, **45**, 885–890.

Ripley, B.D. (1987) *Stochastic simulation.* Wiley, New York.

Rossky, P.J., Doll, J.D. and Friedman, H.L. (1978) Brownian dynamics as smart Monte Carlo simulation. *Journal of Chemical Physics*, **69**, 4628–4633.

Rubin, D.B. (1988) Using the SIR algorithm to simulate posterior distributions (with discussion), in *Bayesian statistics 3* (eds. J.M. Bernardo, M.H. DeGroot, D.V. Lindley, and A.F.M. Smith). Oxford University Press, 395–402.

Rust, J. (1993) Using Randomization to Break the Curse of Dimensionality. Technical Report JR1, Department of Economics, Yale University.

Schmeiser, B. and Chen, M.-H. (1991) General Hit-and-Run Monte Carlo Sampling for Evaluating Multidimensional Integrals. Technical Report, School of Industrial Engineering, Purdue University.

Smith, A.F.M. and Gelfand, A.E. (1992) Bayesian statistics without tears: a sampling–resampling perspective. *American Statistician*, **46**, 84–88.

Tierney, L. (1994) Markov chains for exploring posterior distributions. *The Annals of Statistics*, **22**, 1701–1762.

Vidyasagar, M. (1997) Statistical learning theory and its applications to randomized algorithms for robust controller synthesis. *Plenary Lectures and Minicourses, European Control Conference* (eds. G. Bastin and M. Gevers), 161–189.

7

Complexity Management via Biology

Blaise Morton and Tariq Samad
Honeywell Technology Center

7.1 Introduction

Some of the best examples of control are to be found in natural systems. Mundane activities that we and other biological organisms perform unerringly are well beyond the capabilities of our best artificial control systems. Walking on two (or more) feet, riding a bicycle, catching a fly ball (or a fly (Lettvin et al., 1959)) are feats that we take for granted but that we have not been able to realize in the machines of our design.

The automation systems that human ingenuity has produced are remarkable in their own right. Aircraft fly higher, faster, and with greater payload than any life form, for example. Yet there is a critical biological element in such marvels of the technological age too. Virtually all human-engineered flight requires biological systems by way of pilots and/or ground control. We have been successful in automating flight under normal, nominal conditions, but we have no recourse other than to fall back on the human element when unforeseen situations arise. For example, today's flight management systems can fly aircraft automatically from takeoff to landing, without pilot input, as long as the vehicle experiences no major failures and the flight plan remains fixed. Other engineering achievements in automation and control technology are similar in this respect. An oil refinery, which can contain more than 10,000 individual control loops, is under automatic control for normal operation at or near steady state conditions. In the absence of equipment degradation, changes in the production mix, new crude shipments, changing weather conditions, and so on, it would run on 'autopilot'. In fact there are hundreds of technical staff on site, including operators, maintenance personnel, and control engineers, testifying to the irreplaceability of the human element today.

Human operators are far better than machines in identifying unexpected problems and appropriate responses. In many applications this is the primary reason the human remains in the loop. Autonomy, in the true sense of the word, remains the exclusive preserve of biological systems. The drive toward increasing autonomy in complex systems will only succeed if some of the capabilities of biological systems are incorporated into our engineered systems. This does not necessarily imply that designs for autonomous engineered systems must be driven by detailed biological analogies. Achievements of biological systems can, in many cases, be duplicated with unrelated mechanisms: airplanes and birds represent two markedly different solutions to the problem of moving heavier-than-air bodies through the air. Yet given that we have little idea today on how to engineer autonomous systems, it surely behooves us to study the biological solution to complexity management in some depth.

Our emphasis on autonomy above needs a qualification. Even if we had the requisite technology in our hands today, it is inconceivable that we would be operating pilotless commercial transports. The goal of autonomy is a long-term, and perhaps solely an intellectual, one. From a pragmatic viewpoint, any movement toward greater autonomy represents progress for technologists and their sponsors. Semi-autonomous systems are of at least as much interest. In many cases, increasing the degree of autonomy of a system will require subsuming under automation functions that are currently being performed by people.

We focus here on biological central nervous systems (CNSs). One might wonder why we have selected just this one component of biological systems to study rather than the organism as a whole. Biological sensors and actuators are also superior for some applications to anything we have been able to construct. We still rely on trained dogs for tracking missing humans and intercepting illegal drugs, and in general high-performance olfactory sensing is still the more or less exclusive preserve of biology. Similarly, muscles are hard to beat for certain types of actuation. We believe there is much for us to learn from biology in many respects. We limit our focus partly for pragmatic reasons and partly because our interest here is primarily in the decision making 'intelligence' exhibited by biological systems that we would like to incorporate within our designed automation and control systems. A discussion on biological sensors appears in Chapter 8.

We should note also that certain features of central nervous systems are not desirable for engineering applications. For example, CNSs are hard to design and fabricate to specification, they make mistakes, are hard to monitor, difficult to repair, are not ruggedized, and they require sleep. Our objective is to find ways to design engineering systems that realize the benefits of biological systems, but avoid their drawbacks.

7.2 The Central Nervous System

In higher animals, the nervous system, the endocrine system, and the immune system are the three main control divisions of the body. In a healthy individual, these three divisions work together for the homeostasis and survival of the organism as a whole. All three monitor and help regulate internal states of the body, but only the nervous system concerns itself with the external (extracorporeal) environment: it is the CNS that is associated with sensing and reacting to the external stimuli.

The role of the CNS as the primary sensorimotor control system is generally accepted as scientific fact. The basic control system features of the CNS and their sensorimotor functions are described in standard texts in neuroanatomy (e.g., Kandel *et al.*, 1991); there are neuronal theories of sensory inputs, data processing, and effector outputs to muscles and glands. However, although the control system structure of the CNS is generally accepted, the analysis of the CNS is usually not performed from a control-theoretic perspective. CNS experts tend not to be control experts.

7.2.1 Neurons – the building blocks of the brain

If we look at the brain from the 'bottom up', the building block is the nerve cell, or neuron. There are roughly 10 billion such cells in the human brain, along with an order of magnitude more glial cells that play a supporting role. The nerve cell can be analyzed as an information processing element, with inputs to it arriving via dendrites, being communicated through the dendritic structure to the cell body, or soma, and output signals transmitted over the cell's single axon to synapses that connect (in an electrochemical sense) to dendrites of topological neighbors. Communication between neurons across the synaptic gap thus takes place via neurotransmitters that are expelled from the axon of the source neuron and are absorbed by receptors on the surfaces of the dendrites. Individual cells can be coupled to thousands or more of other cells, so information processing in the brain is very much a parallel operation. Overall, the human brain contains something like 10^{11} neurons and 10^{15} synapses.

Generally speaking, the CNS is bicameral; it is symmetric with respect to reflection through the sagittal plane (i.e., the left and right halves are close to mirror images of each other). Most sensory inputs and motor control outputs cross the sagittal plane at some point in the communication path. So sensory signals from, and action signals to, the right half of the body are associated with the left brain. The cross-lateral connection isn't total, however; for example, part of each eye's input is shared with both sides of the brain during visual processing.

Given the richness of interconnection within the brain, it is no surprise that much of its volume is taken up by axons and dendrites rather than by the cell bodies. In fact, the 'gray matter' of the brain consists largely of the cell bodies, and the white matter is the interconnection structures. In the mammalian brain, a distinctive part of the white matter is the corpus callosum that forms a communication bridge between the right and left halves. In humans, the corpus callosum is estimated to contain 300 million axons.

7.2.2 Some features of biological nervous systems

Several general features of the central nervous system architecture are worth noting and contrasting with our conventional computing technology. First, as already remarked, the CNS has a fully distributed architecture. Not all of the 10 billion neurons in the human brain are operating independently in parallel, but any stimulus or any cognitive activity is associated with large populations of neurons. Similarly, memory and representation in the brain is distributed; individual concepts are not mapped to individual neurons. Neurons in the cerebral cortex tend to fire collectively in activated fields approximately four square centimeters in area.

Unlike digital computing, the elementary computational elements in the brain are slow. Whereas today's logic gates in microprocessors have processing delays that can be measured in nanoseconds, neurons operate at time scales between 10 and 100 milliseconds. This makes some of the sensing and control achievements of biology all the more incredible. Consider a baseball batter hitting a 90+ mph fastball more than 400 feet: the control system, including the sensing and actuation, has to complete its functioning in less than half a second. Our engineered systems are not even close to achieving this level of performance, but any attempt we might make in this direction would require substantial computing power – millions of processing steps would be involved. By contrast, the batter's brain requires perhaps a few dozen sequential computational 'steps'. There is much that is still mysterious about this level of performance, but the parallelism of the brain from a computational perspective is certainly one key enabler.

Neurons are analog computational elements. Interneuron communication is primarily through synaptic transmission of neurotransmitter molecules. In most cases, the influence of one neuron on another is a continuous quantity (of limited precision), not an all-or-none effect. Digital computing brings many benefits, but a notable disadvantage is energy efficiency. As might be expected, taking a real-valued quantity such as the output voltage of a transistor and treating it as equivalent to just one bit of information is a waste of energy. A 'back-of-the-envelope' calculation performed by one of the pioneers of today's digital computers led him to declare that 'The brain is a factor of 1 billion more efficient than our present digital technology' (Mead, 1990).

The building blocks of semiconductor-based computing are extremely reliable, but computing architectures are brittle. Damage to a few transistors or a few wiring shorts can render a microprocessor inoperable. Biological brains present quite a contrast. Neurons atrophy continually. A human infant is born with virtually its full complement of neurons, and the adult human is estimated to lose about 50,000 neurons per day. Yet performance, whether intellectual, sensory, or motor, is increased or maintained for decades. Ultimately, degradation is usually gradual and not catastrophic.

Finally, biological organisms are able to adapt to their environments automatically. Adaptation occurs both at the individual level, where the organism, through trial and error, mimicry, explicit instruction (in the case of some species), or other 'nurture' means, changes its behavior, and at the species level, through the mechanism of natural selection. This latter is critical for all taxa – no biological system is born as a complete blank slate, and theories that postulated a minimal role for genetic determination (e.g., behaviorism) have long since been discredited. At the individual level, adaptation is not exhibited at the lowest rungs of the biological ladder, and as a general rule, the higher the level of intelligence exhibited by an organism – that is, the greater its ability to effectively manage its existence in a changeable, complex world – the more prominent the role of learning over the lifetime of the individual. This observation correlates with the increasingly helpless nature of infants of higher species.

Table 7.1 summarizes the differences in computational characteristics between biological and computational information processing.

Table 7.1 Some differences between conventional and (biological) neural computation

Conventional	*Neurobiological*
Serial	Parallel
Fault intolerant	Fault tolerant
Fixed	Adaptive
Digital	Analog
Fast, reliable components	Slow, failure-prone components

The state of neither our knowledge nor our technology permits modeling the brain with high fidelity. We can, however, develop computational methods that are brain-inspired. The hope is that some aspects of biological computation that are essential to the sophisticated control behavior it helps realize can thereby be captured. Arguments of vitalism aside, it is difficult to dispute the notion that more brain-like computational models should help produce more brain-like performance in areas such as controls.

7.3   The CNS as an Intelligent Control System

When we view the CNS as a control system architecture, we find the treatment in most texts incomplete in two respects. First, the description is incomplete because researchers in the neurosciences have not yet determined all the features of the CNS architecture. This problem will keep researchers busy for years to come, but steady progress is being made by those studying features of the physical architecture of the CNS (identifying types of neurons, neurotransmitters, synaptic circuits, etc.). Secondly, the description is incomplete because some important architectural features are simply not addressed. To some extent, this second issue can be viewed as a consequence of the first, but there are architectural issues that are crucial for designing engineering systems. These issues should be addressed even though, in the absence of a more complete picture at the level of the physical architecture, speculation will be necessary. Examples of these issues include the following:

- What are the functional requirements of the CNS? These requirements need to be articulated hierarchically, starting at the level of the brain as a whole.

- What are the input-output signal requirements, in terms of accuracy, resolution, timing, and throughput? Again, these need to be discovered at all hierarchical levels. We have a reasonably good understanding of signal requirements at the neuronal level and for some cortical elements (e.g., visual cortex).

- What are the principal operational modes of the brain? Candidate modes might be sleep, resting, alert.

- How is failure management accomplished?

As a general modeling methodology, we recommend the HIPO (hierarchy, input, process, output) method (Yourdon and Constantine, 1979). In this method, which is often used for describing engineering system architectures, one starts with a data-flow diagram depicting the entire system as a network of boxes (representing functional processes) interconnected by labeled arrows. The arrows entering a box represent its input elements, whereas the arrows leaving a box represent its output elements. The inputs, functions, and outputs of each box are described, but no specifying description of the algorithm in the box is needed. Instead of specifying the algorithm in each box, one represents it with another data-flow diagram. This second data-flow diagram sits on the second level of a hierarchy that is constructed recursively: each box in a data-flow diagram on the nth level becomes a data-flow diagram on the $(n+1)$st level. The hierarchy grows until every process has been expanded (hierarchically) into data-flow diagrams, all of whose processes are primitive operations (i.e., nondecomposable). The physical architecture and the architecture dynamics have representations on each level of the hierarchy.

7.3.1 Sensory and motor pathways

One structural feature to note is the lengthwise division of the CNS into a posterior region for sensing and an anterior region for acting (Fuster, 1997). This dichotomy is evident all the way from the spinal cord up to the cerebral cortex. In primates, the division between these two regions in the cerebrum is the central sulcus. On the sensory side of the central sulcus, sensory data is fed to the neocortex by thalamic neurons that project to the primary sensory areas. The processing of sensory data in the cortex is distributed and occurs in stages, along functionally distinct neural pathways. At low levels the processing is unimodal (there is no blending of different senses); at higher levels the processing transitions to a multimodal, more abstract form in the sensory association cortex.

On the motor side of the sensorimotor divide, a similar, but reversed, hierarchical flow takes place. The highest level of associative motor cortex is in the prefrontal area of the brain – this area is responsible for temporal organization of behavior critical in the planning of nonimmediate event-triggered responses. The processing of motor data (commands) proceeds from the top level down, flowing from the prefrontal cortex to the primary motor cortex in the precentral gyrus. The primary motor areas represent the basic building blocks of action.

All behavioral responses do not require traversals by signals of the full sensor and motor hierarchies, however; such a rigid strategy would result in time lags and inefficiencies in many situations where higher level processing was not necessary. Instead, the hierarchical pathways are complemented by others that feed sensory data at intermediate levels of processing to appropriate levels in the motor cortex (Figure 7.1). Thus, primary sensor connects to primary motor and unimodal association connects to premotor. In this way, the perception-action cycle can be executed without the involvement of higher levels of the hierarchy when such a response is appropriate. Each level of the hierarchy can inhibit the lower level responses by feedback when higher level processing is appropriate.

7.3.2 Algorithmic processing

Discovering the biological algorithms of cerebral processing remains an active area of research (Koch and Davis, 1994). One of the more intriguing theories in this area is Mumford's (1994; 1996) pattern theory. The basic idea of pattern theory is the hypothesis that the brain synthesizes internal firing patterns of neurons for comparison with the firing patterns produced by outside world stimuli (inputs from the thalamus). Errors between the internally and externally generated signals (modulo warping) are computed in low-level areas and fed back to high-level areas for assessment. If the error is not sufficiently small, a modified pattern is generated and the process is iterated. Mumford's approach is especially promising from our point of view, since it provides a mechanism for command generation and decision-making processes, and as it is close to the standard control-theoretic systems approach. Practical questions remain concerning how a pattern theoretic algorithm might be implemented in the CNS (e.g., where and how are patterns generated and compared, where is simulation performed, what role might a region like the hippocampus play for learning or regenerating new patterns). Perhaps the rapidly growing area of brain imaging (see Roland, 1993) will provide some insight into these matters.

Pattern theory does not address all the aspects of intelligent systems, and there are other features we want to consider. One of these is learning. When speaking of cognitive learning, it is clear that memory should be considered as well, given the conceptual link between storage and retrieval. It is now well established that cognitive learning requires the presence of the hippocampus, although the activities required in the learning process have not been identified. Procedural learning, the learning of motor tasks, is even less well understood. It seems to involve motor cortex, cerebellum, basal ganglia, ventrolateral thalamus, red-nucleus, substantia nigra, and pontine nuclei. Neuroimaging methods have been used to monitor brain activity during learning studies, but the results obtained from different experiments do not allow a simple explanation.

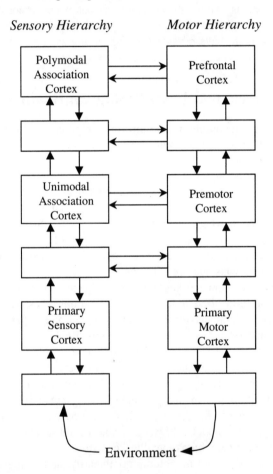

Figure 7.1 Cortical regions and associated signal pathways involved in sensing and action. Unlabeled boxes represent intermediate areas. From Fuster (1997).

Another aspect of the CNS as an intelligent control system should be mentioned. If we consider motor functions, we find that the neocortex does not work alone in controlling voluntary motion. There are two major subsystems closing loops on the voluntary motor-output side: the cerebellum and basal ganglia (see Figure 7.1). Although these two feedback loops undoubtedly have many functions, their better documented features can be inferred from the neural circuitry (see Chapter 35 of Kandel *et al.*, 1991). In a healthy animal, the cerebellum helps correct errors between muscle commands and responses (closing inner loops, in control jargon) while the basal ganglia work on coordinating more complex motions (closing outer loops). There is good evidence that the basal ganglia directly influence cognitive functions as well, so a cognitive theory will need to include the dynamic effects of the closed cerebro-striato-thalamic loop.

Finally, we should briefly mention the role of motivational or emotional subsystems. Certainly the CNS includes such a system, which is associated with the limbic and autonomic nervous systems. Although our emphasis has been on cortical functions, in all mammals these functions exist on top of a foundation provided by the older brain. Emotions and affective displays might have a value for engineering systems, although at our current state of understanding a focus on neocortex seems appropriate.

7.4   Lessons Learned from Biological Brains

Our call for a brain-inspired computing and information processing technology is hardly the first one. It is instructive to review some previous developments that have arisen out of motivations similar to ours.

Above we contrasted the digital nature of today's computers with the analog processing of the nervous system. In fact, one of the progenitors of the digital age was a pioneering mathematical model for biological neurons. In 1943, Warren McCulloch and Walter Pitts published a classic paper in which they laid out a theory for 'nervous nets' in which the model neurons operated according to an 'all or none' law of nervous activity (McCulloch and Pitts, 1943). Today, we would call their neurons logic gates. McCulloch and Pitts demonstrated that their nervous nets were Turing equivalent.

Interest in nervous nets, connectionist models, and neural networks – all terms used for parallel, distributed, brain-inspired models of computation – has waxed and waned ever since. Today, 'artificial neural networks' are an accepted and valuable tool in the signal processing, controls, and related communities (Haykin, 1994). There are many differences between today's models of nerve cells and the first ones. Analog processing, instead of digital, is perhaps the most important. Artificial neural networks have not lived up to the hyperbole that they were generating a decade and a half ago, but they have been successful in solving challenging problems in many domains. They are used generally as nonlinear function approximation (generalized, multivariable curve-fitting) and nonlinear classification or pattern recognition models. Most artificial neural networks rely on simplified formulations of neuron operation, typically consisting of a weighted summation of (dendritic) input signals, a smooth monotonic nonlinearity, and a single-valued output that can then serve as input to other artificial neurons (see Figure 7.2). In some cases, first order dynamics are also incorporated within the processing element.

A key reason for the progress in digital computing is the availability of theories and techniques for reliably designing digital circuits and associated firmware and software. Similar technology for other models of computation is simply unavailable today, and this state of affairs is unlikely to change in the foreseeable future. Some researchers have taken the tack of looking to biology again for the solution to the design problem. Optimization methods modeled after biological evolution have been used for this purpose (see also Chapter 13). Genetic algorithms are the most popular of these methods (Holland, 1975).

Their application to artificial neural network design is illustrated in Figure 7.3 (Harp and Samad, 1991). A population of networks is represented by a coding scheme, analogous to genetic chromosomes. The chromosome for an individual network can be instantiated into an actual neural network, which can be trained and tested on different stimuli. A fitness measure for the network can then be calculated based on various application-specific criteria, including the network size, its learning rate, and its accuracy on training and test data. Once fitness measures for the whole population are computed, the population is sampled and selected 'parents' generate 'offspring' networks. Genetic operators modeled after biological mechanisms of crossover and mutation allow the offspring nets to inherit attributes of their parents while adding diversity to the gene pool.

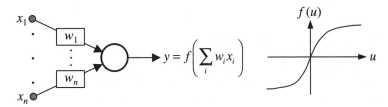

Figure 7.2 Typical computing element for artificial neural networks

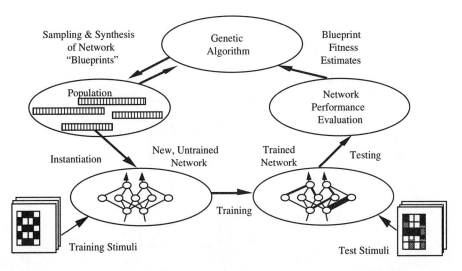

Figure 7.3 Genetic synthesis of neural network architectures illustrating an application for pattern recognition

Inspiration from biology has been drawn not just at the neural level but also at the cognitive level. Expert systems and fuzzy logic systems represent attempts to capture some of the intriguing higher level reasoning capabilities of humans (Kandel, 1991). Again, many useful applications have resulted from this line of research, which can also be considered reasonably mature and established today, even if the grand visions of the original enthusiasts have not been realized.

We note that biologically motivated models of computation have exploited our understanding of both neural- and cognitive-level brain operation. What we believe is missing are significant attempts to capture the architecture of brains.

7.5 Architectural Outlines

In this section we discuss vertebrate and invertebrate architectures for biological information processing. These preliminary architectures are presented, not as solution approaches that in themselves are likely to be practically useful, but rather as examples of the sorts of investigations that can be undertaken to better understand CNS designs for potential engineering applications.

7.5.1 The vertebrate CNS architecture

Figure 7.4 is an oversimplification intended to represent some gross features of the information flow in a vertebrate CNS. The names of the various parts of the brain in the figure are those used for primates, but an analogous picture applies to vertebrates in general. A more complete architecture can be found in Butler and Hodos (1996).

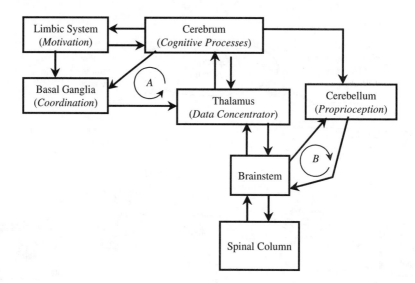

Figure 7.4 Simplified architecture of the primate central nervous system

As illustrated, signals to/from the body pass through the spinal column from/to the brainstem, which consists of the medulla, pons, and midbrain. Those signals from the body destined for the forebrain enter by way of the thalamus, whereas other signals from the body are directed to the cerebellum. Sensory data sent to the cerebrum are processed, fused, and interpreted according to cognitive processes in the cerebrum. The cerebrum is the seat of the motor cortex, which initiates the control signals sent to the muscle effectors, but in many cases a response to a stimulus is not immediately sent. Instead, feedback loop A, involving the basal ganglia and (perhaps) the limbic system, is engaged. The limbic system can help determine an appropriate response based on memory (via the hippocampus) and/or on emotions such as fear (amygdala). The basal ganglia appear to help coordinate complex responses involving concurrent motions of multiple muscle groups. The decision whether to respond immediately to a stimulus or wait for the result of feedback path A appears to be a function of the prefrontal cortex in the cerebrum (Fuster, 1997).

While feedback loop A is involved entirely with the generation of an appropriate response (planning/scheduling), feedback loop B appears to be involved entirely with control of that response (execution). A large bundle of neurons sending signals from the cerebrum to the cerebellum are presumably relating the commanded responses desired from the skeletal muscles – throughout the body. These commands can be compared with proprioceptive feedback indicating the actual motions of the body to generate corrective commands to the muscles through feedback loop B.

As a final observation concerning this diagram, it is worth noting that the reticular formation of the brainstem plays an important role in controlling the state of the higher brain functions. Different nuclei in the reticular formation are capable of moderating the level of activity in selected regions of the higher brain. The level of control ranges from putting those regions to sleep to rousing them to peak levels of activity. The criteria used by the reticular formation to change levels of brain activity are no doubt dependent in part on signals provided by the sensory cranial nerves, many of which enter the CNS through the brainstem. That would explain why sleeping animals wake quickly after a loud noise.

The previous discussion is highly simplistic and ignores many other functions and signals involved in the vertebrate CNS architecture. It is intended only to give nonexperts some idea of how the CNS functions.

7.5.2 The invertebrate CNS

In contrast with vertebrates, invertebrates have much more primitive CNSs, even in the more advanced species. For example, the crayfish CNS contains fewer than 200,000 neurons (Bullock and Horridge, 1965). A crude depiction of a crustacean CNS is shown in Figure 7.5. The brains of crustacea are fairly uniform in terms of internal gross anatomy. The brain directly controls functions of the compound eyes and antennae and is the site of visual and olfactory processing. It is also the origin of higher level motor commands passed to the remaining CNS, as in the case of the vertebrate brain.

Caudal to the brain is the stomatogastric system, which contains the esophogeal ganglion, the commissural ganglia, and the gastric ganglion, each of which contains fewer than 1000 neurons. These ganglia control portions of the stomach and esophagus (as the name suggests) and also appear to provide some primitive vestibular sense of spatial orientation.

The subesophageal ganglion of a crayfish (actually, a fusion of several ganglia) contains perhaps 7000 neurons. It contributes to the control of the:

1. Esophogeal smooth muscles;

2. Mandibles, maxillules, and maxillae (specialized appendages used for ingesting food);

3. Stomach and heart muscle.

The five thoracic ganglia of a crayfish contain around 9000 neurons, whereas the six abdominal ganglia contain fewer than 4000. The thoracic ganglia control the five pairs of walking legs. The last abdominal ganglion of a crayfish is usually larger than the others because of its control function for the tail. Both thoracic and abdominal ganglia help regulate digestive/excretive functions as well.

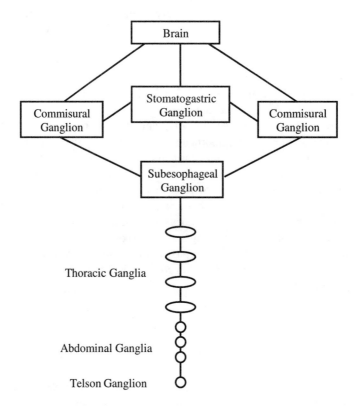

Figure 7.5 Simplified architecture of the crustacean central nervous system

As illustrated above, there are important differences between invertebrate and vertebrate nervous systems, such as the following:

- The invertebrate CNS is relatively simple compared to the highly complex brains of vertebrates.

- As a consequence, individual intelligence is low for invertebrates. On the other hand, they tend to be considerably more populous – the group survival rate is thus high.

- Invertebrate CNSs exhibit much greater specialization and hence less flexibility and adaptability. The neurons responsible for motor commands are often tailored for a single motion with little allowance for context-sensitive variation. In vertebrates, learning enables a wide range of behavior – but at the expense of greater vulnerability prior to learning.

- In invertebrates, the CNS is not localized in a particular anatomical area such as the skull. The control of the organism is therefore distributed. If a significant part of the CNS is destroyed, some functions can still be maintained.

The information processing capabilities of invertebrates may be primitive compared to higher organisms, but invertebrate behavior is also remarkable in its ability to deal with complex environments. We have much to learn from the lower reaches of the biological world as well.

7.6 A Philosophical Controversy

The notion of developing engineered sensors or actuators, or even low-level models of computation, that are based on biologically gleaned principles is uncontroversial. Embodying higher level cognitive capabilities in computational systems, however, is another matter. There are philosophers (and some scientists) who argue that some human capabilities cannot even in principle be realized by the sorts of machines we are contemplating. The levels of autonomy, intelligence, and adaptability exhibited by humans are thereby excluded (the argument goes) from realization in engineered systems.

Two theoretical limitations of formal systems appear to be driving much of the controversy – the issue under debate is whether humans, and perhaps other animals, are not subject to these limitations. First, we know that all digital computers (as the term is used today) are 'Turing-equivalent' – they differ in processing speeds, implementation technology, input/output media, and so on, but they are all (given unlimited memory and computing time) capable of exactly the same calculations. More important, there are some problems that no digital computer can solve. The best known example is the halting problem – we know that it is impossible to realize a computer program that will take as input another, arbitrary computer program and determine whether or not the program is guaranteed to always terminate.

Secondly, by Gödel's proof, we know that in any mathematical system of at least a minimal power (even arithmetic is covered) there are truths that cannot be proved and falsehoods that cannot be disproved (Nagel and Newman, 1958). The fact that we, as humans, can demonstrate the incompleteness of a mathematical system has led some to conclude that Gödel's proof does not apply to humans.

The concept of consciousness lies at the center of this controversy. We take it as given that human-like performance by a machine requires the machine to have something akin to consciousness – an ability to reason about and reflect on its own behavior, not just 'blindly' follow preprogrammed instructions.

In analyzing the ongoing debate on this topic, it is clear that a number of different critiques are being made of what we can call the 'computational consciousness' research program. In order of increasing 'difficulty', these include the following:

- Biological information processing is entirely analog, and analog processing is qualitatively different from digital. Thus sufficiently powerful analog computers might be able to realize autonomous systems, but digitally based computation cannot. Most researchers do not believe that analog processing overcomes the limitations of digital systems (as uncovered by Gödel, in particular); the matter has not been proven, but the

Church-Turing hypothesis (roughly, that anything computable is Turing-machine (i.e., digitally) computable) is generally accepted as fact. A variation of this argument, directed principally at elements of the artificial intelligence and cognitive science communities, asserts that primarily symbolic, rule-based processing cannot explain human intelligent behavior.

- Analog computers can, of course, be made from nonbiological material, so the above argument does not rule out the possibility of engineered consciousness. Assertions that the biological substrate itself is special have also been proposed. Being constructed out of this material, neural cells can undertake some form of processing that, for example, silicon-based systems cannot. Beyond an ability to implement a level of self-reflection that, per Gödel, is ruled out for Turing machines, specifics of this 'form of processing' are seldom proposed, although Penrose's (1989) hypothesis that the brain exploits quantum gravitational effects is a notable exception. (It is worth noting that no accepted model of biological processing relies on quantum-level phenomena.)

- It has also been argued that intelligence, as exhibited by animals, is essentially tied to embodiment. Disembodied computer programs running on immobile platforms and relying on keyboards, screens, and files for their inputs and outputs are inherently incapable of robustly managing the real world. According to this view, a necessary (not sufficient) requirement for an autonomous system is that it undertakes a formative process where it is allowed to interact with the real world.

- Finally, the ultimate argument is a variation of the vitalist one, that consciousness is something extra-material. For current purposes, this can be considered a refrain of the Descartesian mind/body dualist position.

The issue of consciousness in machines has captured the imagination of many as a result of the famous (or notorious, depending on one's take on it) Chinese room thought experiment suggested by John Searle (1980). Searle imagines himself locked inside a room, unable to communicate with anyone outside except through slips of paper passed through a slot in the door. These slips of paper are written in Chinese, a language Searle has no knowledge or understanding of. However, he has been given a voluminous 'script' that details (in English) the algorithmic manipulations he should carry out upon receipt of messages. Some of the messages can have questions written on them, others may describe a story. Searle allows that the script is perfect in that the manipulations result in responses that he can transcribe (the symbols that he reads, manipulates, and writes are meaningless squiggles to him) and pass back to his interrogator. These responses are in fact appropriate in context; to the person outside, Searle must understand Chinese. The point of the Chinese room (thought) experiment is that knowing how the responses were generated, we would not say that Searle 'understands' Chinese. This is a critique of one school of thought that maintains that rule-based algorithmic processing is sufficient for understanding. Variations of the experiment and the argument have since been directed at other types of automated mechanisms.

7.7 Conclusion

In the above discussion we have expressed some views of central nervous systems from a control perspective. The goal was to present facts and theories that could help control engineers expand their viewpoint toward more autonomous applications. We believe this is a worthwhile and very challenging direction for future control research.

In a presentation on this subject, the discussion will unavoidably be more philosophical than analytical. The problem lies in the nature of biological systems, which are the products of haphazard evolution rather than engineering design. We cannot say what a CNS was designed to do because it was never designed in the first place. As a result, before we can produce any mathematical model of CNS function, we must first guess how the CNS works, and the more experience one has in this area, the less willing one is to speculate.

Nonetheless, it is remarkable what progress seems to have been made in this direction, and it is worth knowing about ongoing research that may increase our knowledge in this field. The neuronal model does provide a unifying perspective, and biomedical researchers are producing increasingly more detailed models of CNS function based on neuronal systems. It is noteworthy that, if our neuronal theories are accurate, today's microelectronics technology allows the fabrication of digital neural networks more complex than the CNS of some invertebrates (though the human CNS remains out of reach). Still, we have a lot to learn about designing autonomous control systems with even a small percentage of the capability of biological counterparts having comparable complexity.

For this reason, it may be worthwhile for interested control scientists, and others, to learn more about the anatomy and physiology of biological systems. The standard models taught to medical students (see Kandel *et al.,* 1991) form a good source to start with, although they do not always reflect the state of current scientific knowledge. In fact, most of the basic neuroanatomy and many of the clinical observations on which these models are based were established decades ago, so their gross features have not changed much in recent times. The control scientist should be aware that medical school models are primarily designed to help future doctors in their diagnosis and treatment of pathologies in the clinical environment, so controversial scientific issues are often neglected.

A more rapidly evolving area of CNS study is neuroimaging. The brain activation results obtained by PET scan (see Roland, 1993) seem to offer a whole new source of model data based on physiological activity in the living brain. Although the initial results in this area clearly show that brain dynamics are more complicated than we might have hoped, there appears to be enough structure in results to date that future modeling efforts might be successful. Given that these data are the first to reveal which regions of the CNS are most active during CNS functions, there is reason to hope for significant advances based on this type of study.

References

Bullock, T.H. and Horridge G.A. (1965*) Structure and function in the nervous system of invertebrates* (2 volumes). W.H. Freeman and Co., New York.
Butler, A. and Hodos, W. (1996) *Comparative vertebrate neuroanatomy: evolution and adaptation.* Wiley-Liss, New York.
Fuster, J. (1997) *The prefrontal cortex* (3rd ed.). Lippincott-Raven, Philadelphia.
Harp, S.A. and Samad, T. (1991) Genetic synthesis of neural network architecture, in *The handbook of genetic algorithms* (ed. L.D. Davis). Van Nostrand Reinhold, New York.
Haykin, S. (1994) *Neural networks: a comprehensive foundation.* Macmillan/IEEE Press, New York.

Holland, J.H. (1975) *Adaptation in natural and artificial systems.* University of Michigan Press, Ann Arbor, MI.

Kandel, A. (ed.) (1991) *Fuzzy expert systems.* CRC Press, Boca Raton, FL.

Kandel, E., Schwartz, J. and Jessel, T. (1991) *Principles of neural science* (3rd ed.). Simon and Schuster, New York.

Koch, C. and Davis, J. (eds.) (1994) *Large-scale neuronal theories of the brain.* MIT Press, Cambridge, MA.

Lettvin, J.Y., Maturana, H.R., McCulloch, W.S. and Pitts, W.H. (1959) What the frog's eye tells the frog's brain. *Proceedings of the IRE,* **47**, 1940–1951.

McCulloch, W.S. and Pitts, W. (1943) A logical calculus of the ideas immanent in nervous activity. *Bulletin of Mathematical Biophysics,* **5**, 115–133.

Mead, C. (1990) Neuromorphic electronic systems. *Proceedings of the IEEE,* **78**(10), 1629–1636.

Mumford, D. (1994) Neuronal architectures for pattern-theoretic problems, in *Large-scale neuronal theories of the brain* (eds. C. Koch and J. Davis). MIT Press, Cambridge, MA.

Mumford, D. (1996) Pattern theory: a unifying perspective, in *Perception as Bayesian inference* (eds. D. Knill and W. Richards). Cambridge University Press, Cambridge, U.K.

Nagel, E. and Newman, J.R. (1958) *Gödel's proof.* New York University Press, New York.

Penrose, R. (1989) *The emperor's new mind: concerning computers, minds, and the laws of physics.* Oxford University Press, Oxford, UK.

Roland, P. (1993) *Brain activation.* John Wiley and Sons, New York.

Searle, J. (1980) Minds, brains, and programs. *Behavioral and Brain Sciences,* **3**, 417–458.

Yourdon, E. and Constantine, L. (1979) *Structured design.* Prentice-Hall, Englewood Cliffs, NJ.

8

Sensors in Control Systems

J. David Zook, Ulrich Bonne, and Tariq Samad
Honeywell Technology Center

8.1 Introduction

Control is more than information processing; it implies direct interaction with the physical world. Control systems include sensors and actuators, the critical pieces needed to ensure that our automation systems can help us manage our activities and environments in desired ways. By extracting information from the physical world, sensors provide inputs to control and automation systems. We may label our times the Information Age, but it would be a mistake to believe that advances in automation and control are solely a matter of more complex software, Web-enabled applications, and other developments in information technology. In particular, progress in control depends critically on advances in our capabilities for measuring and determining relevant aspects of the state of physical systems.

Technologists tasked with automation and control of systems of ever-increasing levels of complexity, whether as designers, operators, or managers, or in other capacities, thus need to be familiar with sensor technology. The increasing sophistication of sensors and sensing systems, the considerations driving this sophistication, new sensors and uses of sensors in control systems, the increasing reliability of sensors, and the like, are topics whose relevance today is not limited to sensor application specialists.

Our objective in this chapter is to discuss sensors from these points of view. Our focus is on the role of sensors in control systems and the trends and outstanding needs therein. Since excellent reviews of recent sensor developments and current applications already exist (Middelhoek and Audet, 1989; Frank, 1996; Wolffenbuttel, 1996; Soloman, 1994), we give some selected examples of new sensor developments, without any claims at comprehensiveness. Rather, our goal is to point out the benefits of increased sensor sophistication as well as key approaches and areas where more understanding is needed.

In the following sections, we first offer a definition of a sensor along with several examples. Next, we outline the application of sensors in simple control systems and discuss some of the important attributes of sensors, followed by reviews of recent developments in sensor systems, including sensor compensation and inferential sensing. We then focus on the sensing capabilities of biological systems and lessons we can hope to draw from them. We conclude by presenting some visions for how control and automation can help realize a safer, more productive, and more prosperous future, and the role that sensors will need to play.

8.2 Sensor Fundamentals and Classifications

We can define 'sensor' as a device that converts a physical stimulus or input into a readable output, which today would preferably be electronic, but which can also be communicated via other means, such as visual and acoustic. As perhaps the simplest example, consider a keyboard switch – which provides a signal when the associated key is pressed. The keyboard switch has several desirable features as a sensor. It is inexpensive, it has a high signal-to-noise ratio (its on/off impedance ratio), it is compact, and it has low power consumption. Its reliability and ability to operate over a wide range of environmental conditions are also exemplary.

Unlike most sensors, a keyboard switch lacks an analog input range, and its output is binary. Temperature, pressure, and flow sensors are more typical examples. In these cases, the output is not a binary quantity but a value that is sensitive to a range of those physical conditions. Figure 8.1 shows an example of a state-of-the-art sensor, in this case, a mass flow sensor. As evidenced by this example, many advanced sensors today are microstructure devices that leverage the economies of scale and the fabrication technologies of semiconductor manufacturing.

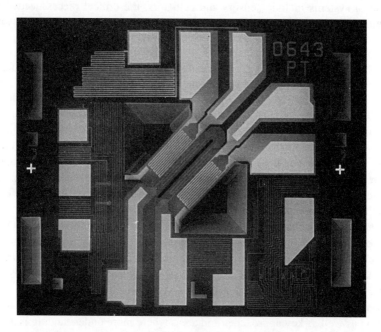

Figure 8.1 A micromachined mass flow sensor die (not packaged)

A semiconductor sensor that provides a wealth of information is the silicon photodiode. A CCD (charge coupled device) array of such devices typically generates on the order of 10^9

bits/second. Yet another example of a near-ideal sensor, it is a very simple device, easily fabricated into arrays with modern solid-state technology, with a very wide dynamic range (the ratio of the maximum to the minimum detectable photon intensity). One characteristic that is almost uniquely ideal is its stability. It has essentially no baseline drift and excellent scale factor stability. The reasons for these ideal characteristics lie in the physics of the device. The silicon photodiode is basically an energy converter – changing light energy (photons) into electrical energy (electrons in higher energy states that can generate current or voltage). When there are no incoming photons, there is no photocurrent – hence the baseline cannot drift. The quantum efficiency is close to unity (meaning one electron per photon) – so the scale factor (photocurrent divided by light intensity) is constant and stable over many orders of magnitude. A well-designed photodiode is linear over eight orders of magnitude in intensity and provides a *primary* standard for light intensity measurement within defined wavelength limits. It is not surprising that photodiodes are at the heart of CCD cameras and are ubiquitous in video information systems.

Still other types of sensors operate on chemical principles and may consist of single molecules. For example, a phenolphthalein molecule (the dye in litmus paper) signals a change in hydrogen ion concentration by changing its color. Similarly, a biosensor may be based on a molecule that interacts with a biological analyte to produce a signal we can pick up with our senses.

The generic block diagram for a sensor shown in Figure 8.2 highlights the role of a sensor as an interface between a control system and the physical world. The detector or transducer converts a physical or chemical phenomenon into (typically) an electrical signal. The signal processor performs one or more of various mathematical operations on the sensed value, such as amplification, rectification, demodulation, digitizing, or filtering. The measured and processed value is communicated to other subsystems (e.g., via a compatible electrical signal) or to a human (e.g., via a display). The sophistication of these functions and of the calibration process varies widely.

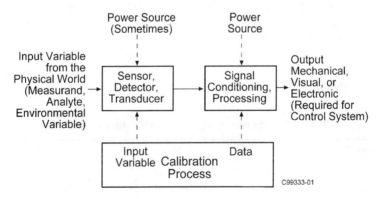

Figure 8.2 Sensor block diagram

The term *smart sensor* implies that some degree of signal conditioning is included in the same package as the sensor. On the more sophisticated end of the spectrum, the 'sensor' unit can include devices or systems with elaborate signal processing, displays, and diagnostic or self-calibration features. Such devices are often referred to as 'instruments', 'analyzers', or 'transmitters' (this last term is common in the process industries), usage which emphasizes that the transducer is but one part of a sensing system in the context of large-scale automation. In this chapter, we will not, however, be making hard distinctions between such sensor categories.

There are many ways to list or classify kinds of sensors. Middelhoek and Audet (1989) and Frank (1996) give several different classifications, based on the form of energy being transduced, and on whether the transduction mechanism is self-generating (like a thermocouple or a piezoelectric material) or is a modulating mechanism (like a thermistor or a piezoresistor). Table 8.1 shows some sensor examples based on a simple classification criterion: man-made versus biological. Either type may be used in an automation system – human operators can be considered part of the system.

A classification focusing on the physical effects that sensors can respond to is the basis of Table 8.2. One may argue about the classification of some examples, in view of their principle of operation being based on mixed effects. For example, the Hall effect depends on an electric current as well as a magnetic field.

8.3 Sensors in Control Systems

The role of a sensor in a simple automation system is depicted in Figure 8.3. The detection and measurement of some physical effect provides information to the control system regarding a related property of the system under control, which we are interested in regulating to within some 'set point' range. The controller outputs a command to an actuator (a valve, for example) to correct for measured deviations from the set point, and the control loop is thereby closed.

Although the control system example of Figure 8.3 is a simple one, it represents a fair number of practical control systems. More complex control systems may have many feedback loops with complex coupling. In especially simple systems, a distinct controller may not be immediately evident. For example, the 'Honeywell Round' thermostat contains a bimetal strip as an analog sensing mechanism that responds to temperature, and the switch attached to it serves as the actuator. This integration of sensor and actuator turns a furnace or other space conditioning device on or off, depending on whether the room temperature is within the set-point differential.

Table 8.1 Examples of engineered and biological sensors

Engineered sensors		Biological sensors	
Sensor	*Sensed quantity*	*Sensor*	*Sensed quantity*
Photodiode	Light intensity	Retina	Light intensity
Psychrometer	Humidity	Cochlea	Sound
Barometer	Pressure	Ear canal	Level, rotation
Thermometer	Temperature	Taste bud	Chemical composition
Phenolphthalein	pH	Skin	Temperature
Timer	Time	Skin and hair	Air flow
Odometer	Distance (inferred)	Olfactory cells	Gas composition

Table 8.2 Classification of sensors based on their sensing principle, with examples

Sensing principle	Examples
Mechanical motion (including mechanical resonance)	Pendulum-clock, quartz clock, spring balance, odometer, piezoresistive pressure sensor, accelerometer, gyro
Thermal (including temperature differences)	Thermometer, thermocouple, thermistor, thermal conductivity detector, transistor built-in voltage, air flow sensors
Optical energy (photons)	Photodiode, CCD camera, Geiger-Mueller tube, color sensor, turbidity sensor,
Magnetic field	Compass, Hall-effect, magnetoresistance, inductive proximity sensor
Electric field	Electrostatic voltmeter, field-effect transistor
Electrical current	Voltmeter, galvanometer, bipolar transistor
Chemical species	Litmus paper, humidity sensor, ZrO_2-based O_2 sensor, SnO_2-based combustible gas sensor, ion-selective electrodes, biosensors
Ultrasonic	Proximity sensors, liquid level, flow sensors
Acoustic	Ears, microphones, optoacoustic gas detectors

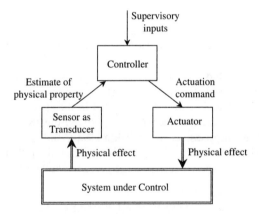

Figure 8.3 Example of a simple closed-loop control system

In general, however, the trend is to incorporate more, not less, information processing with the sensor. The increasing complexity of sensors is in part a consequence of this trend. In many cases, the information processing is being incorporated within the sensor device, blurring the distinction between transducer and processor, and between sensor and instrument.

8.3.1 Desirable sensor attributes

All sensors, whether used individually or as part of a larger automatic control system, need to meet a set of specifications, i.e., performance features. The most common and desirable ones include the following:

- High signal-to-noise ratio
- Reliability
- Safety (including intrinsic safety)
- Accuracy
- Fast response
- Wide dynamic range
- Low cost
- Low power
- Small size, compactness
- Low cross-sensitivity
- Compatible output and display
- EMI immunity.

We discuss several of these features below.

- *High signal-to-noise ratio*: the output signal of a sensor is an obvious measure of the information generated by the sensor. However, the quality of the information depends on the 'unwanted' output, or noise. 'Noise' usually refers to short-term variations in the output signal, which include thermal (or Johnson) noise, shot noise, and sometimes $1/f$ noise (van der Ziel, 1976). But the long-term variations may be the most important property of a sensor. If such drift is not included in the specifications and noise effects are estimated based on short laboratory test periods, the user is likely, in the longer term, to experience surprises, costly recalibrations, and downtimes or even unsafe conditions in the field. An accelerated life test period in the factory and the inclusion of long-term stability in the specifications is therefore recommended.

- *Safety*: demonstrating the safety of a new sensor to potential users can be time consuming. Certain standards have been established so that a manufacturer can obtain safety certificates (such as from Underwriters Laboratory). Safety considerations relate to any of numerous potential hazards: electric shocks, toxic materials, radioactivity (several high-performance sensors rely on subatomic particle emission and detection), acoustic or visual signals, device body temperature, etc. In some applications, assuring safety of human users under normal conditions is not sufficient. Under more stringent conditions of 'intrinsic safety', the sensor is confirmed to not trigger an explosion even under the worst possible failure modes.

- *Low cross-sensitivity and environmental ruggedness*: all transducers are sensitive to properties other than those we are interested in measuring with them. The key point is the extent of this cross-sensitivity and its inclusion in the specifications. For example: with piezoresistors used as strain sensors, their resistance is a function of the strain the device is exposed to. However, resistance does not depend only on strain; it is also a function of temperature. Similarly, a dye whose color changes when exposed to a certain chemical agent will also exhibit color variation with temperature. Pressure, humidity, input voltage, orientation, acceleration, electromagnetic interference – cross-sensitivity with respect to any of these can be a cause for concern.

- *Service life and reliability*: sensors are ultimately physical devices and, as such, have finite service lives and Mean Times Between Failures (MTBF). They can go out of calibration, elements in them can fail entirely, etc. Obviously, control actions taken based on faulty sensors can have catastrophic consequences, so that steps to prevent such events need to be considered: scheduled checks and maintenance, redundant sensors, sensor self-checks, or even sensor 'health monitoring'. Such features add components, cost, software and signal processing and, yes, complexity, with all its ramifications discussed in this and other chapters. The continuing demand for reduction in operating staff in many industrial and government facilities has led to a recent emphasis on automated diagnostics and prognostics (predictions of system degradation).

- *Compact size:* reductions in the spatial dimensions of sensors permit embedding them within systems that are themselves small-scale. Historically, new application areas for sensors have frequently been opened up through advances in miniaturization. A recent example is automotive engine control systems, which now embed sensors for mass airflow, crank angle, ambient temperature, humidity, pressure, and incipient knock. Another important area where compact size is making an impact is biomedical devices. In the past, implantable devices provided periodic stimulation regardless of the state of the system. With miniaturization in sensors, devices today are more 'intelligent': stimulation can now be based on demand, i.e., on accurate measurement of physical parameters. For example, blood sugar of diabetic patients can control insulin release and pressure waves caused by a patient's muscle movement or body motion can be sensed and used to adjust pacemaker stimulation rate (Medtronic, 1997).

- *Compatible output or display:* the output of a sensor has to be communicated to other parts of the automation system – for data logging, operator display, control algorithms, etc. Efficient communication means can greatly simplify sensor integration in a control system. In a process plant, the cost of wiring for communication is often a large component of the cost of sensor installation, especially in large control systems. A possible solution is the use of wireless communication.

- *Low power*: power has to be supplied to the sensor. In many industrial and building applications, the cost of wiring for power can even exceed the cost of the communication channel. Low power consumption opens up other alternatives, such as scavenging power from the communication bus or RF communication using battery power at the sensor. Batteries can be recharged by environmentally powered sources such as solar cells in some installations.

- *Low cost*: most industries that make extensive use of sensors are under perpetual cost-reduction pressures. This issue is not unique to sensors, but it is worth noting here as much as in association with any other aspect of control and automation systems. Informed decisions on cost-benefit trade-offs need to be made to achieve the desired performance and safety at minimum cost. Automation solutions for large-scale, complex systems will be sensor-rich. The developers and operators of such solutions (as distinct from the sensor suppliers) often tend to view sensors as commodities – selection is based on cost; performance differentiation is typically not viewed as significant. Leapfrog technology can, of course, still demand a premium, but in many cases solution and system providers would only check that minimum performance specifications are met.

The above issues are examples that have motivated research and development on sensor technology and on sensor applications, as will be discussed below.

8.3.2 A trade-off between sensor performance and power dissipation

Since a sensor is a provider of information, it would seem natural that the quality of a sensor can be evaluated by the quantity of information it can provide to a control system. Thus it seems reasonable that information theory should provide a measure of sensor quality or performance. One of the benefits gained by adopting an information theoretic perspective is a better appreciation of the trade-off between sensor properties and sensor power dissipation. We discuss the topic briefly and in general terms in this section; a full quantitative explication remains to be developed.

Information generated by a sensor can be considered to be a series of messages related to some characteristic of the physical system whose attributes are being measured. The information reduces our uncertainty as to what state the system is in. Information theory provides a measure of the amount of information in any message called the information theoretic entropy to distinguish it from the entropy of thermodynamics and statistical mechanics. It is simply equal to the minimum number of binary digits (bits) necessary to transmit the message (Pierce, 1980).

It takes energy to generate information. Shannon (1948) showed that a certain energy per bit is required to distinguish signals in an electromagnetic communication channel from the unavoidable background of thermal noise. The energy per bit is given by $kT\ln 2$, where k is Boltzmann's constant and T is the absolute temperature. Improving the maximum signal-to-noise ratio (equivalent to dynamic range) or the response time (related to the bandwidth of the sensor output) of a sensor requires increased power dissipation.

The discussion above is in the context of digitally encoded measurements. However, the transducers used for sensing parameters in control and automation systems are not digital devices that generate coded messages as finite strings of bits. Rather they are analog devices that generate a voltage, current, charge, or change in resistance that must be amplified. It is worth noting here that the scale-up properties of analog communication are substantially poorer than for digital communication. Doubling the dynamic range of an analog sensor output signal requires at least a doubling of the power required for communicating that signal. On the other hand, if the output is expressed in digital form, a doubling requires only one additional bit.

8.4 Sensor Technology Developments

The objective of modern sensor developments is to furnish improvements on one or more of the desirable features listed earlier. For example, if the sensor stability or cross-sensitivity does not meet the level required for an application, the sensor can drift out of calibration, its output can lose its meaning, and the system can lose its ability to control. The addition of means and designs (electronic compensation, signal processing, self-checks, etc.) that prevent such occurrences lead to more sophisticated and complex sensors. Contributing to this complexity is the desire to offload central data processing onto local or distributed processing (i.e., to make individual sensors 'smarter'). Furthermore, to achieve one desired output, a cluster of sensors, a sensor system, analyzer, or instrument may be needed.

Similarly, increasing demands on the performance of systems, such as those providing transportation or chemical processing (greater fuel efficiency, fewer emissions, greater safety, reduced labor content) cause the complexity of the involved automatic control systems to also increase, together with their need for better and/or new sensors.

We discuss some salient trends toward increasing sensor complexity in this section.

8.4.1 Compensation through signal processing

As alluded to above, one of the key problems in developing sensors that are accurate over a range of environmental conditions is cross-sensitivity. Most transducer mechanisms exhibit some influence to variations in properties other than the one they are designed to measure. A flow transducer, for example, may produce a different output as temperature varies, even with the flow remaining invariant. For applications where precision or accuracy are not required, the errors that result from this coupling may be tolerable. In most modern automation systems, however, this is not the case.

Modern sensors correct for cross-sensitivity by incorporating additional transducers for measuring the secondary variables and then employing signal processing and/or statistical techniques to 'cancel out' the undesired influence. As an example, we discuss a temperature-compensated flow sensor for which the sensor output, y, is computed as a function of two sensed signals: the temperature, T, and the flow, x. The form of this function is based partly on our knowledge of the physics responsible for the cross-sensitivity and partly on a formula whose parameters are determined by regression:

$$y = \left(\frac{T}{T_0}\right)^{0.03} \left(a_0 + a_1 x^{p_1} + a_2 x^{p_2} + a_3 x^{p_3}\right)$$

This expression tells us that the temperature effect is fairly small – a ±50°C excursion in T from a nominal temperature T_0 of 288 K only changes y by ±0.5%. The a_i's and p_i's are parameters that are fit using data collected from controlled tests. An example of a fit is shown in Figure 8.4, where the horizontal axis shows the flow sensor output (labeled ΔG) and the vertical axis shows the actual flow rate (V_s).

The S_L quantity also graphed in the figure is the logarithmic sensitivity of the final output to sensor noise, defined by $S_L = |(x/y)(\partial y/\partial x)|$. The figure shows that as flow increases, this sensitivity increases to $S_L > 1$, a consequence of the output signal saturation. The signal noise (not shown) also varies over the flow range, from 10% at low flows to 1% at high flows. Figure 8.4 shows that the saturation negates the benefits of reduced relative signal noise at high flows. In more complex functions, S_L can adopt values of 10, 50, or greater, especially when there are multiple x_i (worst case: an array of similar sensors) and interactions between them (S_L in the multivariable case is a summation of the individual sensitivity magnitudes) (Bonne, 1996). For this reason, the development of sensor arrays to measure complex variables (such as odor or taste) has been fraught with difficulty (such as the need for frequent recalibration).

8.4.2 Inferential and higher-value sensors

The example above illustrates one of the simpler uses of signal processing or statistical techniques in sensors. Over the last few years, the notion of producing a sensor or instrument output that is derived from multiple sensing mechanisms has been extended to a higher level. This concept has reached the point where 'sensors' for parameters that have not been feasible to measure cheaply and reliably can now be developed for broad-based use. The terms 'inferential sensors' and 'higher value sensors' are often used for such systems, but their connection with traditional definitions of sensors is tenuous. In most cases, inferential sensors are not hardware products, unlike most sensors. They are more likely to be software implementations on control system platforms that can access sensor inputs from distributed locations.

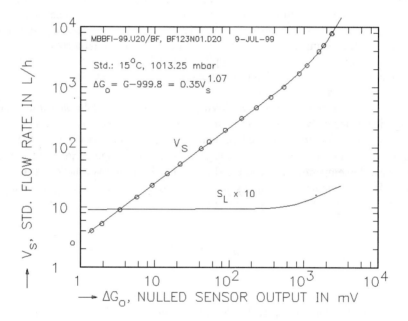

Figure 8.4 Characterization of 200 L/min flow sensor with pure nitrogen: $V_S = 2.8836 \Delta G^{0.93} + 2.9743 \cdot 10^{-7} \Delta G^3 + 2.8886 \cdot 10^{-13} \Delta G^4$. The model has an uncertainty (1-sigma) of ±2.02% and a logarithmic sensitivity range of $1 < S_L < 3$.

One example is a 'comfort sensor'. Devices (other than biological ones!) that directly sense comfort are difficult to even conceive of. However, it seems reasonable to postulate that comfort is correlated with parameters that we can indeed measure. The problem then is a matter of determining the function (or mapping) from measurable parameters to comfort. Formulations of comfort indices have been proposed, with Fanger's (1970) Predicted Mean Vote (PMV) measure having gained some degree of acceptance. PMV is based on six variables: air temperature, mean radiant temperature, relative air velocity, humidity, activity level, and clothing thermal resistance. The International Standards Organization has adopted PMV as its thermal comfort index standard (ISO 7730), and 'comfort sensor' products have since been developed (e.g., see Yamatake, 1999).

In most cases, however, the relationship between the measured properties and the one that is desired to be inferred is unknown. In these cases, an empirical model can be developed. A popular methodology for such modeling is artificial neural networks. A detailed explanation of neural nets is beyond the scope of this chapter; suffice it to note that they provide a nonlinear statistical regression technique that has some attractive properties for high-dimensional problems (Barron, 1993) relative to classical models such as the (low-dimensional) temperature-compensated-flow example shown earlier. Developing an inferential model using neural networks (or any other empirical method) requires first the collection of data from the system. The data must also include measurement of the inferred property value as determined through a temporary sensor or through an offline laboratory analysis – these devices are not needed once the inferential sensor is developed. The development and use of an inferential sensor is outlined in Figure 8.5 for a process industry scenario. Inferential sensors are now widely used in the process industries, such as for estimating octane number in oil refining, smokestack emissions in power plants, and lignin concentrations in wood pulping.

Both this subsection and the previous one highlight the increasingly multidisciplinary nature of sensor technology today: statistics, signal processing, and software are now critical aspects of sensor development.

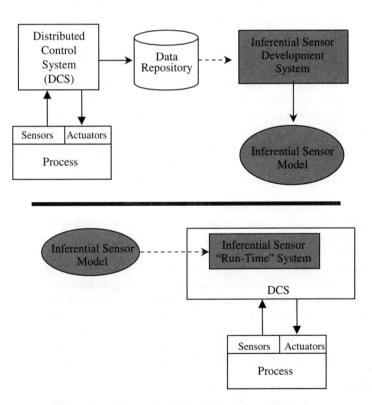

Figure 8.5 Inferential sensor development (top) and operation (bottom)

8.4.3   Self-checking and self-compensating sensors

From the above discussion, it is clear that demand for self-checking and self-compensating sensors is high. Electronic scales today automatically reset their zero settings before objects are placed on them, based on the assumption that most of the time the scales are empty. Even the span or actual weight reading can be easily checked by calibration with a standard weight. Precision pressure transmitters cannot depend on such methods since they are under load most of the time. They must depend on the stability of the piezoresistive elements and of the sensor package to maintain stable baseline and span stability. However, in some cases, auxiliary actuators or valves may be used to enable performing a check on baseline and span for pressure sensors and flow sensors. The following examples illustrate such methods.

8.4.3.1   Chopper as auto-zero device

Infrared detectors that respond to very small temperature differences can be viewed as the classic case where a mechanical actuator is used to provide a reference signal. A lens focuses an element of the scene onto the detector and a rotating chopper wheel is placed in the optical path. The detector alternately sees the scene and the chopper blade, which is always at the same (reference) temperature. A timing signal from the chopper is sent to a switch that alternates the polarity of the output from the infrared detector. Thus the reference signal is subtracted from the signal from the scene. The synchronous signal corresponds to the difference in temperature between the scene and the reference, and the nonsynchronous baseline drift is ignored. It is not unusual for temperature differences of $0.01°C$ (corresponding to $0.00001°C$ at the detector) to be easily seen in a thermal imager.

Chopping a signal and synchronous detection are widely used in optical systems, for example in infrared spectrometers and gas analyzers. The motivation is frequently (but not always) the IR detector drift. In the optoacoustic method for gas concentration, the gas is enclosed in a chamber that acts as a resonator. The chamber is illuminated with IR radiation from a filtered chopped IR source. A microphone is used to pick up pressure changes that occur only if analyte molecules absorbing at that wavelength are present. Chopper-stabilized amplifiers, using switches at the input of an amplifier, are similarly used to convert a dc voltage to an ac voltage to eliminate the effect of baseline drifts.

8.4.3.2   Self-checking flame sensors

Highly reliable optical sensors are needed to monitor the presence of flames in furnaces or boilers so that fuel flow can be interrupted if the flame should go out for any reason. Despite the proven stability of the ultraviolet sensors employed, a further layer of safety can be added to positively check that the sensor's response still meets the 'fail-safe' requirement. A mechanical shutter (chopper) that interrupts the optical path between the sensor and the flame during known intervals is used to perform such checks on flame sensors that are certified to be 'fail-safe' (Honeywell, 1998). Another proven strategy is to use a unique signature or response of the object to be sensed (e.g., the flame-rectified current signal of a flame ionization sensor). In this case, ac interference from faulty wiring appears as a nonrectified ac signal that is readily distinguished from the rectified flame signal.

8.4.3.3   Self-checking oxygen sensors

The classical Nernstian, ZrO_2-based oxygen sensor generates a dc signal that depends on the logarithm of the ratio of unknown/reference oxygen concentrations. One check on the health of such a sensor consists of measuring its impedance: if the value is above a set limit, its health is suspect and it should be replaced. A more sophisticated approach is to force oxygen into and out of a closed ZrO_2 cell (by application of a fixed voltage between reversals) so that set outputs are reached. The 'pumping time' can thus become a linear measure for the absolute oxygen concentration. In addition to checking the cell's impedance, any asymmetry in the pumping time or current profile reveals a faulty sensor (Kroot, 1996).

8.4.3.4 Self-compensated flow sensors

A different calibration problem arises in the case of a mass air flow sensor based on thin-film thermal anemometry (Bonne, 1990). In practice, the output signal depends not on pressure, but on the thermal properties of the gas being sensed, in spite of its mass flow dependence being touted in many textbooks. A theoretically correct compensation can be made if at least the thermal conductivity and specific heat of the fluid are known. Means to measure these online have been developed employing one additional thermal microsensor mounted in a recess, away from the flowing fluid. This approach now enables the online determination of fully compensated (composition, temperature, and pressure) flow of any thermal flow sensor, although it requires additional calibration of thermal conductivity and specific heat sensors (Bonne, 1992).

8.4.3.5 Rebalanced sensors

Another way an actuator is used to eliminate the effects of zero-drift and nonlinearity is to balance the input variable with an opposite effect to keep the input to the sensor at zero. In a force-rebalanced accelerometer, an electromagnetic force balances the force on the proof mass due to acceleration. The rebalancing force is linearly proportional to the actuator current and provides the measure of the acceleration. The proof mass is maintained at its rest (or null) position so that a highly sensitive method such as a Scanning Tunneling Microscope (STM) tip can be used as the sensing mechanism. Such a sensor can easily detect sub-Angstrom motion, and thus is an ideal null detector, even though its output is highly nonlinear (Liu *et al.*, 1998).

8.4.4 Sensor manufacturing

Unlike many other critical automation technologies, sensors are fundamentally 'hardware' devices. Manufacturing, in the traditional sense of the term, thus remains an important consideration for complexity management. Manufacturing for sensors covers both the 'base' detector (increasingly based on microelectronic fabrication processes) and the assembly of the packaged sensor. The software-intensive nature of most of the technologies discussed in this book may lead to the conclusion that manufacturing has been reduced to irrelevance. This would be a mistaken extrapolation. In fact, manufacturing innovations have been at the center of the development and successful commercialization of new sensors – e.g., pressure sensors with automated testing to customize individual compensation algorithms, mass airflow sensors, laser-based gyroscopes, and proximity sensors.

Usually a business plan for a new sensor product is developed in response to either demand from an identified set of customers or a new capability, technology, or sensor design resulting from research and innovation. Potential customers are asked to help define a 'minimum common denominator' specification for the new sensor, and to help identify the potential market size. Based on laboratory and field test results, there must be confidence that the specifications can be met. Before productization can proceed, alternative technologies or manufacturing processes must be examined, the intellectual property position (proprietary information and patents) must be assessed, and possible manufacturing scenarios dependent on volume, labor cost, and location must be considered. However, we have left out one key parameter of the productization and manufacturing process: the human factor. It is the one that significantly increases the complexity of the process, but its discussion falls outside the scope of this text.

8.4.5 Other Drivers for R&D

Several additional factors drive research and development in sensor technology, which we briefly note here:

- *Ease of application*: for sensor data to be meaningful, not only must the sensors be working as assumed, but they must also be positioned properly and installed correctly (i.e., consistent with the control system model). As field installations are labor intensive and costly, ease of installation can be a key competitive edge for a sensor.

- *Sensor-configured system*: the performance of the control system is limited by what we can measure and the feasibility of obtaining measurements. To achieve optimal system performance, we need to have a thorough understanding of the behavior of the system, its controls, and its sensors. We can no longer afford to design a system, add controls at a later stage in its development, and add its sensors at a still later time. A system designer will therefore keep in mind the capabilities of available sensors as he or she designs and configures the system.

- *Rangeability and adaptability*: in line with pressures to reduce the cost of inventory, parts, and installation, sensors have become more flexible and remotely programmable. For example, a lower cost microprocessor with reduced capability may be used if the sensor range can be reprogrammed after installation. In addition, this added flexibility enables a reduction in inventory. Ultimately, such flexibility and reprogramming may be done automatically, by the 'smart' sensor itself, in response to the status of the system.

- *Integration*: further cost savings are achieved by merging the data and signal processing of the sensor and control. Integration of the sensor functions of Figure 8.2 into one monolithic block of a silicon chip is also being practiced, especially where fabrication volumes are high (perhaps more than 1 million/year) and the fabrication processes are compatible. Examples include Honeywell's integrated pressure sensor (Allen, 1984) and Analog Devices' integrated accelerometers, which incorporate both sensing and data processing on one single silicon chip. Further integration will include means for I/O, such as is needed for wireless communication. Additional benefits from such integration are compactness, ruggedness, and low power consumption.

- *Sensor arrays and massive data processing*: it is now affordable to manufacture thousands of silicon-based sensors as compact arrays. This capability has enabled the development of such sensors as the CCD camera noted earlier, and many other 'imaging' sensors. These other examples include scanners and processors for imagery and tomography of signals from infrared, ultrasonic, X-ray, and electron- or nuclear-spin resonance signals.

- *Extreme sensitivity*: the need to detect parts-per-billion and even parts-per-trillion concentrations of trace pollutants, explosives, nerve gases, and bacteria has given impetus to the development of sensors based on solid-state surface effects, biological sensors, and fluorescent chemosensor techniques. The latter have zero baseline and thus (in principle, at least) allow the detection of single molecules (Desvergne and Czarnik,1997).

- *Extreme accuracy*: Some applications require sensor accuracy in the $\leq \pm 0.01\%$ range. Differential pressure sensors with such capabilities are now available and can maintain this accuracy despite changes in temperature and absolute pressure.

8.5 Biological Systems, Chemical Sensors, and Biosensors

Biological systems are an existence proof that low-performance sensors can provide the meaningful data needed to perform sophisticated control functions. Consider the flight control system of a fly – taking evasive action from the fly swatter and landing safely on the ceiling. Or imagine a male moth finding a female several miles upwind. The sensing and control system performance in low power, compact packages that is seen in the biological world is still a futuristic vision for engineers and technologists; the realization of this vision is not imminent.

In Table 8.1, we presented a list of biological sensors and sensing mechanisms. In the area where we have nearly ideal sensing characteristics (electronics and photonics, with CCD imagers providing a particularly good example), we can outperform biological sensors, even the best eyes that have evolved for night vision. For other sensing modalities, chemical and olfactory sensing in particular, we are far behind the performance of biological systems.

8.5.1 Chemical and olfactory sensing in biological systems

A long-standing need exists for chemical sensors – for environmental control, for process control, and for testing almost everything from air and water quality to food and medicine. This is an area in which biological systems have significant advantages over today's manufactured sensors. Biological chemical sensors depend on protein sites on the cell surface, which have a selective affinity for certain molecules. During the last decade, specific olfactory receptors have been identified. Coupled with advances in genomics and cellular biochemistry, such identification provides the beginnings of a molecular understanding of olfaction (Travis, 1999). Adsorption of the odor molecule causes a change in the firing rate of the olfactory neuron. It is believed that moths can detect sex pheromones at sensitivities approaching the level of individual molecules. Humans, although much less sensitive, can discriminate between more than 10,000 odor molecules (odorants) and have about 50 million olfactory neurons that bear about 1000 different types of odorant receptors in four so-called expression zones. Within each zone the odorant receptors are distributed randomly, presumably to provide redundancy in case of damage in a small area.

The evolution of odor receptor proteins has not solved the cross-sensitivity problem discussed above, but in fact makes use of it. Individual types of odor molecules activate multiple receptors, with each receptor type having a different affinity (and therefore a different proportionality between concentration and firing rate) for different odorants. Such a sensor array approach is likely to be a necessary feature of manufactured (nonbiological) chemical sensors for organic molecules, since there is little utility to a sensor that will respond to only one type of molecule. The size of the sensor array is an important issue in chemical sensors. Mammals (and even worms) typically have 1000 types of odorant receptors, whereas fruit flies make do with as few as 100. Since odorants interact with multiple receptors rather than individual ones, the possible combinations exceed by orders of magnitude the number of odors an animal can actually distinguish. Presumably, animals only discriminate between odors that are biologically important to their survival.

Recently, several research groups have developed 'artificial noses' inspired – however loosely – by mammalian olfactory prowess (Nagle *et al.*, 1998). These devices are based on different types of sensor technologies, including metal oxide semiconductors, conducting polymers, surface acoustic wave, and gas chromatography. This is an exciting topic with considerable application potential: food and pharmaceutical manufacturing, environmental monitoring, security systems, and military applications. Although the best of these systems compare poorly with biological noses – they have at most a few dozen sensors and cost up to $100,000 – this research promises to bring a sensory modality that has hitherto been the exclusive preserve of biology within the domain of engineered systems.

An important and highly desirable feature of the olfactory mechanism is that it is apparently unresponsive to odorless molecules, whereas manufactured sensors tend to be sensitive to oxygen and water vapor, in particular. They also respond to odorless combustible gases such as hydrogen and methane. An exception to this is the use of chemical sensors based on conjugated polymers (Desvergne and Czarnik, 1997). These have demonstrated high specificity together with the high-sensitivity characteristic of biological sensors, although they are based on quite different principles. The transduction strategy is to synthesize polymers that are poly-receptor assemblies that act collectively. The electrical conductivity of conjugated polymers is a well known example of a property determined by low concentrations of defects. Fluorescence is another example, where a single defect can significantly quench the fluorescence of the entire chain. Swager (1997) has demonstrated chemosensors for TNT that can detect concentrations in the parts per billion, without significant interference from environmental constituents such as water vapor and CO_2. This continues to be an exciting area of development (Perkins, 2000).

The biological process has some undesirable features as well. It can have a nonlinear response, which can be misleading. For example, a low concentration of indole has a floral scent, while a high concentration has a putrid odor. These features are not unexpected for a system developed by evolution since there is no apparent survival advantage to being able to identify strong odors. The signal processing method (as realized in the olfactory cortex) that evolved along with the chemical receptors did not get "trained" on high concentrations.

8.5.2 Auditory sensors

An example of a nonolfactory biological sensing system that we can learn from is the biological sonar systems used by bats (Suga, 1990). The fact that bats use the 'chirp radar' technique was well known long before it was rediscovered during World War II. Bats emit complex ultrasonic signals and listen to the returning echoes for orientation information and for hunting flying insects. There are nearly 800 species of bat, and probably all of them use echolocation. Range information is conveyed by echo delay, whereas velocity information is carried by Doppler shift. Frequency is detected by the stimulation of hair cells and is coded by the anatomical location of a spiral array of ganglion cells. The rate and pattern of nerve impulses express the amplitude, duration of signals, and time between signals. The signal processing systems use delay lines, multipliers (or AND gates), and frequency tuners that are tolerant of wide variations in signal level. Signal processing in the auditory system is parallel-hierarchical. All of this sophisticated sensing and signal processing takes place in a very small volume with low energy usage. If similar sensing systems could be produced by micro-machining and integrated circuit techniques, they would undoubtedly find application in security and surveillance systems.

8.5.3 The role of sensors in biological control systems

The role of sensors in biological systems is illustrated in Figure 8.6 and is similar in many respects to the role of sensors in engineered control systems. The biological senses provide the interface between the organism and its environment. Many biological organisms constantly probe their environment and compare the information from different senses, providing a means of calibrating the sensory mechanisms.

The building blocks of living systems are cells, and each cell has evolved its structure and its control systems to improve its survivability. Even the simplest single-cell organism has amazing information processing, sensing, and control system mechanisms with complex feedback networks, all tuned to ensure survivability in its environment (Margulis and Sagan, 1997). Natural selection provides the mechanism for producing systems that optimally use resources and avoid threats. The billions of generations of evolution have not only produced optimum response to the environment, but because of the history of changing environments, they have produced systems that can adapt and operate effectively (i.e., survive) in the face of changing environments – changing resources, changing threats, etc. The incredible success of living systems suggests that we have a lot to learn from them, despite their complexity.

Our knowledge about the information pathways and feedback mechanisms in biological systems is meager. We know that information is stored in the sequences of nucleic acids in RNA and DNA and communication takes place by electrochemical pulses and complex biochemical reactions. Biological sensors use the same mechanisms. Since sensing and information processing are both critical to the control system, it is a safe bet that they co-evolved and are closely tied together.

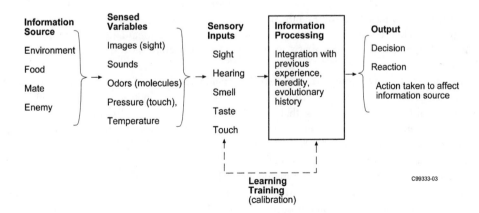

Figure 8.6 Schematic depiction of biological sensing

An evident aspect of biological evolution is the increase in complexity of the system. Cells became more complex and evolved into multicell organisms. Reproduction mechanisms have evolved, as well as sophisticated mechanisms of communication between organisms. In fact, evolution can be viewed as the accumulation of useful information (Ayres, 1994). As such, we may expect that control systems will continue to evolve as well. Here we are also learning from biological systems. The concepts of neural networks and genetic algorithms were inspired by biological systems.

We can make several inferences from a comparison between biological systems and today's control systems:

- Increased reliability and survivability (robustness) are achieved by increased redundancy, which means increased numbers of sensors and interconnections, and therefore increased complexity.
- Adaptability and flexibility go hand-in-hand with increased complexity.
- In terms of complexity, today's control and automation systems are in their infancy compared with even the simplest biological systems.
- As we understand biological systems better, we learn from them. Examples presently in use are neural networks and genetic algorithms.

8.6   Sensor-Enabled Visions for the Future

We have noted a number of trends in the use of sensors for automation and control. These are driven by the desire to have machines replace humans in performing tasks wherever it is economical and safe to do so. A familiar example is the increased use of sensors and computer controls in automobiles. Computers and sensors not only control the engine, but also the brake system, the airbags, the windows and door locks, and even the radio antenna. Some of these uses are simply for convenience, some are for safety or efficiency, and some are required for environmental protection. In all cases, they add to the cost and complexity of the vehicle, but clearly the benefit is perceived to be worth it – the automotive industry operates in a highly competitive, consumer-driven, and regulation-driven market. In the future, we can envision automobiles that warn us of other cars or of traffic problems ahead, that suggest an alternative route, and that even let us know when we are approaching the next turnoff. This requires the development of low-cost navigational and collision avoidance sensors. These are all devices that can be expected to become affordable during the next decade.

In the aviation enterprise, we can also expect increased use of sophisticated navigation and collision avoidance sensors. We can envision the total elimination of airplane collisions, and the ability to choose routes that avoid turbulence or severe weather. In this case, hardware developments are needed in the area of radar and laser scattering to detect clear air turbulence and nearby aircraft. The reliability of the aircraft would be improved by using sensors to detect bearing wear and structural fatigue and to predict when parts are about to fail so that they can be scheduled for maintenance before they do so.

Sensors can also be used to improve the safety and reliability of our basic infrastructure – the essentials we take for granted in the developed world, such as electricity, clean water, and clean air. With widespread environmental monitoring, pollution sources can be identified sooner and breakdowns in the distribution systems (for water and electricity) can be detected and bypassed. As resources become more scarce, sensors and controls will be increasingly used to improve efficiency and regulate consumption.

Health care is an area where the use of diagnostic tests and instruments is growing rapidly. Improvements in capabilities are becoming possible through the use of DNA sequencing to identify genetic defects and diseases at the very early stages, when they can be cured or prevented. At present, diagnostics are offline, but we can imagine that sensors on the wrist could tell us when we are overstressed or overindulgent. With instant feedback we would be reminded to close the loop and change our actions to optimize our health. Although the cost of diagnosis and testing is increased, the overall cost of health care will be decreased.

An area that is not so obvious, but that also has great potential for productivity improvement, is asset management. Significant time and resources are spent in industries, hospitals, and offices looking for equipment or other assets that are not where they are presumed to be. Inexpensive tags and sensors that can remotely detect the tags can locate and track assets. At the extreme end of the spectrum, the homeowner could use his home computer to keep track of items that may get misplaced in the home, such as keys or even eyeglasses. This may seem like overkill, but the technology would not have to be much different from that used to keep track of clothing in a department store.

The above examples suggest that the possibilities for increased use of sensors in automation and control systems are endless. Improvements in the performance of existing sensors, improvements in reliability and robustness, the development of new sensors, and further integration of sensors within control systems are some of the trends in evidence today. As always, research and innovation is driven by application demands. Critical process control and the control of new processes such as biologically based manufacturing of pharmaceuticals are some of the performance drivers. However, many of the applications being envisioned do not require high-performance sensors and transmitters that generate high-quality data. The long development times and high costs of sensor development and maintenance lead us to conclude that cost reduction of existing sensors and finding new ways of using them will continue to dominate the field of sensor development.

References

Allen, P. (1984) Sensors in silicon. *High Technology*, September, 43–50

Ayres, R.U. (1994) *Information, entropy, and progress – a new evolutionary paradigm.* American Institute of Physics, Woodbury, NY.

Barron, A.R. (1993) Universal approximation bounds for superpositions of a sigmoid function. *IEEE Transactions on Information Theory*, **39**(3), 930–945.

Bonne, U. (1990) Flowmeter fluid composition correction. U.S. Patent No. 4,961,348.

Bonne, U. (1992) Fully compensated flow microsensor for electronic gas metering. *Proceedings of the International Gas Research Conference*, **III**, 859, Orlando, FL.

Bonne, U. (1996) Sensing fuel properties with thermal microsensors. *Proceedings of the SPIE Conference on Smart Electronics and MEMS*, paper no. 2722-24, San Diego, CA.

Desvergne, J.P. and Czarnik, A.W. (eds.) (1997) *Chemosensors of ion and molecule recognition.* Kluwer Academic Publishers, Dordrecht, The Netherlands.

Fanger P.O. (1970) *Thermal comfort.* McGraw-Hill Book Co., New York.

Frank, R. (1996) *Understanding smart sensors.* Artech House, Boston, MA.

Honeywell Inc. (1998) Shutter-Check Flame Detectors, Models C7012E, C7012D, C7024E, C7024F, C7061A, C7076A and C7076D. *Tradeline Catalog,* Publ. No. 70-6910, Honeywell Home and Building Control, Golden Valley, MN, 753–763.

ISO 7730 (1994) Moderate Thermal Environments – Determination of the PMV and PPD Indices and Specification of the Conditions for Thermal Comfort (2nd ed.). International Standards Organization, Geneva, ref. no. ISO 7730:1994(E).

Kroot, P. (1996) A new oxygen sensor for cleaner, more efficient combustion. *Scientific Honeyweller,* **15,** 29–33.

Liu, C.H., Barzilai, A.M., Reynolds, J.K., Partridge, A., Kenny, T.W., Grade, J.D. and Rockstad, H.K. (1998) Characterization of a high-sensitivity micromachined tunneling accelerometer with micro-g resolution. *Journal of Microelectromechanical Systems,* **7,** 235.

Margulis, L. and Sagan, D. (1997) *Microcosmos.* University of California Press, Berkeley, CA.

Medtronic (1997) Medtronic rapid technological growth. http://www.medtronic.com/corporate/early.htm.

Middelhoek, S. and Audet, S.A. (1989) *Silicon sensors.* Academic Press, London.

Nagle, H.T., Schiffman, S.S. and Gutierrez-Osuna, R. (1998) The how and why of electronic noses. *IEEE Spectrum,* **35,** 9, 22–31.

Perkins, S. (2000) Eau, Brother! Electronic noses provide a new sense of the future. *Science News,* **157,** 125–127.

Pierce, J.R. (1980) *An introduction to information theory – symbols, signals and noise* (2nd rev. ed.). Dover Publications, New York.

Shannon, C. (1948) A mathematical theory of communication. *Bell System Technical Journal,* **27,** 379–423, 623–656.

Soloman, S. (1994) *Sensors and control systems in manufacturing.* McGraw-Hill, New York.

Suga, N. (1990) Cortical computational maps for auditory imaging. *Neural Networks,* **3,** 3–21.

Swager, T.M. (1997) New approaches to sensory materials: molecular recognition in conjugated polymers, in *Chemosensors of ion and molecule recognition* (eds. J.P. Desvergne and A.W. Czarnik). Kluwer Academic Publishers, Dordrecht, The Netherlands.

Travis, J. (1999) Making sense of scents: scientists begin to decipher the alphabet of odors. *Science News,* **155,** 236–238.

van der Ziel, A. (1976) *Noise in measurements.* John Wiley & Sons, Chichester, U.K.

Wolffenbuttel, R.F. (ed.) (1996) *Silicon sensors and circuits.* Chapman & Hall, London.

Yamatake (1999) Integrated environmental comfort control. http://www.yamatake.co.jp/innovat/iecc.htm.

Part 3

Software and Complex Systems

Engineering systems are complex not only because of the sophistication of their mechanical, electrical, chemical, electronic, or other physical equipment. Increasingly, complexity lies in the sophistication of information processing, as implemented by control and automation software. Software is both an indispensable tool for managing the complexity of physical systems, in that it provides the only way we have available of implementing advanced automation algorithms, and a significant embodiment of complexity itself. Software is thus simultaneously a solution technology and an application domain for complexity management.

Part 3 of this book contains three chapters that address both sides of the software coin. Chapter 9, by Jonathan Krueger, is concerned with managing the complexity of software engineering. The next two chapters focus primarily on describing novel software-based solutions for managing the complexity of engineering systems.

Although concerned with software systems generally, Krueger focuses in particular on real-time automation and control systems. The growth in complexity in control systems is just as great as in any other computational arena. Today's process control systems consist of networks of nodes, each node hosting a 64-bit microprocessor and capable of handling thousands of input and output variables. The complexity of software arises not just from the sophistication of its intended behavior; 'excess complexity' results from suboptimal design, short-sighted bug fixes, and isolated and superficial automation. Functional errors, longer development cycles, and higher costs are some of the adverse consequences. An underlying cause for the rampant complexity of software-intensive systems is the number and diversity of dependencies – dependencies between components of software, between the various functions implemented by it, and between decisions made during the process of software development. The key to managing dependencies in software is abstraction, and Chapter 9 discusses three broad abstraction strategies. First, the decision space can be structured and streamlined through domain analysis, explicit articulation of software architectures, and virtual machines. Secondly, dependency management can be automated, primarily with model-based metatools. Thirdly, the development process itself can be managed through the use of priority-oriented project planning, design reuse mechanisms, and process maturity frameworks. Examples of developed applications and tools illustrate the general remarks.

Chapter 10, by Ricardo Sanz, argues that software agents is the preferred technology today for developing complex control systems. The applications for such systems that are noted include process plants, robotic cranes, and electric power systems. In analyzing those aspects of system complexity that are especially relevant and problematic for control engineering, Sanz highlights intercomponent couplings and uncertainty. Given that no ultimate solution for complexity management exists today, a divide, conquer, and integrate approach is recommended. Control systems technology presents several algorithmic techniques, each well suited for some particular class of problems. Agents, which represent the next step in the modularization of software after object orientation, provide an effective implementation means for the heterogeneous software solutions that are required for complex control systems. For automation and control systems, the most relevant attributes of agent models are encapsulation (an attribute shared with object oriented models), autonomy (the ability to execute methods without being triggered by humans or other agents), and negotiation (the ability to reach acceptable agreements with other agents). The chapter concludes with a description of ICa, an agent-based 'meta-architectural approach to complex controller engineering'. ICa has been used for several applications, including global risk management in a chemical plant, and this application is also discussed.

Another application domain for complex software systems, system health management, is the topic of Chapter 11, authored by George Hadden and several collaborators. System health management includes such activities as fault detection and identification, incipient failure prediction, maintenance scheduling, and remediation. Although a recent field in its own right, system health management is already considered crucial for next-generation automation systems. Some daunting challenges await it. Some of these include: How can we detect and predict failures that have not been seen before in a system? How can the complications associated with multiple co-occurring faults be resolved? How can root causes be uncovered amidst the profusion of alarms? How can the commonalities between different systems be exploited while retaining sensitivity to the differences? Most of this chapter describes in detail an application that has recently been developed for naval fleets and that is now installed on a U.S. Navy hospital ship. This application is for condition-based maintenance of on-board centrifugal chillers and features a novel software architecture that integrates a number of different components. These components include an object-oriented ship model, a knowledge fusion module, and four diagnostic/prognostic algorithms based on artificial intelligence techniques.

9

Managing the Complexity of Software

Jonathan W. Krueger
Honeywell Technology Center

9.1 Introduction

Developing software for control systems is not what it used to be. When digital computers were introduced into control systems, their instruction sets, memory, and I/O were quite limited by today's standards. For example, the first process controllers had a few K-words of memory and just over a hundred inputs and outputs (Stout and Williams, 1995). With the introduction of the integrated circuit, processor sophistication, memory size, and I/O capacity rapidly increased. Sensors and actuators also became more complex, and specialized communication channels entered the scene. Today, process control systems such as the Honeywell Total Plant System can be configured as a network of control nodes, with each node powered by a 64-bit microprocessor and capable of handling thousands of I/O points. Just one of its sensors can hold an order of magnitude more software than an entire system four decades ago.

The rise in hardware capacity was more than matched by the increase in software complexity. Squeaky hardware costs usually got the grease, since software costs were harder to hear and easier to ignore. Software became the proverbial rug covering up the sweepings left behind from hardware simplification. Even if a hardware platform initially offered relaxed software accommodations, software upgrades during the relatively long life-cycles of control systems rapidly consumed the headroom. Requirements for software-borne functionality sooner or later (usually sooner) had to squeeze into tight resources, demanding tedious allocation processes and creating additional layers of software interdependency.

So it is no wonder that many consider software inherently complex (Brooks, 1987). The costs associated with this excess complexity of software are by now widely familiar: unanticipated and undesirable behavior, slower user acceptance, longer development cycles, higher product costs, higher maintenance costs, premature obsolescence.

Software complexity is, almost by definition, a multidimensional beast, much described and lamented in the literature. One of the most significant factors in software complexity seems to be the sheer number and diversity of dependencies between decisions (Figure 9.1). All systems arise from long cascades of decisions. One requirement impacts several specifications, each specification leads to several designs, each design has one or more implementations, and an implementation is made up of interdependent components. Pull on any one of these and a web of relationships reverberates.

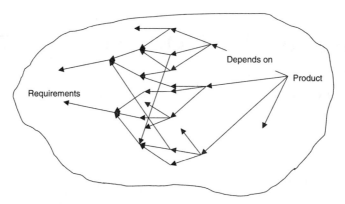

Figure 9.1 A product results from a large web of dependencies between decisions

The decision space of a software development project begins with the conception of the problem and culminates in a set of instructions to a machine. Each decision is based on assumptions and/or the implications of other decisions. Eventually the fan-out subsides and the final few decisions settle into artifacts that put the product out the door. Each dependency can be affected by a change in any one of its conditions, and the impact of such change ripples outward based on the connectivity of the dependencies. Propagating changes across known dependencies is hard enough; changes that impact unknown dependencies can be disastrous.

Software development increases the complexity by adding new kinds of dependencies, many of which are not immediately apparent. For example, the decision to employ a certain binary representation for a variable (e.g., fixed point vs. floating point) may have implications on memory allocation sizes, the timing of math operations, overflow thresholds, exceptions raised, and compatibility with other software.

Software-related dependencies also have great potential for nonlocal impact; that is, the ability of a change in one part of the system to significantly upset an apparently unrelated part of the system, or even another system. For example, most programmers have first-hand experience with the 'off by one' overshoot of a loop or undershoot of a memory allocation, or the failure to update a variable. Such errors can easily ruin a calculation, crash a program, or even cause injury (Leveson, 1995). Software's nonlocality stems from its metaphysical nature; it is a specification interpreted by highly nonlinear hardware. The specification must necessarily make assumptions and approximations about the state of the machine and its context. If for some reason an assumption is violated or an approximation is too far off center (or there is a logical flaw in the specification), the variance is rapidly magnified, causing the system to 'spin' out of control.

Having a loose grip on dependencies between key decisions raises the risk of being wrong when predicting system behavior and when estimating project costs. It can also set up a spiral effect. Changes in a poorly understood area of the decision space are likely to have unforeseen consequences. One change to fix a 'bug' can introduce more bugs, requiring more changes to fix. After a while the software becomes so brittle from repair-induced decay that it must be frozen (no more changes) or totally replaced.

Abstractions serve to organize the decision space so that certain dependencies are made more visible and others are rendered irrelevant. Although beneficial for design, abstractions do not get along well with implementations that must meet stringent performance requirements, since each layer of abstraction introduces another level of translation from the design to the implementation. They escalate the risk of lower performance and more bugs, creating backpressure on the use of abstractions. The abstractions needed for control systems software must equip engineers to manage the increasing load of dependencies while achieving the performance their applications demand.

Software engineering employs three interrelated abstraction strategies to manage dependency spaces. The strategies are common to all systems development, although the tactics are often different because of software's metaphysical nature. The first strategy structures the decision space to minimize the number of dependencies and the potential impact of unknown dependencies. Secondly, tools and methods are deployed to manage the decision space by assessing constraints and propagating changes. Finally, engineering processes are set up to ensure that appropriate thought is given to each key area of decision making.

9.2 Structuring Decisions

9.2.1 *Domain analysis*

Decisions that establish requirements provide a principal anchor for dependencies. The degree of success or failure of many projects can be traced to how they dealt with requirements. Since the requirements are at the top of the dependency food chain, a change there can cause huge tremors throughout the rest of the project. As control systems are pushed into new functional territory, their requirements have become more elusive.

One form of requirements analysis, called domain analysis, distills the general and particular dependencies across a family of systems, and provides a basis for structuring them. Domain analysis grew out of a desire to cut down the number of decisions that are 'reinvented' each time a new product is developed within a family. In domain analysis

> 'We try to generalize all systems in an application domain by means of a domain model that transcends specific applications. [Domain analysis] is thus at a higher level of abstraction than systems analysis. In domain analysis, common characteristics from similar systems are generalized, objects and operations common to all systems within the same domain are identified, and a model is defined to describe their relationships' (Prieto-Diaz, 1987).

A popular flavor of domain analysis called Feature-Oriented Domain Analysis (FODA) is divided into three phases: context analysis, domain modeling, and architecture modeling (Kang *et al.*, 1990). Context analysis gathers relevant documents and draws the initial boundaries of the domain to be modeled (usually a product family). The modeling step establishes the terminology and inter-relationships of the problem space for the chosen domain, usually in object-oriented terms. An example of a small domain model for home automation is illustrated in Figure 9.2. Finally, architecture modeling puts in place one or more architecture definitions (assemblages of components, interfaces, connections, and constraints) that meet the demands of the domain and establish design patterns for applying the architecture(s).

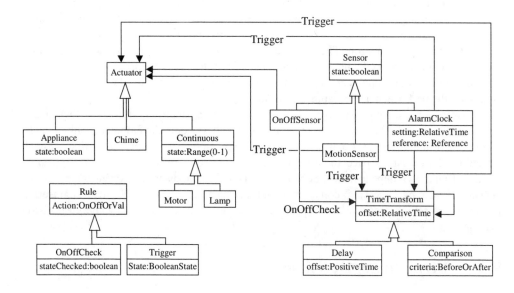

Figure 9.2 A domain model for home automation

To call domain analysis a form of requirements analysis is perhaps a bit misleading. Certainly it does establish a common terminology and a set of requirements that apply to a system domain, but it goes much further. The domain model describes the different parts of a system and their roles. In one sense, the domain model provides generalized designs that can be reused and specialized for new applications that fit the same architecture.

9.2.2 Software architecture

A software architecture specifies how certain kinds of components can be connected to form an application. The components linked together by a software architecture may perform calculations, manage a sensor, or interface to an operator. The connections may specify the flow of data, the timing of dispatches, and the allocation of resources. A software architecture constrains the decision space by identifying the types of dependencies that are allowed in the system, and by prescribing rules for evaluating those dependencies.

All software has an architecture, but it is not always explicit. If it is to be a significant aid to managing software complexity, software architecture needs to be explicit. One way of making software architecture explicit is with block diagrams, as is commonplace for many types of systems. One of the problems with such diagrams, though, is that, although they illustrate components and connections, determining the meanings of the various graphical icons is usually left to the reader, whose interpretation may be different from what the author intended. Augmenting the diagrams with text improves the situation a bit, but the process is tedious and the diagrams are difficult to maintain.

A new breed of software languages called architecture description languages (ADLs), is beginning to address these problems. An ADL standardizes architectural elements in a particular domain, and provides standard syntax for expressing the coarse structure and behavior of an application. It does not describe the algorithmic details of the components; other tools are used for that.

MetaH, developed at Honeywell, is an example of an ADL that specializes in architectures of embedded, hard real-time control systems. (Note: Hard real-time systems place strong ('hard') time constraints on selected computations. Soft real-time systems place high value on quick responsiveness but do not impose hard constraints on execution time.) It has essentially no algorithmic vocabulary, but describes systems in terms of *connected components*. A component is a collection of *attributes* and *implementations* bound together with an *interface,* and can represent software, execution platform, or pure abstraction.

An interface is made up of *interface elements* that serve as endpoints for connections. Connections tie interface elements together to form an *architecture*. The most common kinds of interface elements are *event ports* and typed *data ports*. An event port indicates that the component is capable of raising the specified event. When the event is raised, it changes the schedule of software running on the system depending on what the event port is connected to. The change may be transient or permanent, and may affect one module or the entire architecture. Data ports represent the supply of, or demand for, data. Each data port carries a specific type of data (e.g., integer, array, record) that restricts the set of data ports it can connect to.

In addition to event and data ports, MetaH allows nearly any component to be part of an interface for another component. Usually this means that the interface component is simply a visible and required part of every implementation of the parent component. If Subprogram S appears in Process P's interface, then S must appear in every implementation of P.

An implementation specifies the refinement, or internal structure, of the parent component. The implementation specifies how the component's inputs are consumed and the outputs provided through the interconnection of other components. MetaH allows each component to have multiple implementations that can be employed in different situations. For example, different implementations might be available for different levels of processor horsepower.

Attributes can range from simple scalar values to highly structured information and are used to supply detailed information about a component. For example, a 'processor' object has attributes that specify clock rate, clock jitter, the time it takes to transfer a block of data from one memory location to another, and the time it takes for the executive to dispatch a process (execution thread), among others.

Figure 9.3 depicts the communication pattern among three processes (threads of control) that together implement a parent component. The interface for the parent component appears as one incoming data port, two outgoing data ports, and one outgoing event port. In addition to the 'wires' connecting the ports, Process_B and Process_C share a monitor (protected data structure) through which they communicate.

Several of the MetaH component types correspond to actual objects found in many real-time, embedded systems. These include subprograms and processes for software and processors, memories, channels, and devices for hardware. A process is a group of subprograms tied together in a single thread of control that is dispatched either periodically or in response to an event. Ideally, each process runs in its own protected address space and has a limited execution time, though some execution platforms are not able to enforce these constraints.

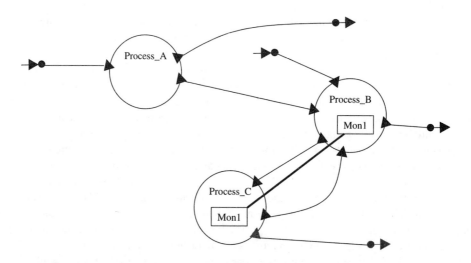

Figure 9.3 An example MetaH architecture description

One type of abstract component, called a mode, partitions the architecture into (potentially overlapping) sets of active processes. A process is active if it can be dispatched for execution. The set of active processes also implies a set of active connections. Only one mode can be active at any time, and transitions between them are triggered by events; therefore, each mode must have one or more out events unless it is a terminal mode. When a transition (event) occurs, the processes that are not part of the new mode are stopped and any new processes in the new mode are started. Any process that was in both the old mode and the new mode is left untouched. The interprocess communication pattern is also updated to reflect the new set of active processes. Figure 9.4 shows a mode structure in MetaH.

Modes are hierarchical, permitting the definition of submodes, or modes within modes. Processes are not hierarchical, however, so it is not (yet) possible to define multiple threads of control within a process.

Two abstract component types support hierarchical organization. The *macro* permits an arbitrary grouping of software components (including macros). Its cousin the *system* groups hardware elements (including systems) together.

In a real multiprocessor embedded application, each process must ultimately run on a processor. Each data transfer connection must ultimately go through some sequence of channels and processors. Each subprogram that is shared between processes must ultimately be bound to a particular memory that can be accessed by the processors involved. The MetaH language allows an explicit specification of these bindings, if desired, or a looser specification that allows a tool to decide (within limits) how to allocate the resources.

Structuring Decisions

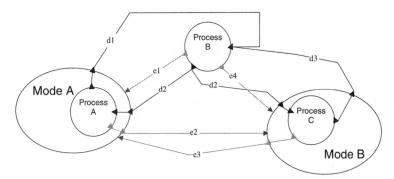

Figure 9.4 Example mode structure in a MetaH software architecture

A MetaH architecture description is more than just a set of pictures. After filling in a few timing-related attributes, the architecture description is ready for schedulability analysis. This form of analysis determines if the computations (including infrastructure overheads) fit into the schedule prescribed by the architecture. The first few turns of the analysis can use budgetary estimates for the execution times of the computational processes. Later on these can be replaced with empirical or theoretical worst case values to yield more precise analysis results. Other forms of analysis can also be done on the architecture, including reliability estimation and safety impact analysis.

In addition to formal analyses, tools can also be used to transform the architecture description into a customized executive that dispatches each computation on schedule and carries data from producer to consumer at the right times, among other things. Only the executive needs to 'know' the architectural specification; the individual components (e.g., algorithms) can be developed without knowledge of those dependencies. The computational logic can come from various sources, including code generators, reused from previous systems, or developed fresh by hand.

ADLs such as MetaH enable the automation of many kinds of dependency management between hardware and software by explicitly representing key software/hardware interfaces in a unified way across platforms, by automating decisions about resource allocation, and by automatically constructing an architecture-specific virtual machine (VM). ADLs for embedded systems greatly simplify the effort to get on the embedded hardware and to evolve the hardware and software within tight performance constraints. The structure they impose on dependencies can also improve the verification and validation picture by limiting the impact of change.

ADLs focus on providing a way of relating interfaces somewhat independently from their implementations. Software interfaces have many dimensions. Even a relatively simple interface to a subroutine is usually more than just a list of values passed in and out. Other dimensions may be just as important, including the order of the values, their type (integer, string, float, structured, etc.), valid ranges, whether the actual values are used or pointers to them, the bit order (most significant bit first or last), correlation between the values, possible error conditions that may be raised, and the list goes on. Hard real-time systems open up other dimensions of interface specification, such as the time at which outputs are available, the subroutine's frequency of execution, and its execution time limit.

Each ADL has its own set of strengths and weaknesses. MetaH does a good job of presenting a view that highlights decisions about communication patterns, roles of independent threads of execution, and the relationship between the software and the execution platform. Other ADLs provide alternative views, such as expectations on the order of events (Luckham *et al.*, 1995) and user interface structure (Taylor *et al.*, 1996). One common goal is the automated translation of the architecture description into code that preserves the properties of the architecture. To do this successfully requires making certain assumptions about the execution platform, in essence requiring a VM.

9.2.3 Virtual machines

An application is built from a set of components and is based on a VM that provides services and controls the overall execution of the application. A VM encapsulates a portion of the decision space and depends on a restricted set of dependencies conveyed through its interface. A currently popular example is the Java Virtual Machine, or JVM. The JVM executes an application that has been compiled into a series of 'bytecodes', where each bytecode is an abstract instruction that does not depend on the processor type, memory addressing, or I/O structure. When certain common libraries of functions are added to the JVM (in essence enlarging its scope), Java applications (e.g., applets) achieve a remarkable level of independence from the hardware they run on.

The browser/HTML/applet/Internet combination illustrates the potential impact of broadly standardized VMs. To the application developer, standardized infrastructure means reduced configuration complexity (many of the development decisions are dictated by the standard), which can make a major difference in development costs. Since the appearance of these standards, the number and accessibility of information management applications has exploded. The same thing is beginning to happen for control system applications.

The consumers and suppliers (clients and services) in control systems pair up at all levels of a system, and any given piece of software may play both roles. At some point each consumer must be *bound* to the resources it requires; that is, the execution platform must execute a concrete set of instructions on a particular processor.

In hard real-time control systems, the stringent performance constraints usually dictate that such bindings are 'hard-wired'; that is, the decisions about how every client accesses its services are made when the software is compiled and linked. Process P requires subprograms Q, R, and S, so the code for those subprograms is linked with the code for Process P. Calls to the subprograms are supported by very efficient instructions on the target processor.

Client software with less severe performance constraints such as supervisory or analytical functions may be bound more loosely to their services, relieving the design process of some of those decisions. In that case, one of the functions of the VM is to first locate the service that is being requested by the client. Taken to the extreme, the VM may encompass two or more physical machines and provide complete *location transparency* between the client and its services. Although the timing constraints may be less severe for some hard real-time applications, the hard constraints dictate a careful approach to implementing the VM. One example of this can be found in real-time CORBA systems (Schmidt *et al.*, 1998).

VMs provide a high degree of portability for the software using them. As long as the behavior does not deviate from the stated expectations, the VM and anything it encapsulates may change without causing further disruption. A device may move to a new address, be replaced by a combination of devices, or be completely removed. Communication between components may take different routes. CPUs may be changed, added, or removed. Services may be reimplemented in alternative programming languages. All of these kinds of changes can be hidden by a robust VM interface.

9.3 Automating Dependency Management

As we have seen, focusing on a particular domain by applying domain analysis and software architecture and through the development of VMs organizes the decision space, streamlining it so that decisions for a class of products are made more efficiently and with fewer dependencies. In any realistic application, however, there is still a vast space of decisions and dependencies to manage.

Of course, architecture is but one aspect of a system, and architecture-oriented development is only one example of a domain-specific software methodology. Software architecture relies on the inputs from other domain specialties (e.g., control algorithms) to provide computational modules and infrastructure.

The formalization of a domain into a domain-specific language such as an ADL enables what is generically referred to as *model-based development* (Bruno, 1995). Model-based development is being increasingly applied in a variety of software domains. It has long been in use for control algorithms, signal and image processing, mode logic, and cockpit display design. Software architecture is a more recent introduction but by no means the least or the last. In each case, the designer employs a specialized – often graphical – language to state decisions and dependencies of his design.

Model-based tools offer much potential, but they have traditionally been very expensive to develop. This, coupled with the fact that highly specialized domains also have small markets, makes it even harder to justify their costs. A similar dilemma faced programming language enthusiasts back in the late 1970s. Developing compilers for new languages was an arduous and error-prone task until the advent of compiler-compiler technology (Aho and Johnson, 1974). The increasing demand for model-based tools is bringing about an analogous suite of tool-building tools, or metatools, for model-based development. The result is essentially model-based development applied to itself.

The primary input to a metatool is an explicit model of the product domain, one of the outputs of domain analysis, discussed earlier in this chapter. In DoME (Krueger *et al.,* 1999), for example, a methodologist records the domain model using a graphical, object-oriented notation. The notation includes features that allow the methodology to indicate how the various objects should be displayed on-screen. Based on the tool specification, DoME automatically constructs a graphical editor that allows a user to create and edit models in the domain. Figure 9.5 shows an example model in a graphical editor produced from a domain model similar to the one shown in Figure 9.2.

The example model depicts a set of 'reflexes' for a particular home. During night-time hours, motion in the living room or opening the front door will turn on a light with a 5-minute inactivity timeout. The living room curtains open at sunrise and close at sunset, and the outlet to the coffee maker turns on at 6:30 a.m. and shuts off at 8:00 a.m. (to prevent fire hazard and excess coffee consumption). This model can be automatically converted into a series of event/action rules that execute on a VM. The VM takes care of the details of sensing the events, commanding the devices, recovering from power failure, maintaining a log, and so forth.

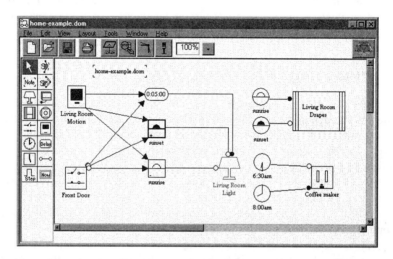

Figure 9.5 An automatically generated home automation tool with example design model

Special purpose tools can be developed and evolved quickly using metatools. A single engineer built the one described above, which included a code generator, from scratch in less than three calendar days. Most of the time was spent getting the domain model right, and once that was in place, the graphical syntax and code generator were completed in a few hours. Animated simulation was added afterwards in just one day.

Model-based tools follow a domain model to capture, organize, and maintain many of a project's dependencies. Many types of change can be automatically propagated through the product models instead of relying on error-prone manual propagation. Model-based tools streamline development in a variety of ways, including consistency maintenance, model analysis, and the generation of artifacts (documentation and code).

9.3.1 Consistency maintenance

To better understand model-based development, it is often helpful to compare it to traditional document-based development (Figure 9.6). In document-based software development processes, product information is stored as a collection of human-readable packages (i.e., documents) that are tenuously linked and with little or no automated means of managing the dependencies between them. Someone may change the name of a function or restrict the range of valid inputs to a module. The document-based process relies on people to track down and update the documents (or code) affected by a change, regardless of how simple it is.

In the model-based development paradigm, an integrated, machine-readable model serves as the central representation for all aspects of the system. The paradigm employs automation to manage the model's dependencies. There are two broad categories of dependency that model-based tools need to handle: derivations and constraints.

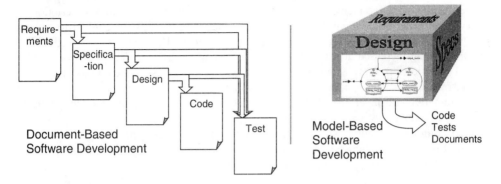

Figure 9.6 Document- vs. model-based software development

A derivation brings one set of objects into a new context after subjecting it to a transformation. Derivations come in all shapes and sizes. A reference (or hyper-) link is the simplest form of derivation whose transformation is trivial and occurs quite often in models. A common use of a reference link is to propagate the name of one object into another. If the name changes, a tool can use the link to update the name of the object linked to it. A more complex derivation may compute a value based on attributes distributed throughout the model. An example of this in a real-time architecture model would be the number of processes at each priority level.

A constraint specifies criteria for validity and, like derivations, may relate several objects and their attributes. One of the most pervasive forms is a type constraint that restricts the set of values that an attribute may have (e.g., a numeric range). A slightly more interesting constraint from the software architecture domain would be the restriction that a process period (the amount of time between invocations of the process) must be an integral multiple of the processor clock period.

Different strategies exist in model-based tools for handling constraints. At the low end, the tool provides commands to the user that cause the constraints to be checked, and the user receives a report of the violations. One difficulty with this approach is that a single problem can raise multiple 'alarms' (violated constraints), requiring extra work to diagnose the problem. At the high end, a tool checks the constraints continuously and adjusts the user interface to prevent, if possible, actions that would cause violations. This avoids the problem with redundant 'alarms' but may frustrate the user who wants to temporarily violate some constraints (e.g., to move the model along a 'short cut' to a new configuration).

Most tools employ a mix of these strategies, but the trend is toward continuous constraint enforcement.

9.3.2 Analysis

Ensuring that one's model (and hence the system) has the right 'ilities', a thorny issue for many control systems, involves the application of a variety of formal and semi-formal methods. Although there are no universally accepted criteria for when a method is considered 'formal', such methods typically exploit abstract representations of the product based on mathematical and logical principles. At one extreme, a formal method rigorously defines all the primitive elements and operations that are used to build up a description of the system. VDM is a formally defined textual notation that has been applied to several software domains, including air traffic control (Hall, 1996). Petri nets are more graphical and are gaining increased attention from control system designers (Moncelet et al., 1998). Formal specifications serve as input to a set of tools developers use to attempt to show that a system meets (or violates) key properties. Formal system modelers must often be content with simulation, however, because the implications of the model are too numerous to exhaustively navigate.

One common difficulty with strictly formal methods is that the sheer size of most industrial problems overwhelms them. There are two reasons for this. First, many methods were developed for generic applications and therefore require a lot of basic, conceptual 'brick making' before even a single course of structure can be laid. It is a bit like trying to specify organic chemistry using basic principles in physics. Projects often run out of patience by the time the first few bricks are ready for mortar. Secondly, the notations for the formal methods are virtually opaque to all but the few who make them the object of their life's work. An interesting round-table discussion of the (lack of) industrial use of formal methods in software development is presented in Saiedian (1996).

Specifying the entire system may be intractable for large systems, but partial specifications can still yield interesting insights while consuming fewer resources. The trick is to provide ways of integrating the parts. Techniques for combining partial specifications have historically been driven primarily by the telecommunications industry, but real-time systems are beginning to receive some attention in this vein as well (Avrunin et al., 1998).

The second problem can be alleviated with more careful attention to language design, resisting the desire to make a formal language completely general and employing graphical metaphors whenever possible (Moser et al., 1997). Model-based tools provide significant potential here because they are already domain specific, and often supply graphical user interfaces for accessing the model. In MetaH, for example, even a partially completed architecture model can yield useful quantitative answers about the real-time schedule, system reliability, and adherence to software safety standards (Krueger et al., 1998).

Stepping back towards the other end of the spectrum, semiformal methods employ less rigorous definitions and yield more qualitative results. The Unified Modeling Language (UML) is currently one of the most popular semiformal methods for describing software systems of all types. It consists of half a dozen interrelated graphical notations that are used to construct views of the system. Standard UML (for which the Object Management Group (OMG) is establishing a standard) supplies well defined semantics at the higher levels of abstraction, such as class structure, states and transitions, and operation signatures. However, many details require the human interpretation of arbitrary strings of text and geometric relationships in diagrams. Efforts are under way to add semantics to UML to better support the specification of real-time systems (Selic, 1999).

Analysis on semi-formal methods is usually limited to various forms of consistency and completeness as guided by the domain model behind the method. Tools look for consistency in the usage of names and key words and in references to related objects. Completeness ranges from having all placeholders filled in, to adhering to preordained structural patterns (e.g., every input is connected to some output of the same type).

9.3.3 Generating artifacts

There are three primary outputs of interest from a software model: documentation, tests, and executable software. Not all models are capable of supporting all three. Of these, documentation adds the least value, since it reflects what is already in the model that is supported by tools which can provide richer access to the information. Some traditional documentation is usually required, however, to satisfy project requirements and perhaps certification authorities. In some cases, the generation of a document is used as a check that the model is complete. In other cases, accessibility to the information is at issue; document formats tend to be less proprietary and more accessible than model formats.

The generation of tests and executable software requires different information from the model, but the information overlaps and is heavily interdependent. Tests focus primarily on allowed inputs and expected results, whereas executable software requires behavioral descriptions. Ideally, a module's expected results are stated independently from its behavioral specification so that the test is actually providing a completely independent validation of the implementation. Some dependency is usually necessary, however (Poston, 1994). An even more desirable course is to validate the tool that generates the software so that its output is trusted (similar to a compiler) and a layer of tests can be eliminated.

As mentioned earlier, a VM can have a tremendous impact on the complexity of the software applications it hosts. Similarly, a software generator has a much easier job if there is a good 'impedance match' between the domain model and the target VM. Ideally, the VM is designed concurrently with the domain model (using other model-based tools), although this is not always possible. If the conceptual distance between the domain model and the (preexisting) VM is too great, it may be more cost-effective to first wrap the VM to narrow the gap. The wrapped VM effectively constitutes a new VM tailored specifically for the software generator.

Artifact generators are, of course, yet another kind of software and have been the subject of model-based development tools in their own right.

9.4 Software Process

Perhaps the greatest complexity in any software development is found in the process itself, chiefly because it requires people to execute it. The 'normal' project management complications are multiplied by the proliferation of dependencies between decisions. There have been many attempts to bring more structure to the people side of software projects, most notably the waterfall, spiral, and RAD models of development. Within process steps, reapplying (reusing) decisions and objects from previous projects helps to reduce costs, provided there are no hidden dependencies. A process engineer's job is never done, however, so measures of maturity help to gauge room for improvement.

9.4.1 Software process models

The waterfall model is so called because it divides tasks into distinct phases, with the results of one phase cascading into the next with little feedback. 'Get it right the first time' was the mantra for waterfallers. Unfortunately, unmanaged risks got the better of many such projects, and the waterfall model fell into disrepute. The waterfall model in effect increased complexity by forcing many decisions to be made too early, before there was enough data to make the decisions wisely, and raised the costs of revising them.

The spiral model (Boehm, 1988) was specifically designed with software development risks in mind. The spiral model spun the waterfall model into a series of analysis, prototyping, and specification loops. Each successive loop focuses on the most significant risks so that, by the last loop, all of the big risks have been taken care of and the product can be polished off predictably.

The spiral model works well for many projects, but persistently vague requirements and increasingly intense schedule pressures have motivated another process formulation: Rapid Application Development, or RAD. The RAD process views requirements as moving targets and the end product as fluid until it goes out the door. In RAD, a review board consisting of representatives from the user base, the customer, and the developer prioritize the functional requirements and allocate a maximum amount of development time to each function. As development proceeds, if it looks like a function's budget will be exceeded, the function is scaled down or jettisoned according to the plan agreed on ahead of time (Boehm, 1999).

9.4.2 Software reuse

Modern software engineering is absolutely dependent on reuse, and the dependence is growing. Applications routinely reuse software intellectual property in the form of operating systems, code libraries, user interface widgets, servers of various kinds, and a host of other reused implementation artifacts. Mass market software components lower development costs by amortizing the development costs of a component over many uses.

One facet of component reuse that is often overlooked in project plans, however, is the process of specifying and choosing the components that satisfy the needs of the application. Having a good specification of the requirements for a component is essential for performing a search. Given a lot of alternatives, deciding which one best satisfies the project's many dependencies may be nearly as complex as developing a custom solution. Standardizing application 'look and feel' and architectural style – interface and interconnection rules – helps ensure a complete component specification, simplifying the selection process. Sometimes, however, an exact fit is nowhere to be found, and some adaptation is necessary. Software tends to be so brittle that if much more than 10% of a potential component requires adjustment to fit the application, it may be more economical overall to build a new component from scratch.

Reuse advocates have proposed organizational approaches to cultivating software component reuse (Ahrens and Prywes, 1995). They typically separate development into two primary activities: system development and component development. In the extreme case, the two activities take place in two separate organizations, although both fall under a common product family. System development personnel go through the 'normal' requirements analysis, architecture, and design, but the components of the system have to come from the component developers. To obtain a desired component, the system developers must supply a specification to the component developers, who in return commit to supply the component. If one does not already exist that meets the specification, they modify an existing one or create a new one, and in so doing add to the knowledge base. Incentives should be set up so that it is in the system developers' best interest to avoid supplying a specification that says more than it needs to, and it is in the component developers' best interests to supply a satisfactory component as quickly as possible.

Reusable software components have improved the economics of many kinds of software, and yet they focus on a relatively small portion of the overall software: implementation. More significant potential for reuse lies in the requirements, design, test, and maintenance phases. Intellectual assets in these earlier phases are more difficult to identify, characterize, and codify. One approach originally came from a very nonsoftware perspective: building architecture (Alexander et al., 1977). Design patterns, originally introduced to foster qualities for towns, buildings, and construction, focus on structuring situational design knowledge for reuse in new but similar situations. For example, one of Alexander's patterns suggests that when building a small public square within a neighborhood, it should be no wider than 50–60 feet across. His absolute upper limit is 70 feet, based on studies that the edge of comfortable communication for two people with healthy vision is about 75 feet. A pattern language consists of a set of interrelated design patterns.

A design pattern usually has a name (for reference purposes) and describes a specific problem, its context (common constraints, motivation), a solution, its applicability, consequences, and related patterns. Patterns often come in related groups and can be hierarchically decomposed.

Pattern languages are used to establish common structure and terminology among a collection of patterns. The Portland Form pattern language of the Portland Pattern Repository is an example of this. As described there, a pattern

> '... places itself within the context of other forces, both stronger and weaker, and the solutions they require. A wise designer resolves the stronger forces first, then goes on to address the weaker ones. Patterns capture this ordering by citing stronger and weaker patterns in opening and closing paragraphs' (Cunningham et al., 1994).

Since the reinvention of patterns, they have been the subjects of intense study by a growing number of process-minded computer scientists who are attempting to extend the usefulness of patterns to, among other things, code generation and architecture generation. There is even a set of patterns for developing pattern languages (Meszaros and Doble, 1995).

Model-based tools provide a natural home for design patterns. The computer-readable design record can be used to initiate and narrow searches for applicable patterns. Once the designer selects an appropriate pattern, the tool could automatically apply some aspects of it and track the application of the rest. The applied patterns automatically become part of the design rationale and a means for assisting the design in subsequent steps. Such pattern cognizant tools are few and far between, and those that exist are still in their infancy.

9.4.3 *Software process maturity*

In an effort to establish objective criteria for judging the merit of an organization's software development capability, the Software Engineering Institute (SEI) developed and published the Capability Maturity Model (CMM) for software (SEI, 1994). The CMM defines five levels of maturity for an organization. The first level, appropriately called Chaotic, describes organizations that develop software almost entirely by the seat of their pants and the heroic performance of highly competent people. Larger businesses especially dislike such volatility. Project planning, tracking, quality assurance, and configuration management move chaotic organizations to the next level, called Repeatable.

An organization operating at the Defined level (Level 3) has written down the process it follows, takes steps to make sure the practitioners understand and follow it, and fosters coordination between groups. The Managed level (4) adds metrics to the mix of tools used to gain deeper insight into and control of the software development enterprise. The last rung on the ladder is the Optimizing organization, wherein continuous improvement based on quantitative data supports ongoing, incremental incorporation of new ideas and technologies.

The CMM is more of a framework than a specific approach for handling complexity. It encourages a software development organization to recognize the complexity of what they do and to put process steps in place for managing it better. It is interesting to note that other disciplines besides software are looking for similar frameworks; there are 'maturity models' for systems engineering, project management, and people management. The improvements necessary to ratchet up CMM levels often apply directly or indirectly to other quality frameworks such as TQM (Total Quality Management) and ISO-9000.

9.5 Conclusion

Software is inherently complex because of the high-dependency fan-in and fan-out of decisions. The trend for control system software to become more complex is accelerating, and nothing seems to be able to stop it. The beast cannot be killed or tamed, but it can be trained. A disciplined synthesis of methods, tools, and processes will allow us to create, organize, and maintain the mind-numbing web of dependencies necessary for the production of modern software systems. Domain analysis, software architecture, model-based development, and software process discipline offer important, synergistic benefits. Domain analysis focuses our creative energies on value-added decisions in a product family. Software architecture defines taxonomies of components, connections, and analytical methods for a family of applications. Model-based tools provide automation that assists with the expression, analysis, and evolution of product requirements, architecture, and design and their transformation into deliverable product. Software processes foster collaboration between individuals and between organizations by prescribing, in effect, interface specifications for each step. Control system software is a lot different than it used to be, but the very technology that has forced us to look at systems in new ways has given us the ability to build tools that amplify our thinking.

References

Aho, A. and Johnson, S. (1974) LR parsing. *ACM Computing Surveys*, **6**(2), 99–124.

Ahrens, J. and Prywes, N. (1995) Transition to a legacy- and reuse-based software life cycle. *Computer*, **28**(10), 27–36.

Alexander, C., Ishikawa, S. and Silverstein, M. (1977) *A pattern language*. Oxford University Press, Oxford, U.K.

Avrunin, G., Corbett, J. and Dillon, L. (1998) Analyzing partially-implemented real-time systems. *IEEE Transactions on Software Engineering*, **24**(8), 602–613.

References

Boehm, B.W. (1988) A spiral model of software development and enhancement. *Computer*, **21**(5), 61–72.
Boehm, B.W. (1999) Making RAD work for your project. *Computer*, **22**(3), 113–114.
Brooks, F. (1987) No silver bullet. *Computer*, **20**(4), 10–18.
Bruno, G. (1995) *Model-based software engineering*. Chapman & Hall, London.
Cunningham, W. *et al.* (1994) The Portland Pattern Repository. http://c2.com/ppr.
Hall, A. (1996) Using formal methods to develop an ATC information system. *IEEE Software*, **13**(2), 66–76.
Kang, K.C., Cohen, S.G., Hess, J.A., Novak, W.E. and Peterson, A. (1990) Feature-Oriented Domain Analysis (FODA) Feasibility Study. CMU/SEI-90-TR-21, Software Engineering Institute, Pittsburgh, PA.
Krueger, J. *et al.* (1999) The DoME guide. http://www.htc.honeywell.com/dome, Honeywell Technology Center, Minneapolis, MN.
Krueger, J., Vestal, S. and Lewis, B. (1998) Fitting the pieces together: system/software analysis and code integration using MetaH. *DASC 98 Proceedings,* IEEE.
Leveson, N. (1995) *Safeware: system safety and computers*. Addison-Wesley, Reading, MA.
Luckham, D., Kenny, J., Augustin, L., Vera, J., Bryan, D. and Mann, W. (1995) Specification and analysis of system architecture using Rapide. *IEEE Transactions on Software Engineering*, **21**(4), 336–355.
Meszaros, G. and Doble, J. (1995) MetaPatterns: a pattern language for pattern writing. http://hillside.net/patterns/Writing/pattern_index.html.
Moncelet, G., Christensen, S., Demmou, H., Paludetto, M. and Porras, J. (1998) Analysing a mechatronic system with coloured petri nets. *International Journal on Software Tools for Technology Transfer,* **2**(2), 160–167.
Moser, L., Ramakrishna, Y., Kutty, G., Melliar-Smith, P. and Dillon, L. (1997) A graphical environment for design of concurrent real-time systems. *ACM Transactions on Software Engineering Methodology*, **6**(1), 31–79.
Poston, R. (1994) Automated testing from object models. *Communications of the ACM*, **37**(9), 48–58.
Prieto-Diaz, R. (1987) Domain analysis for reusability. *COMPSAC 87 Proceedings,* 23–29.
Saiedian, H. (1996) An invitation to formal methods. *IEEE Computer*, **29**(4), 16–17.
Schmidt, D., Levine, D. and Mungee, S. (1998) The design of the TAO real-time object request broker. *Computer Communications*, **21**(4), 294–324.
Selic, B. (1999) Turning clockwise: using UML in the real-time domain. *Communications of the ACM*, **42**(10), 46-54.
Software Engineering Institute (SEI) (1994) *The Capability Maturity Model.* Addison-Wesley, Reading, MA.
Stout, T. and Williams, T.J. (1995) Pioneering work in the field of computer process control. *IEEE Annals of the History of Computing*, **17**(1), 6–18.
Taylor, R. *et al.* (1996) A component- and message-based architectural style for GUI software. *IEEE Transactions on Software Engineering*, **22**(6), 390–406.

10

Agents for Complex Control Systems

Ricardo Sanz
Universidad Politécnica de Madrid

10.1 Introduction

Control system applications in industry are reaching complexity frontiers that are hampering further progress in automation. It is time to revisit our current control technology with the objective of identifying what new developments are needed that will allow us to cross these frontiers and how these needs can be satisfied. This technology problem is not exclusively related to the mathematical theory of control systems, but also with such wider issues as software engineering, project management, and controller design reusability.

Information processing technology is the key to the future of systems automation. Control systems are nothing but information processors; the more complex the task, the more complex the processor that performs it. Independently of the core disciplines – such as dynamical systems theory – control engineers need to master computer science, communications, and software engineering that are basic tools for complex controller engineering.

This chapter focuses on the use of the agent-oriented approach to software-intensive control system construction. It tries to show the proximity of the agent-based perspective to control systems engineering. It will stress the need for incorporating within the controls discipline some of the methods and ideas that are under development in this area of research. This need is bidirectional; people working in agent-based systems lack the control engineer's experience with building autonomous entities. In a sense, the clearest examples of agent-based systems are to be found in the realm of control systems. All controllers, complex or not, are agent-rich.

10.2 Examples of Complex Control Systems

As stated earlier, this chapter deals with the construction of complex controllers using agent technology. But, *What is a complex controller?* I will present some examples to let the reader grasp the concept before proposing definitions.

10.2.1 Chemical process control

Chemical plants (see Figure 10.1) are composed of a number of different interacting units that perform certain basic processes to realize, overall, a complex processing structure.

Control systems in these plants are usually based on distributed control technology, employing conventional and advanced control algorithms to steer the various units. Distributed control systems in process plants can manage up to several thousand measured and/or controlled variables to keep units at set points or recalculate set points for optimization. They also perform alarm management and some form of limited diagnosis of plant faults. These plants operate in high-availability regimes, stressing the requirements of control systems.

Figure 10.1 Chemical plants such as the acrilonitrile reactor in the petrochemical complex of Repsol S.A. in Tarragona, Spain, are very complex systems that can employ state-of-the-art controllers

Top-level, plantwide strategic control is usually done by humans. They base their decisions on previous experience or on supervisory and optimization packages. One of the main problems control engineers must confront in these plants is the abstract or summary nature of some of the top-level objectives. Although automated management of the plant demands online measurement of these factors, it is not easy to establish the safety status of a plant at a specific moment.

Besides the abstract nature of some of the objectives, they are strongly coupled in intricate ways. The development of a global model to resolve the control task is therefore difficult. We can say that present production conditions are always near some form of criticality (see Figure 10.2) due to the nature of the process, the market circumstances, and/or political regulations. Control of chemical processes is getting harder due to this trend toward extreme operating conditions.

Figure 10.2 Top-level objectives in plant management go beyond simple production or quality set-point control. A network of interacting objectives complicates the task of global decision making.

For example, cement production plants commonly use coal to heat the kiln. This is the main cost in a cement plant (apart from people), and so its reduction is quite important. Classical cost reduction was based on cutting the amount of coal burned. This led to kiln operation at a lower temperature, near the critical temperature where the chemical reaction does not complete, consequently producing a bad product.

Modern cost reduction is based on the use of alternative, cheaper fuels. Examples of these cheaper fuels are coke, fungi mycelium, and peach (yes, peach!). These fuels reduce the production cost by sacrificing other objectives. Coke has a large sulfur content that contaminates the environment and spoils the quality of the product (sulfur is a poison for cement). Fungi mycelium and peach have to be desiccated and have low calorific power, reducing the control capability of the kiln.

10.2.2 Large robot control

New trends in service robotics are leading to the construction of new types of robots for applications beyond manufacturing. Examples include robots designed for building construction. Automated construction is an important objective because of the large number of human deaths in industrial accidents in the construction sectors.

These robots need to be large, with wide spans and high load capacity. Figure 10.3 shows one of these robots designed in our laboratory (Barrientos *et al.*, 1997). The robot has hydraulic actuators to reach the required load levels, due to the handled object (a brick in the photograph) or to the weight of the robot itself. Controllers must also take into account the flexibility of the robot links due to their length.

Figure 10.3 Complex controllers do not need to be distributed controllers for large plants. The photograph shows one of the construction robots developed in the ESPRIT ROCCO project that used a complex heterogeneous controller.

10.2.3 Electric systems control

Electricity generation and distribution systems are composed of a heterogeneous collection of subsystems of varying complexity that can span a whole continent. Although electrical systems are very well understood and can be modeled quite well, the number of components and their interactions lead to extreme complexity. Market pressure is also a factor to take into account; cost reduction is a basic objective for these plants.

An example is the trend toward total reduction of plant personnel in hydroelectric or wind power plants. These are the simpler generation plants, with very high availability levels. The present trend is to reduce in-plant personnel, including operators and maintenance personnel, to zero, operating the production plants from a remote site (see Figure 10.4).

The distributed control system for these superplants spans hundreds of kilometers instead of hundreds of meters. Wide area broadband networking is employed to provide a communication infrastructure capable of handling all the information needed to properly operate the plant.

One major problem with autonomous operation is the reduction of safety due to physical separation. Operators cannot physically observe the real plant to validate digital information. Redundancy in control systems is used to minimize the failure risk, using information systems replication and/or complementary measures, for example, real-time video, to provide operators assurance regarding plant state.

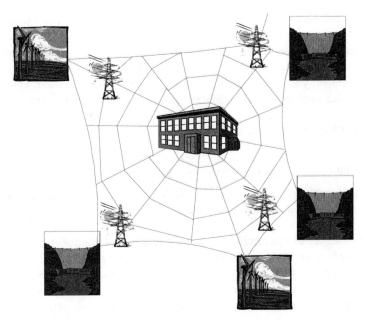

Figure 10.4 Remote operation of electric distribution and generation plants from a single operation center can be done due to the availability levels of these types of plants

10.3 Control Systems Complexity

10.3.1 The nature of control systems complexity

We can identify three basic types of complexity in control systems:

- *Primary complexity* due to system size. For example, a thermal power plant can have 20,000 measured variables.
- *Secondary complexity* due to couplings between subsystems. A paradigmatic example of secondary complexity is nonlinearity.
- *Tertiary complexity* due to uncertainty. Examples of uncertainty are faults or chaotic behavior of systems.

We must take into account that we need to manage not only the plant, but also the controller-plant system, and that complexity affects not only the plant but the controller itself. This perspective is becoming increasingly important as controllers reach functionality levels that were unthinkable some years ago: distributed, nonlinear, robust, fuzzy, reconfiguring, fault-tolerant, and so on.

Primary complexity (size) should not, purely by itself, be a problem in a controller today, because a disciplined systems engineering approach and the rising power of computers can cope quite well with it.

Secondary complexity (coupling) is a very big problem. Traditional control system technology has been focused mainly on a mathematical theory based on linear differential equations (i.e., *linear control theory*). This is not the ultimate approach, because systems of interest mainly fall beyond the analysis capabilities of this theory. For example, physical systems tend to be nonlinear. Inner couplings between plant subsystems are created when trying to optimize plant operation (e.g., when doing co-generation). Plant-controller models are getting to be too complex to solve analytically or even numerically.

Tertiary complexity is also of special relevance for the control systems community. The main sources of uncertainty when building or operating a control system are

- The lack of knowledge about the plant and its environment that appears in the form of poor models;
- The uncertain nature of perception (noisy sensors, hidden variables, etc); and
- The unpredictable behavior of the controlled system due to bad controller design, hardware malfunction, unhandled events, or software errors.

In fact, we can say that uncertainty is the *raison d'être* of control engineering (as illness is for doctors or hunger for farmers). In software engineering, the term *future-proof* is used to refer to the capability of a software system to accommodate future changes. Control engineers make systems that are inherently future-proof (i.e., they can adapt themselves (the controller + the plant) to events in order to meet some design objective).

Perhaps other dimensions of complexity can arise when we introduce new factors into consideration. For example, the type of modeling methodology used (continuous, discrete, hybrid, rule-based, fuzzy, etc.), the type of control technology employed (in relation to the type of model), the platforms where the controller is to be deployed, the programming languages or operating systems used, the development team structure and expertise, and so on.

It should be clear that the conventional two-phase approach of control systems engineering (model-system/design-controller) needs to be expanded to take into account systems that cannot be easily modeled using conventional technologies. The current situation leaves much to be desired. There remains an outstanding need for the Great Unification Theory (GUT) of control systems.

10.3.2 Tackling control complexity

In the meantime, those of us who build complex controllers should concentrate our efforts on a realizable objective such as reducing the difficulty of complex controller construction using the available technologies. There are several approaches to tackling complexity. Some examples are shown in Table 10.1.

Table 10.1 Complexity attacks

Name	Approach	Requires	Example
Brute force	Plain work	Computing power	Supercomputing
Approximation	Simplification work	Man hours	Poor man's approach: linearization
Divide and conquer	Engineering	Experience	Modularization
Change of view	Theory	Deep insight	Conceptual revolution

Brute force approaches are focused on applying techniques that work for simple problems to complex ones. In most problems, complexity growth is nonlinear in relation to problem size; it grows disproportionately. Examples can be found in the use of optimizing packages based on ordinary differential equation models or expert systems for large plant control. In the first case, the complexity of a whole plant can lead to models with one million equations that can hardly be computed in the standard way. In the second case, inference engine performance decreases with knowledge base size, especially since in most cases a clear methodology for systematic construction of large knowledge bases is lacking.

Approximation approaches try to use simplification to attain solvability. To achieve this objective they sacrifice precision. The effort here is put into determining what are the simplifications that minimize the loss of precision. Examples of this approach are linearization of nonlinear systems, time-scale focusing in simulations, or knowledge base focalization in rule-based systems. Parts of the model are not used – sacrificing precision – so that known solution methods can be applied.

Divide and conquer can be considered a kind of brute force approach. The core idea is to decompose the problem into smaller problems without loss of quality. As stated earlier, problem complexity tends to grow faster than linearly, and decomposition tries to approach linear scale-up. In complex systems, linearity is not achievable since interaction effects imply expenditures of effort for the integration of partial solutions. The utility of divide and conquer relates directly to the ease with which the inevitable interaction factors can be minimized.

The last approach is the most desirable and the most difficult. Some problems can be simplified by a change in the solution model. System complexity is a measure also of our difficulties in understanding the system, or to be more precise, the difficulties in grasping it, of having it in our minds. Models (paper-based, computational, or mental) reflect our understanding of systems, and complex models show that we do not have easy ways to capture our knowledge of the systems.

These *Change of view* attacks on complexity provide conceptual tools that simplify the models we use (mental or external). This is what many people are seeking in new control systems technology. Examples are passivity-based control, neural network learning control, and control based on genetic algorithms. It is not easy to predict what theories will be successful enough to be considered a conceptual revolution. In technology at least, the label 'revolutionary' is bestowed by hindsight.

My opinion is that we have not yet identified the 'right' solution to tackling complexity, even though many contenders are being promoted (see Figure 10.5). Complex controller engineering tends to be done using a mixture of all these approaches, with divide and conquer as the main guideline for global problem structuring.

This is the focus of this chapter: divide the problem, build partial solutions using available technologies, and integrate them to make the whole solution.

10.4 Heterogeneity and Integration

In the present situation, the recommended approach for advanced controller construction is to use the best technology available for specific parts of the system under construction. Decomposition strategies (Alarcón *et al.*, 1994) are valuable assets, but unfortunately, in their current instantiations, they do not have the extension and depth necessary for wide application. We must confront the problems of decomposition, construction, and integration.

Figure 10.5 The Control Wars: What is the ultimate technology for complex control?

10.4.1 Heterogeneity of control systems technology

Typical examples of control technologies that address problems beyond the capability of classical control system theory (theory *à la* the *IEEE Transactions on Automatic Control*) are those grouped under the flag of intelligent control: expert systems, neural networks, fuzzy systems, genetic programming, and so on.

It is not easy to demonstrate the stability of a rule-based expert controller for a cement kiln, and a control theory purist may dismiss the approach as a consequence. This is not a pedantic position, because behavior certainty is important for us. However, we cannot – and do not want to – renounce these approaches because these technologies are the single, most promising alternative for complex control system construction. In fact, behavior certainty cannot be achieved using classic approaches, because even when you can demonstrate the stability (or performance) of a specific controller, in practice this is an artificial result, and what you get is artificial confidence in it.

Real controllers control real systems, and this means that they are far from the entities used in the theoretical proof. The real system does not exactly behave the same way as its model and the real controller is not the theoretical one for several reasons, including physical limits of actuators, rounding errors in computation, or programming errors in the code that implements the controller.

Soft systems technologies excel at some problems. Examples of their use can be found in any area of control system technology:

- *Expert systems* can exploit knowledge acquired by persons during years of learning effort, leading to expert controllers that can out-perform human operators.

- *Artificial neural networks* can be trained to model plant subsystems better than physically grounded models in cases where the latter are unavailable or infeasible to develop. Further, in some applications they can be retuned online.

- *Fuzzy technologies* are adept at uncertainty management, having demonstrated their capabilities in linguistic knowledge transfer with humans or management of perception uncertainty.

- *Genetic algorithms* can explore design regions that escape conventional methodologies, finding, for example, efficient operating conditions for thermal power plants when working far from their design set point.

Until the Control GUT arrives, complex controllers will be built using heterogeneous technologies because the problems in a real plant are also heterogeneous. (Unification approaches do in fact exist, mainly centered around integrated modeling methodologies and model-based algorithms (virtual engineering), but the difficulties they confront are extremely tough: the need for extensible modeling formalisms, the problems of combinatorial explosion of algorithms based on deep knowledge, etc.)

10.4.2 People heterogeneity

Heterogeneity does not only manifest itself in the systems we build. If we consider the business structure of the control systems engineering market, we see several roles for people:

- *Final users*: they are scattered into niches with limited interaction and with a wide and heterogeneous collection of requirements, from soft requirements in noncritical systems to strict requirements in safety-critical applications. Even in a specific application the collection of users can be heterogeneous, from operators to plant managers or maintenance personnel.
- *System integrators*: they design solutions to users' problems and build systems integrating products usually developed by others.
- *System developers*: they create new information and control systems for specific purposes (complete systems and/or components).
- *Tool developers*: like system developers, they construct systems, but with a wider market perspective. They build metasystems for those who build systems. The tools they construct help apply their designs to several applications.
- *Researchers*: they focus their interests toward the exploration of ideas rather than the exploration of real plant needs. Their products are often solutions in search of problems.

Each of these populations has specific objectives for the control system and expresses different requirements for it. From our perspective, we need a common control system framework that leads to a wider integration of the visions of all these stakeholders.

10.4.3 Integration technology

One of the key words in the area of complex systems engineering is *integration*. Integrated approaches to complex systems construction are based on the identification of three broad categories of technological entities:

- Base technologies
- Integration architectures
- Integration technologies.

Base technologies are those employed in the construction of specific components. In the case of complex controller construction, they are basically classic control technologies (robust control, model predictive control, etc.), soft technologies (expert systems and relatives), or information management technologies (user interfaces, real-time databases, etc.). Integration architectures are global, reusable application designs that make possible the integration of components implemented using the base technologies. Integration technologies are those which make possible the integration of the components. In the case of complex controllers, software integration is one of the major problems these technologies must solve.

Drawing a parallel with the construction of buildings, base technologies are the bricks and the windows, the integration architecture is the building framework, and integration technologies are cement and soldering guns. Or, to be more precise, base technologies do parts of the work, architectures define how parts build the whole, and integration technologies deal with interfacing between components. The modern way to achieve integration in software-intensive activities such as complex control systems is through agent technology.

10.5 Objects, Components, and Agents

In the very beginning, subroutines were built to conserve memory in the computer. Although not done with reuse in mind, it was in fact the first example of reuse. What served initially to maximize the use of machine resources ultimately improved human effectiveness (Clements, 1995).

The object-oriented approach to software system construction promised reusability by means of encapsulation without sacrificing adaptability (by means of inheritance). The paradigm of reusability in modern software construction is component-based software engineering.

Object technology used to be considered a poor choice for control systems due to the difficulties of complying with real-time requirements (for example, uncontrolled chains of virtual function calls). But now the vision has changed with the realization that object orientation can facilitate mappings from conceptual domain models to software implementations, and that it is well suited for implementing systems that have a strong connection with real objects.

Object technology offers a good foundation for component-based software engineering due to the encapsulation of behaviors of components. Syntactic specification of interfaces is, however, not enough for real-time systems; semantic specification of class functionality and constraints is usually done in natural language, and this approach does not permit formal analyses of the correct behavior of the reused component.

The next step in this chain of modularization of software components is agent technology, about which there is much confusion and misunderstanding. We assert that agent technology is the technology of choice for complex control systems construction. In fact, this is not a radical change in control systems construction, because as we will see, we have been building agents for years.

10.5.1 Agent technology hyperbole

The classic paper by Franklin and Graesser (1995) introduces a common theme for discussion: *What is the nature of a software agent?* Questioning the meaning of the term is natural in the context of agent research due to the confusion originated by the overuse of a buzzword.

Agent technology is not a software technology but a system organization technology. This means that the scope of agency concepts goes beyond pure software systems. Much of the confusion that appears in the agent technology literature is due to the appropriation of terms that is so widely seen in information technology. People want to use the word *agent* to distinguish their work from others. Needlessly restrictive definitions for the concept of *agent* can be found everywhere. (For example, 'an agent is a Java piece of code that can be moved to be executed in a remote computer'.)

When I confront these situations, I like to read dictionary definitions. (As an exercise, try to get a definition for *complex system* using a dictionary. You may discover that both words have the same meaning!) *Agent* conveys two main meanings:

- *The one who does something.* This is the main interpretation in the case of multiagent systems.
- *Doing something on behalf of another.* This is the preferred interpretation in Internet agents.

Adjectives have been used to modify the undefined core meaning of the word agent, the two most overused being *autonomous* and *intelligent*. 'Intelligent agents' is virtually a tautology. Autonomous agents warrant some discussion because autonomy is that elusive objective of control systems engineering.

From my point of view, autonomous agents are programs – no less, no more. If you read the definition of *autonomous* given by Jennings *et al.* (1998):

'... be able to act without the direct intervention of humans (or other agents), and ... have control over its own actions and internal state'

you will reach the same conclusion. I also suggest reading the beginning of the first chapter of the classic book of Abelson, Sussman, and Sussman (1985). In this book about basic programming, the beginning of the first chapter describes what is a *computer process*.

So if agents are programs, what is the reason for the hyperbole about agents in information technology? They have new journals, new symposia, research groups, and a lot of funding. For those of us involved in the development of automation and control systems, it is difficult to understand the reasons behind this wide interest in a technology that hardly seems new.

My conclusion is simple: agents are associated with the marvel of creation of intelligence. Intelligence is the capability of exploiting information to do things. This is what agents do. This is also what controllers do. This is why agent models are the natural way of building controllers.

The agent model of building systems is based on getting global behavior through the interaction of entities (the agents) that pursue small-scale, self-owned objectives. Agencies are the systems built following this model. A major part of research and development in agent technology is the development of agent frameworks, that is, software environments where agents can be built and executed easily.

Agent-based design is not a matter of the technology you employ but the mental model you have about the systems you are building. It is not at all necessary to build an agency using an agent infrastructure, even when frameworks can simplify the work. You can build agents even in Visual Basic!

10.5.2 Aspects of agency

The classic view of agency (Jennings *et al., 1998*) incorporates a collection of properties that may or may not appear in any specific implementation. The classic property set for an agent is the triad *situatedness, autonomy,* and *flexibility.* Situatedness means that the agent is placed in some environment, senses the environment, and acts on it; autonomy means that the agent is able to do its work without help from humans and other agents; and flexibility is the capability of being *responsive, proactive,* and *social.* This definition matches almost any piece of real-time code running on a computer (any running process) with the possible exception of proactiveness.

Other agent definitions incorporate more aspects to better profile their agency concept. Examples are capabilities of *negotiation* (reaching an acceptable agreement with other agents), *mobility* (being able to move the execution to another support system, for example another host), *coordination* (communication with other agents to avoid interference between activities), *facing* (having a graphic presentation that represents the *emotional* state of the agent), and *cooperation* (collaboration with other agents to reach a common objective).

For me, the most relevant issues in the agency model are encapsulation, proactiveness, and negotiation. Encapsulation is the basis of the object-oriented model, and agent systems have nothing new to say about it. Proactiveness means that the agent can initiate activities (execute methods) based on its own state and sensory input interpretation. This differs from the classical message-based object model in the sense that there will be agent activities that are not requested by other agents.

Agent-based systems can be viewed as collections of active objects; that is, objects that have methods invoked by the object itself and not by external requests. This was not contemplated in classic object models such as Smalltalk, but today it fits perfectly within new paradigms such as CORBA. This change is due to the change to concurrent object models based on distribution or on multithreading.

From my point of view, the only different issue of relevance in agency models is negotiation. Negotiation implies that the agency interaction structure is no longer static. Agents can not only respond to predefined requests from predefined agents, as in static agencies, but they can negotiate with other agents to reach an agreement that is satisfactory for all agents. This leads to a distinction between static and dynamic agencies. In the case of complex control systems, dynamism is of immense practical importance to cope with uncertainty in the state of the agency. Negotiation opens the door to optimization, reconfiguration, fault tolerance, hot-swapping, extension, and adaptation.

Negotiation (or object service dynamism) introduces an extra degree of uncertainty, however. It will be very difficult in general to completely determine the state of the agency at any time in the future with enough precision to guarantee real-time performance. This is the reason why most control agencies are static (i.e., the interactions between agents are predefined).

10.5.3 Agent literature

I recommend five starting points from the agency literature for more details:

- The article 'A Roadmap of Agent Research and Development', by Jennings, Sycara, and Wooldridge, in the first number of the journal on *Autonomous Agents and Multi-agent Systems*, published by Kluwer Academic Publishers in 1998.

- The series of books titled *Intelligent Agents N,* written by various authors and published by Springer. They are collections of papers presented at various symposia.

- The book *Readings in Agents* edited by Huhns and Singh and published by Morgan Kaufmann Publishers in 1998. This is a collection of classic papers on agents.

- Numbers 1–2 of volume 27 of the journal *Robotics and Autonomous Systems*, dedicated to agent-based manufacturing systems.

- The book *Computational Theories of Interaction and Agency* edited by Agre and Rosenschein and published by MIT Press in 1996. It is a reprint of volumes 72 and 73 of the journal *Artificial Intelligence*.

Although agents and control systems are very strongly related, little systematic work has been done in the specific use of this technology in control systems. Examples are scattered in various publications: Bonasso *et al.* (1997), Fischer (1999), Hayes-Roth (1995), Nielsen (1995), Occello and Demazeau (1997), Sanz *et al.* (1998), Sheremetov and Smirnov (1999), Wang and Wang (1997), Wittig (1992), and Wooldridge *et al.* (1999).

10.6 Agents for Complex Control

10.6.1 The ICa approach

Our approach to complex control systems construction is based on the modular composition of agent-based subsystems over an integration framework. We call this approach ICa, which stands for Integrated Control Architecture.

ICa is not, strictly speaking, an architecture. It started, as the name suggests, as an architecture for integrated controller construction for the process industry, but it went *meta* quite early, so we can say now that ICa is a meta-architectural approach to complex controller engineering (Figure 10.6). ICa promotes architecture-based development of control applications and relies on domain engineering to produce reusable assets.

In ICa we distinguish three main parts:

- *The ICa Generic Agents:* an extensible object framework that lets developers exploit previous work.

- *The ICa Glue:* a modular object request broker based on CORBA specifications that serves as integration middleware for the agents.

- *The ICa Methodology:* a reuse-based methodology that focuses on the use of design patterns to maximize design effectiveness.

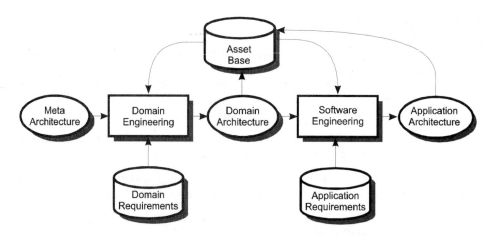

Figure 10.6 The ICa meta-architectural process promotes an architecture-centered process of complex controller development. One of the main steps is the definition of the application architecture based on specific domain architectures and the reverse process of architectural asset construction.

10.6.2 Patterns as controller designs

ICa bases its methodology on the reuse of pieces of code (the generic agents) over an integration middleware (the glue) according to domain architectures for complete applications or parts of applications. We document these designs as patterns (Buschman *et al.*, 1996; Lea, 1994, Aarsten *et al.*, 1996, Kuikka *et al.*, 1999).

The ICa pattern language provides a collection of design patterns ranging from elementary controllers to information-technology-specific aspects such as brokering or database federation (Sanz *et al.*, 1999c).

10.6.3 The proper size of an agent

As with any other fashionable technology, agent solutions are being force-fit into inappropriate application areas. An example is the time-to-contract factor associated with negotiation. Static agencies have predefined contracts between agents, so they are quite predictable in behavior because negotiation is not needed.

Other problems are related to agency efficiency. In the words of Kersten and Noronha (1998), human 'negotiators more often than not reach inefficient compromises', and this is a clear risk in negotiation-based agencies. Although having a compromise – a solution – is valuable, inefficiency can no longer be accepted in industrial systems if alternatives exist.

The conclusion is that the need for specific agent behaviors is determined by the application or domain requirements. For example, to provide fault tolerance to a critical database application, hot replication can be used. This is obviously a burden for the agency and for the database agents themselves. The proper size of an agent is the size needed by the application and not the size required by the agent framework. Agents should provide the functionality needed for the application and not more, but if you take agents from a repository and put them into a framework, you will get a minimum size and functionality (for example, negotiation capabilities or autoreplication) depending on the ideas of the designer of the framework. Thus the basic ICa agent does not provide any method and only needs space for two pointers (used to map the agent's name and a transport to interact with other agents).

The true reusability of a component is not based on providing broad functionality but on reducing to a minimum the design compromises of the component. Design decisions partition the design space for the component, limiting its reusability. Component designers must provide methods to adapt the components to the domains where users intend to apply them (see Figure 10.7).

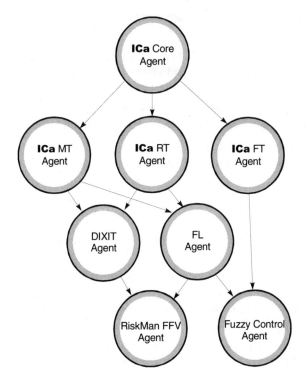

Figure 10.7 The RiskMan Fuzzy Filtering and Validation Agent was built by inheritance of behaviors from generic agents that provided specific functionalities, such as the DIXIT agent (distributed strategic control), the FL agent (fuzzy inference and knowledge base management), or the ICa MT agent (multithreading) (Sanz *et al.*, 1998).

10.6.4 Development methodology

The ICa development methodology is based on domain engineering concepts. The basic idea is *progressive domain focalization* (Sanz et al., 1999a). We do not make a strong distinction between domain engineering and application engineering because we consider application engineering to be subdomain engineering (i.e., components built or tailored for a specific application can be reused in a domain that is included in the previous one). This forces developers to *think generic* even when building application-specific things. This process matches the idea expressed in Figure 10.7 in relation to the generic agents.

The ICa methodology proposes a base list of activities:

- *(Sub)domain analysis:* to provide support for cohesion in different control systems, subdomain analysis must be fused with the domain analysis previously done. This phase should include classical requirements engineering phases.
- *Control tasks identification:* this is related to domain analysis but focuses on control aspects more than plant behavioral aspects that are mostly addressed in the domain analysis phase. Both individual and group control tasks need to be identified.
- *Control technology selection:* the more suitable technologies for solving control tasks are identified.
- *Controller architecture design:* the base architecture is defined based on domain architectures. Control task partition and grouping follow compositional criteria (speed, technology, layering, etc.). This means agency design.
- *Detailed controller design:* depending on the degree of reusability of controller assets, the need for redesigning and reimplementing controllers will be diverse.
- *Controller implementation:* isolated control modules are constructed based on true reuse and parametrization or on complete reconstruction of the controller. This means agent implementation.
- *Integration:* this is done following architectural guidelines. Middleware support simplifies this process.
- *Test and validation:* system testing, verification, and validation are also included in the activities.
- *Maintenance:* both classical maintenance (defect correction, incorporation of new functionality, continuous operation) and controller maintenance are implied. The latter includes retuning, continuous learning, knowledge enhancement, controller monitoring, etc.
- *Make it generic:* reusable assets can be obtained from application-specific developments. Re-engineering of specific components may be required.
- *Asset base maintenance:* the incorporation of new assets into the asset base and the reengineering of obsolete assets.

10.7 Some Examples of Agents in Control Systems

There are many developments in agencies for control applications. Here I comment on just three examples that can serve as introductory illustrations.

10.7.1 ARCHON

ARCHON (Witting, 1992) is a classic application of agent-based systems for distributed process control. Developed in an ESPRIT project, the final result was a framework and a methodology for the construction of coarse-grained agent applications in industry. Examples of its use are applications in electricity transmission systems and particle accelerators.

Any ARCHON agent is structured into four main components:

1. A High-Level Communication Module (HLCM) for interagent communication.
2. A Planning and Coordination Module (PCM) to control agent actions.
3. An Agent Information Management (AIM) module, a world model manager.
4. An Intelligent System (IS), which is the core of the agent (the task performer).

HLCM, PCM, and AIM are the portions contributed by the framework that are used to wrap the IS. The IS can be a preexisting system or a new component developed for the application.

10.7.2 RiskMan

RiskMan is an application developed for *global risk management* in a chemical plant complex (see Figure 10.8). The scope of RiskMan includes:

- *Risk reduction*: enhancing practices for minimizing hazards
- *Risk prevention*: by means of online fault detection and isolation
- *Risk control*: by means of emergency management in accordance with safety regulations, operating policies, and constraints of the plant.

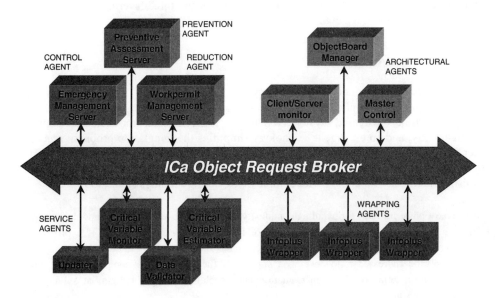

Figure 10.8 RiskMan is an application for plantwide risk control (prevention, reduction, and management) in a petrochemical complex. It uses heterogeneous technologies such as expert systems, fuzzy filters, and neural network models to implement a strategic decision support system for plant personnel. Agent interaction is done through the ICa object request broker. Non-agentified components (such as user interfaces) and helper applications interact using specific integration mechanisms and have been suppressed from the figure.

The RiskMan application was developed around three subsystems that were devised for the three main risk management tasks above:

1. A *workpermit management system* to handle all works that are performed in the plant. This subsystem tries to minimize the risk of each work by itself (by automatic handling of work-specific safety measures), the risk of each work in a specific plant condition (workpermits are issued taking into account plant state), and interwork interactions.
2. A *preventive system* to identify faults and plant state deviations from safety conditions.
3. An *emergency management system* to help people handle emergencies when they occur. The system is designed to disseminate real-time information about an emergency situation, coordinate the activities of persons, interact with external entities, and predict emergency evolution.

The application was built using the ICa framework and consisted of ICa agents, legacy applications, wrappers, helper components, and independent non-agentified components.

10.7.3 Control of smart airplane structures

Sahasrabudhe and Mehra (1998) present an architecture for smart structure control in airplanes that is composed of three layers: a reactive layer for real-time control (RCL); a supervisory layer (SCL), and a hierarchical multiagent cooperative action layer (DAL). Agency is introduced in the uppermost layer, whereas real-time properties can only be assured in the lowest one. This is a typical example of the three-layering concept used in agent implementations (Sanz *et al.*, 1999b).

10.8 Conclusions

Agent technology (active object technology) is the natural way to apply information technology in real-time applications that interact with physical systems. It features reduced development time, increased system reliability, reduced cost, modular development, reconfigurabilty, and so on. Agent technology simplifies the integration problem, rendering easier the development of modular systems with evolving requirements (as is the normal case in complex controller engineering). In control systems engineering it is an especially promising technology for coping with complexity. Quoting Jennings *et al.* (1998):

> 'The most powerful tools for handling complexity in software development are modularity and abstraction. Agents represent a powerful tool for making systems modular. If a problem domain is particularly complex, large, or unpredictable, then it may be that the only way it can reasonably be addressed is to develop a number of (nearly) modular components that are specialized (in terms of their representation and problem solving paradigm) at solving a particular aspect of it'.

CORBA can be used in real-time industrial applications to provide an effective platform for communication that constitutes a foundation for enterprise integration architectures reaching even the sensors. Systems built using this technology are open, evolvable, and adaptable; it is quite simple to incorporate other types of advanced features within an existing system (for example, adding a new application or making a system fault tolerant).

As a final comment, I can say that building agent-oriented applications is not a question of using the proper tools and infrastructures. It is just a cultural attitude – a design stance that you can use to think about your application. The terms employed can be somewhat provocative at times: 'the agent thinks that ... ', 'the controller believes that ... ', 'the sensor feels ... ', and so on. Some people may claim that you are anthropomorphizing machines, but the correct interpretation is that you are putting the terms *think, believe,* and *feel* in their proper meaning.

Minds are control systems, and engineered control systems are artificial minds we build to make our machines intelligent enough to achieve their design objectives. As Minsky (1985) said, our control agencies are *societies of mind*.

Acknowledgments

We would like to acknowledge the funding for this work provided by the European Commission, the Spanish Comision Interministerial de Ciencia y Tecnología, and the European Science Foundation through several research projects.

References

Aarsten, A., Brugali, D. and Menga, G. (1996) Designing concurrent and distributed control systems. *Communications of the ACM*, **39**(10), 50–58.

Abelson, H., Sussman, G.J. and Sussman, J. (1985) *The structure and interpretation of computer programs*. MIT Press, Cambridge, MA.

Alarcón, M., Rodríguez, P., Almeida, L., Sanz, R., Fontaine, L., Gómez, P., Alamán, X., Nordin, P., Bejder, H. and de Pablo, E. (1994) Heterogeneous integration architecture for intelligent control. *Intelligent Systems Engineering*.

Barrientos, A., Gambao, E., Saltaren R., Balaguer, C. and Aracil, R. (1997) Hierarchical control architecture for large range robots with static deflection correction. *Proceedings of SYROCO'97*.

Bonasso, R.P., Firby, R.J., Gat, E., Kortenkamp, D., Miller, D. and Slack, M. (1997) Experiences with an architecture for intelligent, reactive agents. *Journal of Experimental and Theoretical Artificial Intelligence*, **9**(2).

Buschman, F., Meunier, R., Rohnert, H., Sommerlad, P. and Stal, M. (1996) *Pattern oriented software architecture: a system of patterns*. John Wiley & Sons, New York.

Clements, P.C. (1995) From subroutines to subsystems: component-based software engineering. *The American Programmer*, **8**(11).

Fischer, K. (1999) Agent-based design of holonic manufacturing systems. *Robotics and Autonomous Systems*, **27**(1–2).

Franklin, S. and Graesser, A. (1996) Is it an agent, or just a program?: a taxonomy for autonomous agents. *Proceedings of the Third International Workshop on Agent Theories, Architectures, and Languages*, Budapest, Hungary, August.

Hayes-Roth, B. (1995) An architecture for adaptive intelligent systems. *Artificial Intelligence*, **72**, 329–365.

Jennings, N.R., Sycara, K. and Wooldridge, M. (1998) A roadmap of agent research and development. *Autonomous Agents and Multiagent Systems*, **1**, 7–38.

Kersten G.E. and Noronha, S.J. (1998) Rational agents, contract curves and inefficient compromises. *IEEE Transactions on Systems, Man and Cybernetics – Part A: Systems and Humans*, **28**(3).

Kuikka, S., Tommila, T. and Venta, O. (1999) Distributed batch process management framework based on design patterns and software components. *Proceedings of the 14th IFAC World Congress*, Beijing, China.

Lea, D. (1994) Design Patterns for Avionics Control Systems. Technical Report ADAGE-OSW-94-01, SUNY Oswego.

Minsky, M. (1985) *The society of mind*. Simon and Schuster, New York.

Nielsen, P.E. (1995) SOAR/IFOR intelligent agents for air simulation and control. *Proceedings of the 1995 Winter Simulation Conference*.

Occello, M. and Demazeau, Y. (1997) CELLO: an agent model with real time constraints. *Proceedings of the First International Conference on Autonomous Agents*, 488–489.

Sahasrabudhe, V. and Mehra, A. (1998) A multi-agent control system framework for smart structures. *Proc. AIAA 98*.

Sanz, R., Matía, F., de Antonio, A. and Segarra, M. (1998) Fuzzy agents for ICa. *Proceedings of FUZZY-IEEE 98*, Anchorage, AK.

Sanz, R., Alarcón, I., Segarra, M.J., de Antonio, A. and Clavijo, J.A. (1999a) Progressive domain focalization in intelligent control systems. *Control Engineering Practice*, **7**(5), 665–671.

Sanz, R., Matía, F. and Puente, E.A. (1999b) The ICa approach to intelligent autonomous systems, in *Advances in autonomous intelligent systems* (ed. S. Tzafestas), chapter 4, 71–92. Kluwer Academic Publishers, Dordretch, The Netherlands.

Sanz, R., Segarra, M.J., de Antonio, A., Matía, F., Jiménez, A. and Galán, R. (1999c) Patterns in intelligent control systems. *Proceedings of the IFAC 14th World Congress*, Beijing, China.

Sheremetov, L.B. and Smirnov, A.V. (1999) Component integration framework for manufacturing systems re-engineering: agent and object approach. *Robotics and Autonomous Systems*, **27**(1–2).

Wang, H. and Wang, C. (1997) Intelligent agents in the nuclear industry. *Computer*, **30**(11), 28–34.

Wittig, T. (1992) *ARCHON: an architecture for multi-agent systems*. Ellis Horwood, Chichester, U.K.

Wooldridge, M., Jennings, N.R. and Kinny, D. (1999) A methodology for agent-oriented analysis and design. *Proceedings of the Third Annual Conference on Autonomous Agents*, 69–76.

11

System Health Management for Complex Systems

George D. Hadden, Peter Bergstrom, and Tariq Samad
Honeywell Technology Center

Bonnie Holte Bennett
Knowledge Partners of Minnesota

George J. Vachtsevanos
The Georgia Institute of Technology

Joe Van Dyke
PredictDLI

11.1 Introduction

Although some aspects of system operation, such as feedback control, are by now widely automated, others such as the broad area of system health management (SHM) still rely heavily on human operators, engineers, and supervisors. In many industries, SHM is viewed as the next frontier in automation.

Performance, economics, and safety are all at stake in SHM, and the emphasis on health management technology is motivated by all these considerations:

- Systems are being operated closer to the performance and stability edge. Even small problems with equipment can have major adverse impact. Not only can system performance be significantly degraded, but more important, human health and environmental safety can be compromised.

- System shutdowns are increasingly unacceptable from an economic perspective. In the global competitive environment, profit margins are being squeezed. System health problems that result in the shutdown of the system, or any of its major subsystems, can turn profits into losses.

- There is a cross-industry trend in corporations toward reduction in operational staff at all levels of skill and seniority. This by no stretch means that the human is being put out of the loop; however, the role of the human is changing. People are being faced with broader responsibilities, and automation is needed to support a minimal staff.

System health management has always been a topic of significant interest to industry (Bristow *et al.*, 1999). Only relatively recently, however, have the numerous aspects of health management begun to be viewed as facets of one overall problem. The term itself has gained currency only recently. We now understand SHM as encompassing all issues related to off-nominal operations of systems – including equipment, process/plant, and enterprise. As for the capabilities that fall under the SHM label, the following are particularly notable:

- Fault detection: identifying that some element or component of a system has failed.

- Fault identification: identifying *which* element has failed.

- Failure prediction: identifying elements for which failure may be imminent and estimating their time to failure.

- Modeling and tracking degradation: quantifying gradual degradation in a component or the system.

- Maintenance scheduling: determining appropriate times for preventive or corrective operations on components.

- Error correction: estimating 'correct' values for parameters, the measurements of which have been corrupted.

Technologists are seeking to exploit advances in diverse fields for developing SHM solutions. As might be expected, the variety and complexity of problems that SHM encompasses preclude any single-technology answers. Hardware, software, and algorithmic technologies are all required and are being explored. For highly data-intensive, computation-intensive, or communication-intensive applications, an SHM solution can require a hardware architecture design, integrating sensors, actuators, computational processors, and communication networks. Different algorithmic techniques may be needed for signal processing, including Fourier and wavelet transforms and time series models. Artificial intelligence methods such as expert systems and fuzzy logic can be helpful in allowing human expertise and intuition to be captured. There is also increasing interest in fundamental modeling, especially in failure mode effects analysis (FMEA), a systematic approach for identifying what problems can potentially occur with products and processes. Finally, software architectures are required to manage the multiple devices, data streams, and algorithms. With Internet-enabled architectures, an SHM system can be physically distributed across large distances.

11.1.1 Challenges in system health management

Our successes in capturing common failure mechanisms has resulted in safer, more reliable, and more available systems. An interesting corollary is that we are now seeing failure modes that were rarely seen before. The lack of empirical data or experiential knowledge in such cases renders many diagnostic and prognostic methods unusable. Other types of knowledge must be relied upon, generally based on a human expert's understanding of system operation. Approaches are needed for integrating the diverse types of knowledge that will allow health management to be conducted under all conditions, and for all components of the system.

Another failing with many conventional methods for fault identification is that they assume that faults occur singly. This only covers part of the problem. Surprising relationships can occur among various failure modes. A lightning strike or thunderstorm on a refinery or building, for example, may take out a few pieces of equipment. Or a fault in one device may cause problems in otherwise unrelated machines that depend on it for their input (perhaps separated by several intervening devices). Compound faults often do not have independent symptoms, and predicting or diagnosing multiple faults is not simply a matter of dealing with each separately.

Even when there is a single fault, its symptoms will be masked by any number of additional symptoms generated by logically upstream and downstream subsystems. It is not uncommon for refinery operators to be faced with hundreds of alarms. Ferreting out the root cause remains an unsolved problem, except in simplified cases.

Another implication of system health management for large-scale complex systems is that, as it considers the full spectrum of system operations, SHM must deal with the large differences in the time scales involved. Vibration data from a motor may need to be collected at nearly a megahertz for shaft balance problems to be detectable, whereas flooding in a distillation column is a phenomenon that occurs on a time scale of many minutes. System architectures, and algorithms, that can deal with these extremes of sampling rates are needed and not readily available.

There are also challenges relating to health management procedures. Preventive maintenance is the norm in industry today. In this approach, individual components or larger systems are checked on a regular basis, with a predetermined frequency. Preventive maintenance is generally considered overly conservative, and, as personnel costs escalate to the point where they are dominating the operating budget, alternatives are being sought. The notion of 'just in time' maintenance has been suggested, by which is meant that, ideally, maintenance is done only when it is necessary. Turning this notion into feasible schemes is a challenge for SHM.

We conclude this section by noting one opportunity for improved system health management. Few of the components or systems we deal with are one of a kind, yet the prospects for improving health management of one element based on experiences with other, similar elements have not been pursued. It should be noted that similarity does not mean identity here. All pumps from one manufacturer might have some similar failure characteristics, or all temperature transmitters based on the same sensing technology might all be susceptible to drift under particular environmental conditions, or even all rotating equipment exhibiting third harmonic excess vibrations can expect extremely reduced life. The structure in which these generalities can be organized is itself an exercise in complexity. We need new approaches to capture such generalities and take advantage of them.

11.1.2   List of Acronyms

The following list of acronyms may be of use to the reader:

ADO	Active Data Objects
AMOSS	Airline Maintenance and Operation Support System
API	Application Programming Interface
ASM	Abnormal Situation Management
ATL	Active Template Library
CBM	Condition-Based Maintenance
COM	Component Object Model
DC	Data Concentrator
DCOM	Distributed Component Object Model
DRAM	Dynamic Random Access Memory
EMA	Electromechanical Actuator
FFT	Fast Fourier Transform
FMEA	Failure Mode Effects Analysis
HTC	Honeywell Technology Center
ID	Identifier
KF	Knowledge Fusion
LCD	Liquid Crystal Display
MCA	Machinery Condition Assessment
MPROS	Machinery Prognostic and Diagnostic System
MUX	Multiplexer
NASA	National Aeronautics and Space Administration
NRL	Naval Research Laboratory
ODBC	Open Database Connectivity
OLE	Object Linking and Embedding
ONR	Office of Naval Research
OPC	OLE (Object Linking and Embedding) for Process Control
OOSM	Object-Oriented Ship Model
PCMCIA	Personal Computer Memory Card International Association
PDME	Prognostic/Diagnostic/Monitoring Engine
PSD	Power Spectrum Distribution
RMS	Root Mean Square
RSVP	Reduced Shipboard Manning Through Virtual Presence
SBFR	State-Based Feature Recognition
SHM	System Health Management
WNN	Wavelet Neural Network

11.2   Condition-Based Maintenance for Naval Ships

Although there appears to be no dearth of challenges and difficulties, some ambitious programs currently under way are attempting to make the transition from the piecemeal maintenance and diagnostic practices of the past to a true SHM philosophy. In the remainder of this chapter, we discuss in some detail one such program with which we have been involved (Bennett and Hadden, 1999; Edwards and Hadden, 1997; Hadden *et al.*, 1996; Hadden *et al.*, 1998; Hadden *et al.*, 1999a; Hadden *et al.*, 1999b).

This project, supported by the Office of Naval Research of the U.S. Department of Defense, is focusing on condition-based maintenance (CBM) for naval ships. Condition-based maintenance refers to the identification of maintenance needs based on current operational conditions. In this project, system architectures and diagnostic and prognostic algorithms are being developed that can efficiently undertake real-time data analysis from appropriately instrumented machinery aboard naval ships and, based on the analysis, provide feedback to human users regarding the state of the machinery – such as its expected time to failure, the criticality of the equipment for current operation, and so on. Using these analyses, ship maintenance officers can determine which equipment is critical to repair before embarking on their next mission – a mission that could take the better part of a year.

11.2.1 MPROS Architecture

The development of the ONR CBM system, called MPROS (for Machinery Prognostic and Diagnostic System), had two phases. The first phase had MPROS installed and running in the lab. During the second phase, we extended MPROS's capability somewhat and installed it on the Navy hospital ship *Mercy* in San Diego.

MPROS is a distributed, open, extensible architecture for hosting multiple online diagnostic and prognostic algorithms. Additionally, our prototype contains four sets of algorithms aimed specifically at centrifugal chilled water plants. These are:

1. PredictDLI's (a company in Bainbridge Island, Washington, that has a Navy contract to do CBM on shipboard machinery) vibration-based expert system adapted to run in a continuous mode.

2. State-based feature recognition (SBFR), an HTC-developed embeddable technique that facilitates recognition of time-correlated events in multiple data streams. Originally developed for Space Station *Freedom* (a precursor to the International Space Station), this technique has been used in a number of NASA-related programs.

3. Wavelet neural network (WNN) diagnostics and prognostics developed by Professor George Vachtsevanos and his colleagues at Georgia Tech. This technique, like PredictDLI's, is aimed at vibration data; however, unlike PredictDLI's, their algorithm excels at drawing conclusions from transitory phenomena rather than steady-state data.

4. Fuzzy logic diagnostics and prognostics also developed by Georgia Tech that draws diagnostic and prognostic conclusions from non-vibrational data.

Since these algorithms (and others we may add later) have overlapping areas of expertise, they may sometimes disagree about what is ailing the machine. They may also reinforce each other by reaching the same conclusions from similar data. In these cases, another subsystem, called *Knowledge Fusion* (KF), is invoked to make some sense of these conclusions. We use a technique called *Dempster–Shafer Rules of Evidence* to combine conclusions reached by the various algorithms. It can be extended to handle any number of inputs.

MPROS is distributed in the following sense: devices called *Data Concentrators* (DCs) are placed near the ship's machinery. Each of these is a computer in its own right and has the major responsibility for diagnostics and prognostics. Except for Knowledge Fusion, the algorithms described above run on the DC. Conclusions reached by these algorithms are then sent over the ship's network to a centrally located machine containing the other part of our system – the *Prognostic/Diagnostic/Monitoring Engine* (PDME). KF is located in the PDME. Also in the PDME is the *Object-Oriented Ship Model* (OOSM). The OOSM represents parts of the ship (e.g., compressor, chiller, pump, deck, machinery space) and a number of relationships among them (e.g., part-of, proximity, kind-of). It also serves as a repository of diagnostic conclusions – both those of the individual algorithms and those reached by KF. Communication among the DCs and the PDME is done using Distributed Common Object Module (DCOM), a standard developed by Microsoft.

11.2.2   Why centrifugals?

Our central mission in this project was to design a *shipwide* CBM system to predict remaining life of all shipboard mechanical equipment. However, given our funding level, implementation of such a system in its entirety would have been much too ambitious. In light of this, we chose to illustrate the general principles of our design by implementing it in a specific way on the centrifugal chilled water system. The most obvious result of this philosophy is that occasionally we chose a more general way of solving a problem over a 'centrifugal-chiller-specific' solution.

There were two main reasons for our choice of centrifugal chillers: system complexity and commercial applicability. These air conditioning systems combine several rotating machinery equipment types (i.e., induction motors, gear transmissions, pumps, heat exchangers, and centrifugal compressors) with a fluid power cycle to form a complex system with several different parameters to monitor. This dictated the requirement for a correspondingly complex and versatile monitoring system. Dynamic vibration signals must be acquired using high sampling rates and complex spectrum and waveform analysis. Slower changing parameters such as temperatures and pressures must also be monitored, but at a lower frequency, and can be treated as scalars rather than vectors, as with vibration spectra. All of these monitored parameters and analysis techniques are combined using a versatile diagnostic system. The final product has the inherent capability of diagnosing not just the whole air conditioning system, but each of its parts as well, making it a potentially very useful tool for monitoring any pump, motor, gearset, or centrifugal compressor in the fleet.

Secondly, selection of the air conditioning system as the subject will mean a high probability of commercial applicability of the resultant monitoring system. Numerous industrial, military, commercial, and institutional facilities use large centrifugal-chiller-based air conditioning systems throughout the United States and the rest of the world.

11.2.3   Data Concentrator hardware

The DC hardware (Figure 11.1 shows the HTC-installed DC) consists of a PC104 single-board Pentium PC (about 6 in. by 6 in.) with a flat-screen LCD display monitor, a PCMCIA host board, a four-channel PCMCIA DSP card, two multiplexer (MUX) cards, and a terminal bus for sensor cable connections. The operating system is Windows 95™, and there are connections for keyboard and mouse. Data is stored via DRAM. The DC is housed in a NEMA enclosure with a transparent front door and fans for cooling. Overall dimensions are 10 in. by 12 in. by 4 in. The system was built entirely with commercial off-the-shelf components with the exception of the MUX cards, which are a PredictDLI hardware subcomponent, and the PCMCIA card, which was modified from a commercial two-channel unit to meet the needs of the project.

The MUX cards provide power to standard accelerometers and are controlled using the PC's IO port. Each of the two MUX cards can switch between four sets of four channels, each yielding up to 32 channels of data. Of those 32 channels, 24 can power standard accelerometers. All channels can be configured to sample DC voltage signals. Additionally, all channels are equipped with a root-mean-square (RMS) detector that can be configured to provide a digital signal when the RMS of the incoming signal exceeds a programmed value. This allows for real-time and constant alarming for all sensors.

The four-channel PCMCIA card samples DC and AC dynamic signals. The highest sampling rate exceeds 40,000 Hz, and the length of sampled signals is limited only by the PC's storage capacity.

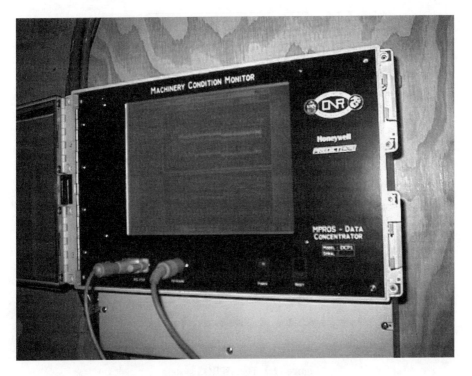

Figure 11.1 Data concentrator installed at HTC

11.3 Details of the MPROS Software

Figure 11.2 shows a diagram of the MPROS system. The PDME consists entirely of software and runs on any sufficiently powerful Windows NT machine. A potentially large number (on the order of a thousand) DCs are installed on the ship and report diagnostic and prognostic conclusions to the PDME over the ship's network. In the following sections, we describe the various software parts of the system.

11.3.1 PDME

The PDME is the logical center of the MPROS system. Diagnostic and prognostic conclusions are collected from DC-resident as well as PDME-resident algorithms. Fusion of conflicting and reinforcing source conclusions is performed to form a prioritized list for use by maintenance personnel.

The PDME is implemented on a Windows NT platform as a set of communicating servers built using Microsoft's Component Object Model (COM) libraries and services. Choosing COM as the interface design technique has allowed us to build some components in C++ and others in Visual Basic, with an expected improvement in development productivity as the outcome. Some components were prototyped using Microsoft Excel, and we continue to use Excel worksheets and macros to drive some testing of the system. Communications between DC and PDME components depend on Distributed COM (DCOM) services built into Microsoft's operating systems.

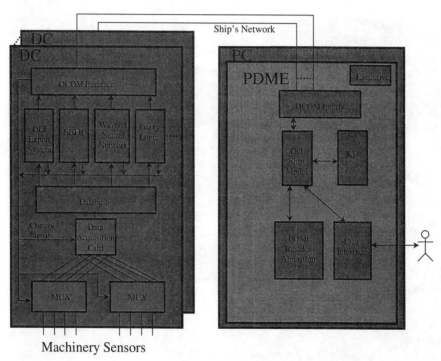

Figure 11.2 The MPROS system

11.3.1.1 User interface

Although MPROS is likely to be embedded in a larger system, we did find it necessary to build a user interface, if only to demonstrate the system's capabilities to others. Figure 11.3 shows the layout of this rudimentary interface. The sample screen shown indicates that for machine A/C Compressor Motor 1, six condition reports from four different knowledge sources (expert systems) have been received, some conflicting and some reinforcing.

Details of the MPROS Software

After these reports are processed by the KF component, the predictions of failure for each machine condition group are shown at the bottom of the screen.

This display is updated as new reports arrive at the PDME and are accumulated in the OOSM.

11.3.1.2 Object-Oriented Ship Model

The OOSM is a persistent repository for machinery state information used for communication between the various prognostic and diagnostic software modules. In addition to diagnostic and prognostic reports, the OOSM also models the physical, mechanical, and energy characteristics of all the ship's components, including the machinery being monitored. Exposing an integrated programming interface to all this information eases the development of new knowledge-based algorithms for diagnostics and prognostics.

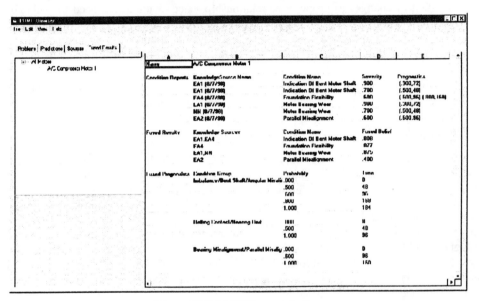

Figure 11.3 MPROS user interface

Object Model We chose an object-oriented design to provide a consistent interface to all the developers using the OOSM to store, retrieve, and monitor changes to information of interest to MPROS components.

Entities in the OOSM are modeled as objects with properties and relationships to other entities. Some of the OOSM objects represent physical entities such as sensors, motors, compressors, decks, and ships, whereas other OOSM objects represent more abstract items such as a failure prediction report or one of the diagnostic or prognostic algorithms. Some common properties include name, manufacturer, energy usage, capacity, and location. Common relationships include part-of, kind-of, connected-to, and energy flow.

As part of easing the use of an object-oriented model for developers, the persistence is entirely managed in the background. In fact, except for retrieving the first object in a connected graph of objects, no understanding of the persistence mechanism is necessary.

Contents We have modeled a portion of the information about the system under observation in the OOSM. This includes information about the motors, compressors, and evaporators in the chillers we are working with. We have also modeled relationships between the failure predictions and the relevant equipment to provide for a future expert system to analyze interactions among equipment subsystems.

Application Programming Interface (API) As mentioned in preceding sections, a consistent API for developers has been defined. Briefly, it consists of functions to retrieve specific object instances, to view the values of properties, to update their properties and relationships, and to create and delete instances. This API, based on COM, has been used from C++, Visual Basic, and Java programs to work with the OOSM.

Events An event model has been implemented for the OOSM that allows client programs to be notified of changes to property or relationship values without the need to poll. This facility is provided by OLE Automation events, making it usable from C++, Visual Basic, and Java. The KF component uses this to process failure prediction reports automatically as they are delivered to the OOSM. The PDME browser also uses events to update its display for users.

Persistence of object state in the OOSM is implemented using a relational database. Object types are mapped to tables and properties and relationships are mapped to columns and helper tables. This mapping approach has helped in system debugging and has proven very reliable in operation.

Implementation The OOSM is implemented in C++ using Microsoft's Active Template Library (ATL) and Active Data Objects (ADO) library. We chose this approach because of the control it offered over object lifetime, performance, and reliability.

11.3.1.3 Knowledge fusion

Knowledge fusion is the coordination of individual data reports from a variety of sensors. It is higher level than pure 'data fusion,' which generally seeks to correlate common platform data. Knowledge fusion, for example, seeks to integrate reports from acoustic, vibration, oil analysis, and other sources, and eventually to incorporate trend data, histories, and other components necessary for true prognostics.

The knowledge fusion components must be able to accommodate inputs that are incomplete, time-disordered, fragmentary, and that have gaps, inconsistencies, and contradictions. In addition, knowledge fusion components must be able to collate, compare, integrate, and interpret data from a variety of sources. To do this, knowledge fusion must provide both inference control that accommodates a variety of input data and fusion algorithms with the ability to deal with disparate inputs.

Knowledge fusion follows this procedure:

1. New reports arriving to the PDME are posted in the OOSM.

2. New reports posted in the OOSM generate 'new data' messages to the knowledge fusion components.

3. The knowledge fusion components access the newly arrived data from the OOSM. They perform knowledge fusion of both diagnostic and prognostic reports.

4. Conclusions from the knowledge fusion components are posted to the OOSM and presented in user displays via the graphical user interface.

Implementation To-date, two levels of knowledge fusion have been implemented: one for diagnostics and one for prognostics.

Our approach for implementing knowledge fusion for diagnostics uses Dempster–Shafer belief maintenance for correlating incoming reports. This is facilitated by use of a heuristic that groups similar failures into logical groups.

Dempster–Shafer theory is a calculus for qualifying beliefs using numerical expressions. For example, given a belief of 40% that A will occur and another belief of 75% that B or C will occur, it will conclude that A is 14% likely, B or C is 64% likely, and assign 22% of belief to unknown possibilities. This maintenance of the likelihood of unknown possibilities is both a differentiator and a strength of Dempster–Shafer theory. It was chosen over other approaches (e.g., Bayes nets) because the others require prior estimates of the conditional probability relating two failures – data not yet available for the shipboard domain.

The system was augmented by heuristically collecting similar failures into logical groups. This facilitates processing and streamlines operation because Dempster–Shafer analysis looks at each failure in light of every other possible failure and is required to produce the likelihood of unknown possibilities. In the MPROS case, this is inadequate because it would assume mutual exclusivity of failures. This is not a realistic assumption. There can, in fact, be several failures at one time, and two or more of them might be independent of one another. Thus, we developed the concept of logical groups of failures. Failures that are all part of the same logical group are related to each other (for example, one group might be electrical failures, another lubricant failures, etc.). Moreover, failures within a group might be mistaken for one another, so any two of them are logically related and should share probabilities when they are both under consideration. Note that this does not preclude multiple failures within a group all being suspect concurrently; it simply ensures that they are tracked and weighted correctly.

The second level of knowledge fusion combines time to failure estimates. Time to failure is represented in our system as a list of one or more time points, probability pairs, called the 'prognostic vector'. For example, the prognostic vector with the single member '((3 months, 0.1))' indicates that the system has a 10% likelihood of failure within 3 months. The prognostic vector '((2 weeks, 0.1) (1 month, 0.5) (2 months, 0.9))' indicates a likelihood of failure of 10% within 2 weeks, 50% within 1 month, and 90% within 2 months.

Our approach to the fusion of prognostics information is to combine the lists, taking the most conservative estimate at any given time period and interpolating a smooth curve from point to point. For example, suppose we have a prognostic for a given component that indicates it will perform well for 3 months, then experience some trouble making it as likely to fail as not by 4 months and almost surely to fail within 5 months. The prognostic vector for this case is ((3 months, 0.01) (4 months, 0.5) (5 months, 0.99)). Suppose further that we need to combine this with another report showing that the same component will experience some small trouble at 4½ months. This prognostic vector is ((4.5 months, 0.12)). Under our current approach, we ignore the second report and stick with the first, which is more conservative. If, however, the second report indicates a much higher likelihood of failure, say ((4.5 months, 0.95)), then this report would dominate and the extrapolation of the curve beyond this point would indicate an even earlier demise of the component than the first prognostic vector.

Interfaces provided One of the goals of the MPROS system is to encourage the incorporation of many diverse expert systems supplying diagnostic and prognostic conclusions based on similar, overlapping, or entirely disjoint sensor readings. At the same time, we recognized that these diverse results must be unified into a meaningful report to the system's users. To this end, a standard protocol has been defined for reporting failure predictions to the PDME for fusion and display.

The general incoming report format may contain the following data fields (not all reports need use all fields):

1. KnowledgeSourceID: the unique MPROS object ID for the instance of the diagnostic/prognostic algorithm generating the report.
2. SensedObjectID: the unique MPROS object ID for the sensed object to which this report applies.
3. MachineConditionID: the unique MPROS object ID for the diagnosed machine condition (usually a failure mode).
4. Severity: numeric value in the range 0.0 to 1.0 indicating relative severity of machine condition to operation. Maximal severity is 1.0.
5. Belief: numeric value in the range 0.0 to 1.0 indicating belief that this diagnosis is true. Maximal belief is 1.0.
6. Explanation: an optional text string providing a human-readable description of the diagnosis.
7. Recommendations: an optional text string providing a human-readable description of the recommended actions to take.
8. Timestamp: the time when this report should be considered effective.
9. Additional Information: an optional text string providing human-readable additional information.
10. Prognostic vector: a vector of time point, probability pairs indicating projected likelihood of failure (as described above).

Diagnostic knowledge fusion generates a new fused belief whenever a diagnostic report arrives for a suspect component. This updates the belief for that suspect component and for every other failure in the logical group for that component. It also updates the belief of 'unknown' failure for the logical group for that component.

Prognostic knowledge fusion generates a new prognostic vector for each suspect component whenever a new prognostic report arrives.

Future directions for knowledge fusion Several high-level control extensions are under consideration for future extensions. First, multilevel data are represented by the OOSM. We are not currently exploiting this fully. For example, we could reason about the health of a system based on the health of a constituent part. Currently, only the parts are tracked. Secondly, spatial reasoning using the OOSM could lead us to fuse information about spatially related components. One example of a spatial relation is proximity. For example, a device might be vibrating because a component next to it is broken and vibrating wildly. Another example is flow. Flows are relationships that represent fluid flow through the system (one component passing fouled fluids on to other components downstream), electrical flow, or mechanical flow of physical energy. Third, temporal reasoning components could be implemented to scrutinize failure histories and provide better projections of future faults as they develop.

Two other future directions for knowledge fusion are the refinement of specific knowledge fusion components for diagnostics and for prognostics. For example, Bayes nets seem to be a promising approach to diagnostic knowledge fusion when causal relations and a priori relationships can be teased out of historical data. Prognostic knowledge fusion could be improved with the addition of techniques from the analysis of hazard and survival data. These approaches scrutinize history data to refine the estimates of life-cycle performance for failures, and the refined inputs to the prognostic analysis should yield better projections of future failures.

11.3.1.4 Resident algorithms

As shown in Figure 11.3, the PDME has the capability to host prognostic and diagnostic algorithms. Some reasons for placing the algorithms in the PDME rather than the DC include: the algorithm requires data from widely separate parts of the ship, the algorithm can reason from PDME-resident components (a model-based diagnostic and prognostic system, for instance, might use only the OOSM), and so on.

Although we provide this capability in our general architecture, our current system does not place diagnostic/prognostic algorithms in the PDME – all of them run in the DCs.

11.3.2 Data Concentrator

11.3.2.1 Scheduler

At the heart of the DC software is an event scheduler. This software component runs the show by organizing all necessary events. For example, the standard vibration test and analysis is executed routinely by the scheduler. To do a standard vibration test, the scheduler first triggers execution of the vibration data acquisition component, and when that operation is complete, the scheduler fires off the vibration analysis component and then triggers communication of the results. In a similar fashion, the scheduler invokes WNN and SBFR routines to collect and analyze process variables.

Behind the scenes, the vibration data acquisition component must control the MUX, extract sensor setup information from the DC database, request several specific PCMCIA card tests, and store the results in the DC database (see next section). Then the processing begins by extracting the stored vibration data, equipment descriptors, and diagnostic criteria from the database. All of this is then sent to an expert system DLL that applies stored rules for each equipment type and derives the diagnoses. The DLL then passes the results back to the DC database.

Each of the components extracts information from and stores data in the DC database, which is configured as a database server and can be accessed by client PCs on the network. In this way, the PDME or any other client can command the scheduler to conduct another test and analysis routine.

11.3.2.2 Database

Central to the operation of the DC is an open architecture, ODBC-compliant, relational database designed to store all the instrumentation configuration and machinery configuration information, test schedules, resultant measurements, diagnostic results, and condition reports. The commercially available database design is already field-tested and proven effective in many industrial facilities. The design was slightly altered to accommodate particular DC hardware settings and scheduling parameters. This database approach allowed the autonomous development of various kinds of measurement and analysis techniques by several of the DC development team members while still using the same database and communications protocol with the PDME.

11.4 Prognostic/Diagnostic Algorithms

MPROS has four sets of prognostic/diagnostic algorithms: the PredictDLI expert system, WNNs, fuzzy logic, and state-based feature recognition. These are all resident in the DC.

11.4.1 PredictDLI expert system

Currently, all standard machinery vibration FFT analysis and associated diagnostics in the DC are handled by the PredictDLI expert system. This system was initially developed for use in the Machinery Condition Assessment (MCA) program in 1988. MCA is a U.S. Navy program on the aircraft carriers that has efficient fleet maintenance as an ultimate goal. Beyond naval applications, this expert system is installed in hundreds of industrial and manufacturing facilities, where it is steadily improved through the introduction of rules and procedures for analyzing industrial machines (such as extruders) typically not found on aircraft carriers. For the Navy, this expert system runs on desktop PCs as part of a system built around walk-around data gathering and analysis. It provides savings in analysis time for contractors doing on-board machine condition monitoring and puts a sophisticated and powerful diagnostic tool in the hands of sailors on the ships. One study found that the system exceeds 95% agreement with human expert analysts for machinery aboard the Nimitz-class ships.

The DC takes this scenario one step further in that it provides this same powerful diagnostic capability without the need for manual vibration measurement collection. All necessary measurements are made and analysis steps are completed in the DC with only the diagnostic results routinely fed up to the PDME for eventual human consumption. Spectra and diagnostic backup data are also available on request but are not primary outputs.

The PredictDLI expert system provides an intermediate level of sensor and knowledge fusion. The frame-based rules application method employed allows the spectral vibration features to be analyzed in conjunction with process parameters such as load or bearing temperatures to arrive at a more accurate and knowledgeable machinery diagnosis. For example, since some compressors vibrate more at certain frequencies when unloaded, the PredictDLI expert system rule for bearing looseness can be sensitized to available load indicators (such as prerotation vane position) to ensure that a false positive bearing looseness call is not made when the compressor enters a low-load period of operation. This kind of knowledge and sensor fusion is again found at the PDME, which ascertains the relative believabilities of the various diagnostic systems and derives a reasonable conclusion. It is, however, advantageous for the vibration expert system to use all available known system responses when analyzing the vibration patterns, because it minimizes the necessary PDME decisions and improves overall system accuracy.

On the horizon, efforts are under way to develop a new system component that will apply certain fundamental time domain vibration analysis and feature-extraction techniques to the data sampled for FFT analysis. This new PredictDLI expert system component will enhance the ability of the DC to more easily and accurately discern various gear and bearing faults and reduce the possibilities of false positive bearing wear diagnoses on water pump systems operating with cavitation or flow noise. This enhancement will be made possible through the combination of time domain features with spectral features in one matrix for rules application.

An elementary level of machinery prognostics has always been provided by the PredictDLI expert system, which since its inception has provided a numerical severity score along with the fault diagnosis. This numerical score is interpreted through empirical methods that map it into four gradient categories – Slight, Moderate, Serious, and Extreme – which correspond to expected lengths of time to failure described loosely as: no foreseeable failure, failure in months, weeks, and days of operation. This approach to prognostics was developed and has been improved upon during the last 10 years and was further refined for the DC and the PDME through the introduction of believability factors for each of the diagnoses. These believability factors are based on PredictDLI's statistical database, which demonstrates the individual accuracy of each diagnosis by tracking how often each was reversed or modified by a human analyst prior to report approval.

11.4.2 Integrated WNNs and fuzzy logic

Components, machines, and processes fail in varying ways depending on their constituent materials, operating conditions, and so on. Failure modes are typically monitored by a sensor suite that is intended, for failure analysis purposes, to capture those failure symptoms which are characteristic of a particular failure mode. Consider, for example, the case of a typical process such as an industrial chiller found on Navy ships and commercial and other facilities that require climate control. Typical failure modes include evaporator tube freezing and decreased seawater flow through the condenser, which are characteristic of process failures, as well as a variety of vibration-induced faults that are affecting mechanical and electromechanical process elements.

It is generally possible to break down the sensor data (and correspondingly, the symptomatic evidence) into two broad categories (see Figure 11.4). One concerns temperature, pressure, level, and so on, whereas the other exemplifies high-bandwidth measurements, for example vibrations and current spikes. Failure modes associated with the first category may develop slowly, and data are sampled at slow rates without loss of trending patterns. High-frequency phenomena, though, such as those accompanying a bearing failure, require a much faster sampling rate to permit a reasonable capture and characterization of the failure signature. Moreover, process-related measurements and associated failure mode signatures are numerous and may overlap, thus presenting serious challenges in resolving conflicts and accounting for uncertainty. This dichotomy suggests an obvious integrated approach to the fault diagnosis problem (Mufti and Vachtsevanos, 1995; Vachtsevanos *et al.*, 1992): Process-related low-bandwidth faults may be treated with an expert system based on fuzzy rules (see Figure 11.5) (Kang *et al.*, 1991), whereas high-bandwidth faults are better diagnosed via a feature extractor/neural network classifier topology (see Figure 11.6). This approach is adopted below in addressing typical chiller failures. Although the chiller is used as the demonstration testbed, the basic diagnostic architecture with low-bandwidth measurements, such as those originating from process variables, is generic and applicable to a wide variety of complex engineered systems and industrial processes.

11.4.2.1 Fuzzy logic – low-bandwidth failure detection and identification

The knowledge base consists of a fuzzy rule set. Typical rules for diagnostic purposes are as follows:

1. If Evaporator Liquid Temperature is Low and Difference between Chilled Water Discharge Temperature and Evaporator Liquid Temperature is Increasing, then Failure Mode is Refrigerant Charge Low.

2. If Evaporator Liquid Temperature is Low and Compressor Evaporator Suction Pressure is Low, then Failure Mode is Refrigerant Charge Low.

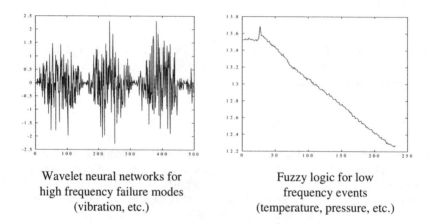

Wavelet neural networks for high frequency failure modes (vibration, etc.)

Fuzzy logic for low frequency events (temperature, pressure, etc.)

Figure 11.4 The two-pronged approach of the diagnostic module

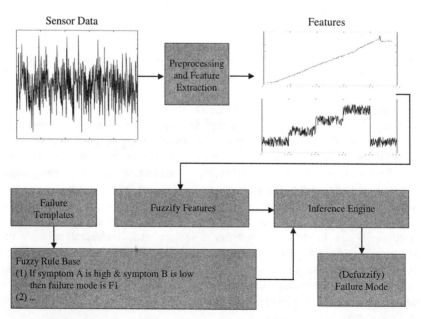

Figure 11.5 Fuzzy diagnostic system layout with feature extraction

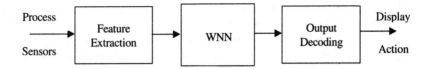

Figure 11.6 A fault diagnostic system based on wavelet neural networks

The linguistic labels Low, Increasing, and the like are assigned appropriate fuzzy membership functions and an inference engine is called upon, given input sensor data (the symptomatic evidence), to declare a fault condition.

The diagnostic system uses Dempster–Shafer theory to return a 'degree of certainty' or belief in the failure mode that has been detected by the fuzzy logic routine. Using results from evidence and possibility theories, mass functions can be extracted from fuzzy input membership functions. A mass function is a mapping from the power set of the set of propositions to the interval, [0,1]. Obtaining mass functions is the key step in determining the belief of a failure mode.

Once the mass functions are obtained for each sensor, Dempster's rule of combination is performed on each to combine the evidence supporting the different propositions. Dempster's rule of combination is as follows:

$$m_{12}(A) = \frac{\sum_{B \cap C = A} m_1(B) m_2(C)}{1 - K}$$

where the A, B, C's are sets of propositions, m's are mass functions, and K is a normalizing constant. The set A could, for example, represent the failure mode, refrigerant charge low, or could be a set of failure modes such as evaporator tube freezing and chiller tube fouling.

After determining the final combined evidence mass function, a belief function can be constructed. Belief functions map the power set of the set of propositions to the interval, [0,1]. Belief functions present a lower bound on probability distributions. Plausibility functions, which can also be created from mass functions, are considered an upper bound on probability distributions. A degree of certainty measure can now be obtained that conveys how much belief there is in a detected failure mode occurring.

Degree of Certainty = m(Failure of Interest) − Belief(not(Failure of Interest))

The combined use of fuzzy logic and Dempster–Shafer theory (Kang et al., 1991) is greatly facilitated by using the same rule base and input membership functions for both (i.e., they share the same expert knowledge base).

11.4.2.2 Vibration analysis – high-bandwidth failure detection and identification

The WNN belongs to a new class of neural networks with such unique capabilities as multiresolution and localization in addressing classification problems (Echauz and Vachtsevanos, 1996). For fault diagnosis, the WNN serves as a classifier for occurring faults. A fault diagnostic system based on the WNN is illustrated in Figure 11.6. Critical process variables are monitored via appropriate sensors.

The data obtained from the measurements are processed and features are extracted. The latter are organized into a feature vector, which is fed into the WNN. The WNN then carries out the fault diagnosis task. In most cases, the direct output of the WNN must be decoded to produce a feasible format for display or action.

For example, the WNN can be used to perform the diagnosis of a bearing fault. Many bearing faults can be classified, such as those on races, rolling balls, and lubrication materials. Here, for simplicity, the focus is placed on the diagnosis of whether the bearing is normal or defective. Through vibration measurements, several vibration signals for a bearing are obtained. Then the peak of the signal amplitude and the peak of the signal's PSD are chosen as the features. (Many other quantities such as the standard deviation, cepstrum, DCT coefficients, wavelet maps, temperature, humidity, speed, and mass can also be selected as the features.) From the vibration signals, a training data set is obtained and used to train the WNN.

Once trained, the WNN can be employed to perform the fault diagnosis. Signals and their PSDs from a normal bearing and a defective one are shown in Figure 11.7.

For the good bearing, features = [0.3960 0.1348]
For the defective bearing, features = [4.9120 9.2182]
[0 1] = WNN([0.3960 0.1348]) → The bearing is good!
[1 0] = WNN([4.9120 9.2182]) → The bearing is defective!

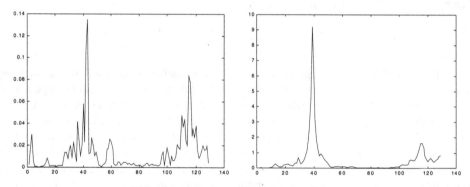

PSDs from good and defective bearings

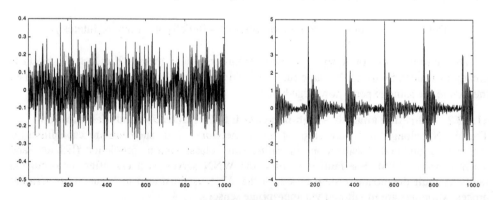

Signals from good and defective bearings

Figure 11.7 Signals and their PSDs from normal (left plots) and defective bearings (right plots)

This example is limited to a binary fault classification, but the approach can be extended to handle multiple fault classes.

11.4.3 State-based feature recognition

SBFR is a technique for the hierarchical recognition of temporally correlated features in multi-channel input (Hadden *et al.*, 1994; Nelson and Hadden, 1994). It consists of a set of several enhanced finite-state machines operating in parallel. Each state machine can be transitioned based on sensor input, its own state, the state of another state machine, measured elapsed time, or any logical combination of these. This implies that systems based on SBFR can be built with a layered architecture, so that it is possible to use them to draw complex conclusions such as prognostic or diagnostic decisions.

Our implementation of the SBFR system requires very little memory (100 state machines operating in parallel and their interpreter can fit in less than 32 Kbytes) and can cycle with a period of less than 4 ms. It is thus ideal for embedding into the DC. We have successfully applied SBFR-based diagnostic and prognostic modules to several problems and platforms, including valve degradation and failure prediction in the Space Shuttle's Orbital Maneuvering System (OMS), imminent seize-up in Electromechanical Actuators (EMAs) through electrical current signature analysis and other parameters, and failure prediction in several subsystems (including Control Moment Gyro bearing failure) in the Space Station *Freedom*'s Attitude Determination and Control System (ADCS).

To give a sense of how SBFR works, we discuss an example from the aerospace domain. The two-state-machine system shown in Figure 11.8 was used to predict a seize-up failure mode in an EMA. EMAs are essentially large solenoids meant to replace hydraulic actuators for the steering of rocket engines. Prediction of this fault was done by recognizing stiction in the mechanism. The two state machines recognize stiction in the following way: Machine 0 recognizes spikes in the drive motor current. Machine 1 counts the spikes that are *not* associated with a commanded position change (CPOS). When the count is greater than four, a stiction condition is flagged, and higher level software (e.g., the PDME) can conclude that a seize-up failure is imminent.

We now consider these machines in more detail. Starting with Machine 1, notice that it has two states labeled **Wait** and **Stiction**. Each state has transitions into and out of it with labels C and A, standing for condition and action, respectively. The condition on one of the transitions out of the **Wait** state is 'Status: $0 \neq 0$ & CPOS unchanged'. This means that the transition will be taken if both of the following are true. First, the status of Machine 0 (the other machine) must be nonzero. This means that a spike has occurred. Secondly, no command has been issued to change the position of the EMA. If both of these are true, the action associated with this condition is executed. In this transition, the action is (1) to set the status register of Machine 0 back to 0 so that it can continue looking for spikes in parallel with the actions of any other state machines, and (2) to increment local variable 1 (Local:1), which represents the number of spikes noticed by Machine 1.

Notice two things: first, a state machine can transition states based on the state of another machine; secondly, each machine can have any number of local variables, one of which, the 'status', is distinguished by being readable and writeable by any of the state machines, including, of course, the one in which it resides.

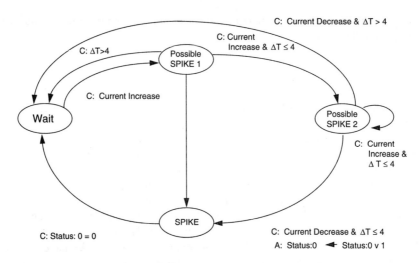

(a) Current spike machine (Machine 0)

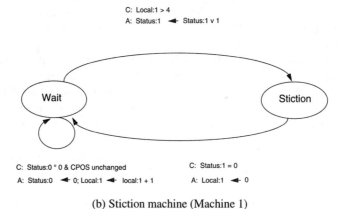

(b) Stiction machine (Machine 1)

Figure 11.8 Two state machines

Another transition out of the **Wait** state occurs if local variable 1 is greater than four (i.e., more than four spikes have been noticed). If this transition is taken, the status register of Machine 1 (the current machine) is set to 1. (Actually, only the lowest bit is set to 1, since we would like to save the option of using other bits for some other purpose.) Machine 1 then enters the **Stiction** state. At this point, some other agent – another state machine or some other software – can recognize that stiction has occurred and take appropriate action (e.g., notify the operator). That agent then has the responsibility to reset Machine 1's status register to 0, allowing the machine itself to set the count back to 0 and start over.

The spike recognition machine (Machine 0) is somewhat more complex; it has four states and seven transitions. Although it is not necessary to cover this machine in detail, a few points should be made. The most important is that this spike recognition machine is relatively noise-free. There are time constraints (ΔT in Figure 11.8) with which changes in the current must be consistent before a spike can be recognized. Intermediate states

(**Possible Spike 1** and **Possible Spike 2**) exist from which the machine can return to the **Wait** state so that it can be sure about whether the fluctuation in the current should be labeled a spike or not (fluctuation could be due to random noise). Notice also that some of the transitions can be taken without any explicit actions.

The **Spike** state has one transition out of it. This occurs when Machine 0's status register (the current machine) is reset to 0. Recall that Machine 1 did this after it had recorded the existence of the spike.

The sizes of the current spike machine (Machine 0) and the stiction machine (Machine 1) are 229 and 93 bytes, respectively. The interpreter that executes the SBFR system in the DCs is about 2000 bytes long.

We chose 12 failure modes to address with SBFR in MPROS. Each of these has an associated set of one or more state machines used to predict that failure mode.

11.5 Installation and Validation

Our system is complete and currently installed in two places: in the basement of the Honeywell Technology Center on a Carrier 19DK chiller and in San Diego on the *U.S.N.S. Mercy's* (T-AG-019) #1 AC plant – a York 363-ton chiller. The system has detected and reported a number of problems, including a bad bearing on one of the seawater pumps.

The *Mercy* was chosen for several reasons:

1. It is in a climate that was more likely to require cooling during the winter shipboard test phase of our centrifugal chiller prognostics and diagnostics system.

2. It was likely to remain stationary and not put out to sea (as, for instance, an aircraft carrier would).

3. It contains pieces of equipment that are the target of our prognostics and diagnostics efforts.

4. We have a good working relationship with the crew.

A question we are often asked is, 'How are you going to prove that your system can really predict failures?' This question, as it turns out, is quite difficult to answer. The problem is that we are developing a system we claim will predict failures in devices, but these devices fail relatively rarely. In fact, for any one failure mode, it is entirely possible that the failure mode (although valid) may never have occurred on any piece of equipment on any ship in the fleet! We have several answers to this question:

- We are still going to look for the failure modes. We have a number of installed data collectors both on land and on ships. In addition, PredictDLI is collecting time domain data for several parameters whenever their vibration-based expert system predicts a failure on shipboard chillers. This gives us data that we can use to test our system.

- As Honeywell upgrades its air conditioning systems to be compliant with new non-polluting refrigerant regulations, older chillers become obsolete and are replaced. We have managed to acquire one of these chillers and ship it to York, and we are now constructing a test plan to collect data from this chiller through carefully orchestrated destructive testing. Two more chillers were likely to have become available in December 1999.

- Faults can be seeded. Our partners in the Mechanical Engineering Department of Georgia Tech are seeding faults in bearings and collecting the data. These tests have the drawback that they might not exhibit the same precursors as real-world failures, especially in the case of accelerated tests.

- Honeywell and its teammates, York, PredictDLI, the Naval Research Laboratory, and WM Engineering, have archives of maintenance data that we will take full advantage of in constructing our prognostic and diagnostic models.

- Similarly, these partners have human expertise that we can tap in building our models.

Although persuasive, these answers are far from conclusive. The authors would welcome any input on how to validate a failure prediction system.

11.6 Conclusions

In the not too distant past, automation was employed largely to manage systems under nominal operating conditions. The realm of automation rarely extended to abnormal conditions – people were expected to handle these. Whether it was equipment failure, severe environmental disturbances, or other sorts of disruptions, the responsibility for predicting and diagnosing faults and returning the system to normal operation rested squarely on human staff. Developers of control systems and their applications were concerned about these issues only to the extent that they needed to provide the appropriate information and decision support to operators, engineers, and supervisors. The actual prognosis, diagnosis, and remedial actions were generally outside the scope of automation.

We have succeeded in our original mission almost too well, and this success has led to a broadening of our ambitions for automation and control systems. This has happened even as the scale and complexity of the physical systems – whether naval ships or commercial buildings or factories – have dramatically increased. Further, requirements for human and environmental safety have become much more stringent.

As might be expected, problem complexity translates to solution complexity. For instance, the more time we have to plan our response before a failure occurs, the better off we are – catastrophic failures can be avoided, human safety can be maximized, repair actions can be combined, and so on. To increase this time, we must find new ways to access data that we have not sensed before. In addition, we have to construct software that derives prognostic and diagnostic conclusions from increasingly subtle correlations among the sensed data.

As another example, different prognostic and diagnostic algorithms have different, but overlapping, areas of expertise. Using more than one is essential as machinery grows more complex, but then we are almost certain to have the problem of conflicting conclusions in the areas of overlap. Moreover, taking advantage of reinforcing conclusions from various algorithms is also desirable. An additional layer of software (and thus complexity) – the Knowledge Fusion component described in this chapter – is required to deal with this effect.

We have described in some detail one specific maintenance and prognostic system that integrates a number of different, mostly software-based, technologies for an application where minimal human staffing is desired, and not just for economic reasons. Our experience suggests that despite the multifarious complications associated with health management for complex enterprises, effective solutions are feasible to build.

Acknowledgment

The authors gratefully acknowledge the support of the Office of Naval Research, grant number N00014-96-C-0373.

References

Bennett, B.H. and Hadden, G.D. (1999) Condition-based maintenance: algorithms and applications for embedded high performance computing. *Proceedings of the 4th International Workshop on Embedded HPC Systems and Applications (EHPC'99).*

Bristow, J., Hadden, G.D., Busch, D., Wrest, D., Kramer, K., Schoess, J., Menon, S., Lewis, S. and Gibson, P. (1999) Integrated diagnostics and prognostics systems. *Proceedings of the 53rd Meeting of the Society for Machinery Failure Prevention Technology.*

Echauz, J. and Vachtsevanos, G. (1996) Elliptic and radial wavelet neural networks. *Proceedings of the 2nd World Automation Congress,* Montpellier, France, May 27–30.

Edwards, T.G. and Hadden, G.D. (1997) An autonomous diagnostic/prognostic system for shipboard chilled water plants. *Proceedings of the 51st Meeting of the Society for Machinery Failure Prevention Technology.*

Hadden, G.D., Bennett, B.H., Bergstrom, P., Vachtsevanos, G. and Van Dyke, J. (1999a) Machinery diagnostics and prognostics/condition based maintenance: a progress report. *Proceedings of the 53rd Meeting of the Society for Machinery Failure Prevention Technology.*

Hadden, G.D., Bennett, B.H., Bergstrom, P., Vachtsevanos, G. and Van Dyke, J. (1999b) Shipboard machinery diagnostics and prognostics/condition based maintenance: a progress report. *Proceedings of the 1999 Maintenance and Reliability Conference (MARCON99).*

Hadden, G.D., Edwards, T.G. and Van Dyke, J. (1996) Condition based maintenance for shipboard machinery. *Proceedings of the 67th Shock and Vibration Symposium.*

Hadden, G.D., Edwards, T.G. and Van Dyke, J. (1998) Shipboard machinery condition based maintenance. *Proceedings of the Maintenance and Reliability Conference (MARCON98).*

Hadden, G.D., Nelson, K.S. and Edwards, T. (1994). State-based feature recognition. *Proceedings of the 17th Annual AAS Guidance and Control Conference.*

Kang, H., Cheng, J., Kim, I. and Vachtsevanos, G. (1991) An application of fuzzy logic and Dempster–Shafer theory to failure detection and identification. *Proceedings of the 30th IEEE Conference on Decision and Control,* Brighton, England, 1555–1560.

Mufti, M. and Vachtsevanos, G. (1995) An intelligent approach to fault detection and identification. *Proceedings of the American Control Conference.*

Nelson, K.S. and Hadden, G.D. (1994) Real-time feature recognition in medical data. *Proceedings of the AAAI 1994 Spring Symposium, Artificial Intelligence in Medicine.*

Vachtsevanos, G., Kang, H., Cheng, J. and Kim, I. (1992) On the detection and identification of axial flow compressor instabilities. *Journal of Guidance, Control, and Dynamics,* **15**(5), 1216–1223.

Part 4

Complexity Management and Networks

As a metaphor for complexity, 'systems of systems' has been superseded by 'networks of systems'. National power grids, telecommunication systems, transportation infrastructures, the World Wide Web – networks have become ubiquitous in our high-technology world. However, virtually any large-scale system can usefully be seen as a network – neither the subsystems nor the interconnections need be necessarily physical elements. Complex software, for example, is a network of functional dependencies.

Observations and analyses of networked systems, both in the abstract and in the particular, are therefore important activities for complexity management as a research undertaking. With this in mind, the final part of this book collects four chapters that cover a broad range of views on networks as complex systems.

Chapter 12, by Steven Green and Joseph Jackson, focuses on the commercial air transportation system in the United States. A dramatic transformation of this system is under way, from one kind of network to another, much more complex one. Commercial aircraft today generally fly along 'highways in the sky', a situation that is within our abilities to manage but which exacts a significant cost in throughput and efficiency. What is being envisioned is a 'Free Flight' environment in which airspace users have considerable freedom to fly routes that are optimized for fuel and/or time (or other criteria). Green and Jackson describe the current U.S. national airspace system in some detail and outline the technology and the procedures that it relies on. Limitations of this system have led to the Free Flight concept. Free Flight involves both new technology – airborne as well as ground based – and new operational procedures. Three phases have been identified through which Free Flight is expected to be incrementally implemented. These phases are to occur in the years 2000–2002, 2003–2005, and beyond 2005. The chapter concludes with a discussion of five long-term research needs for air traffic complexity management.

Next, in Chapter 13 by Martin Wildberger, the focus shifts to electric power systems and to a particular complexity management approach, complex adaptive systems. Like the air traffic system, power systems are also undergoing a profound transformation. The driver in this case is deregulation – the global trend toward an open, competitive, market-driven industry structure. Deregulation represents a sea change from centralized to distributed control of the power grid. Distributed network-based control has numerous advantages in terms of robustness, response time, and efficiency, but it requires a different perspective and new solution approaches. Complex adaptive systems (CASs) are suggested as a natural fit. CASs can be considered an extension of agent-based computing (see Chapter 10) in which the agents are endowed with adaptation capabilities, typically algorithms inspired by biological evolution. The chapter also describes a prototype complex adaptive systems tool that has recently been developed. The tool incorporates agents that represent physical, corporate, and market entities. Its primary objective is to help participants in the deregulated power industry gain strategic insight into its operation and evolution.

Chapter 14, by Massoud Amin, views national infrastructures as complex networks. Examples of such networks are electric power grids, oil and gas pipelines, telecommunication and satellite systems, computer networks, transportation systems, banking and finance, and utility services. The chapter describes a new program being undertaken jointly by the Electric Power Research Institute and the U.S. Department of Defense called the Complex Interactive Networks/Systems Initiative (CIN/SI). The genesis of this program was a report by the U.S. Presidential Commission on Critical Infrastructure Protection which concluded that infrastructure protection requires a cooperative effort between government agencies, infrastructure owners and operators, and the research community. CIN/SI aims to develop tools and techniques that enable large scale and interconnected national infrastructures to self-stabilize, self-optimize, and self-heal. Three specific objectives have been defined. First, modeling is needed to understand the real dynamics of complex interactive networks. Secondly, measurement, visualization, and prediction techniques are needed. Thirdly, distributed decision and control systems need to be developed. Six multi-university consortia are currently being funded under CIN/SI; their research topics, the institutions involved, and long-term and near-term objectives are summarized.

The final contributed chapter in this volume, by John Doyle, suggests that complex networked systems may form the basis of a new science and technology that accords a central place to complexity and uncertainty. Four themes are presented that could form the basis of this new discipline. The first is 'convergent, ubiquitous, pervasive networking', referring not only to the increasing number and role of networks in the world but to a future where there may be a single integrated 'network of networks'. The second theme concerns the multiscale, multiresolution, heterogeneous aspects of complex networks. Third, robustness is seen as the dominant issue in complex systems – in contrast to yesterday's emphasis on information, entropy, energy, and materials. The final theme is that the new science will be a rigorous one, marrying a new mathematics with the engineer's emphasis on real-world systems rather than academic examples. The rest of the chapter discusses several examples of moderately complex systems, including automobile airbags, compact disc players, the Mars Pathfinder, and Formula One racing. These discussions exemplify issues of robustness, uncertainty, and modeling that are associated with all complex systems.

12

Current and Future Developments in Air Traffic Control

Steven Green
NASA Ames Research Center

Joseph Jackson
Commercial Electronic Systems, Honeywell

12.1 Introduction

The air transport industry, traditionally a large consumer of control and automation technology, presents new challenges in complexity management. Much of modern control theory was developed for this industry and its military counterpart. The challenge of controlling a single aircraft has, for the most part, been tackled successfully. Although modern air transports are highly automated, the efficient management of multiple aircraft, and their interactions, is relatively uncharted territory. This is particularly true with respect to the safe sharing of limited airspace under dynamic traffic conditions. Additional complexity stems from airline scheduling practices that attempt to maximize the profitable utilization of resources (equipment and personnel) within regulatory constraints and operational uncertainties (e.g., weather).

The Federal Aviation Administration (FAA) is responsible for providing air traffic control (ATC) and Traffic Flow Management (TFM) services to a wide variety of airspace users (e.g., airlines, military, general aviation) within the National Airspace System (NAS). The FAA ensures safety (aircraft separation) by restricting flight operations. Many of these restrictions have tended to be static in nature and designed into the airspace structure. Although this approach has been extremely successful in the past, future economic growth requires a new approach to achieve even greater levels of airspace capacity/throughput and flight efficiency. Today's air traffic controllers operate at their maximum capacity with greater frequency while airspace users are demanding increased capacity and flexibility in their flight operations. These factors have led to the concept known as 'Free Flight', the principal driver for the development of new air traffic management (ATM) technologies (RTCA, 1995). Without enhanced automation in the form of computer aids and information displays, the desired growth in system capacity and user flexibility will not be feasible.

NAS modernization is particularly challenging because upgrades must take place in parallel with operations (analogous to changing the tire on a moving car). In addition, operations are distributed and non-homogeneous in nature, and the stakeholder community is quite diverse. Stakeholders include the FAA, which provides NAS services; airspace users, who select aircraft capabilities and fleet compositions; airport operators, who determine the airport features of priority to the community they serve; manufacturers (airframe, avionics, and ground system) and information service providers, who develop the equipment that provides operational capability; and the human operators (flight crews, dispatchers, and air traffic controllers) who operate the equipment. Put simply, 'optimization' by a single stakeholder is unlikely to provide the desired system-wide benefits. Instead, a coordinated effort across stakeholders is needed for cost-effective modernization.

The purpose of this chapter is to describe the complex nature and operation of the current NAS and illustrate several automation challenges on the path toward modernization. Although the airspace user community is quite diverse, this chapter will focus on airline operations in the U.S. Issues related to other airspace environments, such as Europe, oceanic, and less developed nations, are not addressed. Some researchers (IEEE, 1993; Kahne and Frolow, 1996; Perry, 1997) offer additional insights into the complexity and automation issues related to ATC.

12.2 National Airspace System

12.2.1 History

The NAS, like the modern airplane, has evolved significantly from its beginning. Initially, airspace operations were truly 'Free Flight'. Early ATC services were developed and provided by the users themselves (Nolan, 1990). In the early 1920s, some of the larger 'air service' operations developed procedures to promote safety through the dissemination of weather information and the separation of flights. These services were first provided by air carrier personnel who were primarily concerned with the safe 'dispatching' of aircraft (e.g., safe weather conditions). By 1935, an interairline air traffic agreement formed an experimental center for the Chicago-Cleveland-Newark corridor, staffed by airline dispatchers. In 1937, the responsibility for 'ATC services' was transferred to the government.

ATC procedures and supporting technologies have evolved over time. Early airline-ATC communications were between controllers and dispatchers via teletype. Advances in radio communications and supporting infrastructure led to the direct communication between ATC and pilots. Early navigation aids (navaids) were sparse and routes were limited. Surveillance was based on pilot position reports, a technique still used today in remote airspace. Flights were separated procedurally by clearing only one flight into a volume of airspace at a time. Given the crude level of supporting technology, these volumes of airspace were quite large by modern standards.

As traffic density, flight speed, and system complexity grew through the middle of the century, the demands on the system revealed bottlenecks in areas such as navigation and surveillance. The deployment of navaids such as Very-high-frequency Omni-range Receivers (VORs), Distance-Measuring Equipment (DME), and Instrument Landing Systems (ILSs) greatly increased route flexibility and airport capacity during instrument conditions. The deployment of radar provided a revolution in surveillance that facilitated new procedures for direct separation of aircraft rather than procedural separation of aircraft from airspace. The corresponding reduction in separation criteria increased *en route* and terminal airspace capacity. Radar also increased the controller's flexibility in managing arrival and departure traffic independent of navaids and routes.

Advances in information technology in the 1960s and 1970s greatly reduced the bottlenecks in ATC coordination due to manual intersector coordination. Growth in traffic and flight range increased the need for flight plan data in downstream sectors to facilitate the planning and coordination of inbound traffic. Radar alone did not provide a complete traffic picture. Controllers had to correlate flight plan data with their radar displays by manually moving 'shrimp boat' data tags across their screens. The implementation of the current network of NAS computer systems, software, and supporting interfacility communications introduced automation to facilitate flight plan processing, track correlation, and hand-offs between sectors and facilities.

In recent years, system-related bottlenecks led to the development of traffic flow management services. Traffic flows have become highly dependent on the inter-relationships between major airports, airline schedules, and weather. Airline schedules, driven by economic considerations, often press airport capacities to their limit during arrival and departure rushes. Dynamic capacity reductions, typically due to weather, may last up to several hours or more. A capacity reduction at one high-density airport can drastically impact the system, with delays rippling to other airports. TFM services have been evolved to mitigate the impact of these disturbances through dynamic initiatives (restrictions), including arrival metering/spacing, traffic flow re-routes, and ground delays.

12.2.2 Airspace structure

This chapter will focus on 'controlled' airspace operations (requiring ATC clearance), which comprise most commercial flight operations. A comprehensive description of airspace classification, flight rules, and ATC operations may be found in (Nolan, 1990; FAA, 1993; FAA, 1998).

Airspace within the continental United States is divided into 20 geographic regions (Figure 12.1), each under the jurisdiction of an Air Route Traffic Control Center (Center). Centers typically control the airspace from near the surface up to flight level FL600 (60,000 feet). Centers are subdivided into sectors (Figure 12.2), as many as 20 or more, with stratification into layers of low altitude (e.g., surface up to FL240), high altitude (e.g., at and above FL240), and sometimes ultra-high altitude (e.g., above FL350). The geographic dimensions are typically designed around 'nominal' traffic flows to distribute workload among controllers and expedite traffic. Each sector, managed by at least one 'tactical' controller, has a discrete radio frequency for air-ground voice communications. During sustained periods of low traffic density, multiple sectors are often combined into one.

Regions of Special Use Airspace (SUA) may be defined as off-limits to civilian flights for reasons of national security or flight safety. Depending on application, SUAs vary in size up to thousands of square miles, with altitudes ranging from the surface to the upper flight levels. Some SUAs are always 'active' (e.g., White House), whereas others are scheduled. When active, SUAs may require significant detours from the most cost-efficient flight paths.

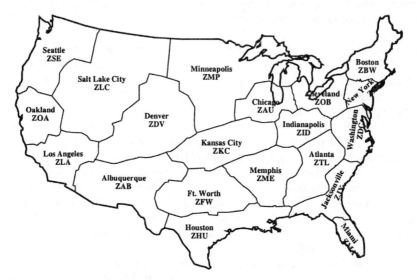

Figure 12.1 U.S. air route traffic control centers

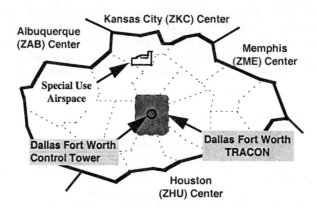

Figure 12.2 Fort Worth Center airspace

Other regions of airspace may be delegated from a Center to a Terminal Radar Approach Control (TRACON) facility. TRACONs are designed around high-density terminal areas (often serving multiple airports) to expedite arrival and departure transitions with *en route* (Center) airspace. TRACON airspace, also divided into sectors, generally extends from near the surface up to about 15,000 feet above the ground. Although Center radar coverage extends over most of the nation, TRACONs are generally served by shorter range radar. The higher TRACON accuracy supports lower radar-separation standards (3 miles vs. 5 in Center). Smaller regions of airspace are controlled by ATC Towers that are responsible for operations near and on the airport surface.

In addition to Tower, TRACON, and Center facilities, the FAA operates the ATC System Command Center (ATCSCC). The ATCSCC does not directly control, or communicate with, any individual flight. Instead, the facility monitors traffic flows and capacity throughout the NAS and coordinates TFM initiatives (between ATC facilities) to relieve excessive traffic congestion. These initiatives are then implemented by the ATC facilities that have control over the affected flights.

Figure 12.3 illustrates the typical sequence of ATC interactions over a typical flight. Prior to departure, the airline submits a flight plan specifying the aircraft type, navigational capability, departure location and time, destination, desired route of flight, cruise altitude and speed, and estimated flight time. If the planned route transits impacted airspace, ATC may modify the route and/or the ATCSCC may assign a predeparture ground delay.

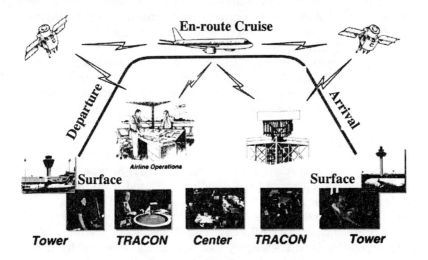

Figure 12.3 ATC phases of flight

Prior to departure, the pilot receives a flight plan clearance (coordinated between the departure Center, TRACON, and Tower). Prior to takeoff, the pilot must receive taxi clearance to the runway. Upon 'release' from the departure TRACON, the Tower controller may clear the flight for takeoff.

After clearing airport traffic, the Tower controller hands the flight over to a TRACON-departure controller. The TRACON provides separation services in the terminal area and facilitates the flight's transition into *en route* airspace. As soon as practical, TRACON controllers hand off the flight to the adjacent Center.

As a flight progresses through *en route* airspace, it undergoes a series of hand-offs between sectors and Centers. Each hand-off involves a formal transfer of 'track control' between sectors followed by a pilot switchover to the new sector's frequency. The controller's primary responsibility is to ensure separation between flights. Workload permitting, controllers also respond to requests for changes in routing and altitude.

During the arrival phase, control is transferred to the destination TRACON. TRACON controllers merge traffic, establish sequences, assign runways, and provide 'vectors' to the final approach. Arrivals are then handed off to the Tower for landing, after which they receive taxi clearances to their appropriate gates.

12.2.3 NAS Automation

Although the current level of NAS automation is considered archaic compared to the modern flight deck, it has enabled tremendous air traffic growth to date. A comprehensive review of NAS automation systems is beyond the scope of this chapter, but several salient functions are presented to illustrate key capabilities.

12.2.3.1 Information flow

Perhaps the most significant capability of the NAS is the underlying information infrastructure (FAA, 1999b). Each Center and TRACON facility uses local computational resources for radar tracking and flight plan processing. The NAS network supports the sharing of this information between sectors and across facilities. When a flight plan is filed/amended, the local ATC facility forwards the information to downstream facilities. Each facility translates the flight plan information into local coordinates and distributes the data in 'flight strips' to affected sectors about 30 minutes prior to the flight's arrival in that airspace. As a flight progresses, each controller updates the planned route (to reflect changes that affect downstream sectors) and the NAS automation updates the estimated time of arrival (ETA) based on radar tracking. Flight strips provide a backup to radar tracks and enable controllers to anticipate hand-offs and coordinate incoming flights.

12.2.3.2 Decision support aids

Examples of decision support aids that are built on top of the NAS information infrastructure include Conflict Alert, trend vectors, the Arrival Sequencing Program (ASP), and the Traffic Situation Display (TSD). Conflict Alert and trend vectors are *en route* (Center) tools used at the radar sector, whereas ASP and TSD are TFM tools.

Conflict Alert is an automatic function that analyzes the three-dimensional velocity vector of each flight to tactically predict any 'loss of separation' approximately 2 minutes ahead. Radar separation is lost when any two aircraft are separated by less than legal minimums (e.g., 5 miles and 1000 feet of altitude). Controllers (and supervisors) receive alerts on their plan view (radar) displays via flashing data blocks (alphanumeric flight information displayed with radar tracks). Controllers normally use procedures and clearances to ensure separation. Conflict Alert provides a safety net to alert controllers of rare occasions when separation may fall below the minimum.

Trend vectors, on the other hand, are simple graphics used to illustrate the future positions of tracked flights on the radar display. Controllers can display each flight's vector (based on the current track position and ground speed) 1–8 minutes into the future. Although the vector does not account for changes in direction or speed, vector comparisons are useful for merging flights and detecting potential conflicts.

ASP is an arrival-metering tool for major airports used by Traffic Management Coordinators (TMCs) within Center facilities. ASP is used to monitor arrival delays by comparing predicted arrival time to a target airport acceptance rate (flights/hour). When predicted delays exceed a target value (e.g., 4 minutes), TMCs initiate metering for arrival traffic when flights are generally within 20 minutes of their meter fix (often located at the TRACON boundary). During metering, sector controllers adjust traffic to conform to ASP meter times displayed directly on the radar display. This time-based approach enables TMCs to distribute delay across sectors, in an equitable fashion, based on the dynamic traffic load.

TSD provides individual facilities (Centers, TRACONs, and the ATCSCC) with a shared awareness of traffic flows. TSD displays a mosaic of aircraft position, velocity, and flight plans that is collected and distributed across the NAS by the Enhanced Traffic Management

System (ETMS). This 'big picture' facilitates strategic planning of traffic flows around airspace that is dynamically impacted by weather or congestion. The monitor alert function alerts TMCs when sector traffic densities are predicted to exceed nominal limits. The FAA also provides limited ETMS data to airspace users to facilitate flight following, and provide for a common situational awareness with airspace users.

12.2.4 Airline enterprise

An airline's main product is its schedule, namely, the airline's ability to deliver the right flights (city pairs) with the right connections, at the right time. The objective of airline operations is to operate the schedule safely and at a profit. This leads to a delicate balance between capital expenditures (for schedule expansion or efficiency improvements) and the cost of daily operations.

Given the industry's high level of capitalization and low profit margins, airline business decisions are highly sensitive to changes in equipment or operating procedures. Airline business cases typically depend on revenue enhancement, cost avoidance, or competitive position (Chew, 1997). Revenue may be enhanced by the addition of new/larger aircraft to increase service (add new city pairs or increase frequency). Cost avoidance may involve upgrades (e.g., avionics) to avoid exclusion from preferred routes/altitudes. Competitive position frequently involves a balance between price and service (meals, amenities, seating, and service). Other operating cost considerations include operations, maintenance, training, and special requirements for certain airspace or airports.

As a result of this economic environment, airlines are cautious when acquiring new aircraft, retrofitting older equipment, or upgrading avionics. This results in a mixture of capabilities within and across airline fleets. When the variation in airline 'missions' is also considered, it becomes obvious that a homogeneous fleet is just not economically viable. A fundamental 'chicken and egg' problem haunts modernization efforts. Most avionics upgrades depend on greater NAS capabilities for return on investment, whereas many NAS upgrades may be dependent on widespread aircraft equipage. Certainly, mixed equipage will continue to be a pacing factor in modernization.

12.2.5 Airline operations

Airline operations are typically orchestrated within an Airline Operational Control (AOC) center. An AOC's primary objective is to operate its flights safely and legally. The secondary objective is to profitably manage the airline's schedule while operating flights as efficiently as possible (FAA, 1997b). Meeting these objectives, within a dynamic ATC environment, requires complex problem solving related to fleet and flight optimization.

An AOC is composed of flight dispatchers and a variety of personnel with specialized functions. Meteorologists analyze weather data such as predicted winds aloft, convective weather hazards, and turbulence. ATC coordinators monitor NAS status (e.g., TFM initiatives) and act as the real-time point of contact for airline interaction with ATC facilities. Maintenance personnel manage scheduled and unscheduled maintenance of equipment to ensure compliance with regulations and dispatchability. Crew representatives manage crew assignments, reserves, and training with consideration for details such as qualifications, duty time limits, and schedule/connections.

Dispatchers play a central role by planning flights prior to departure, monitoring flights while *en route,* and replanning flights as needed. As a focal point for operations, dispatchers support airborne flight crews and coordinate airline resources on the flight's behalf. The modern dispatcher workstation typically includes a networked PC that provides graphical weather information, dispatching and flight planning tools, and a TSD showing flight locations in real time and air-ground communications links.

In addition to the planning, monitoring, and optimization of flights under normal conditions, the AOC is the most critical airline component for managing rare-normal operations. Rare-normal operations occur when significant transients (e.g., weather) disrupt the airline's schedule and network. A missed connection can cost an airline many thousands of dollars in hotel stays for delayed passengers, costs to deliver baggage, and loss of goodwill. However, schedule delays and missed connections also impact flight crew and aircraft/equipment logistics. Disruptions to just one part of the schedule can ripple through the entire network of flights, as delayed flights prevent other dependent flights from departing. The continued buildup of delays may lead to a critical point where the original schedule is impossible to maintain and the AOC's priority is simply to recover.

12.2.5.1 A typical day in an AOC

On a seasonal basis, the AOC builds and publishes a schedule in the Official Airline Guide (OAG). At the start of each day, scheduled flights are distributed to individual dispatchers. Other departments have already assigned a specific aircraft and crew to each flight. Within several hours of each flight's planned departure time, the dispatcher generates a flight plan based on the latest data (e.g., weather). These flight plans are provided to the pilots for acceptance/revision, and filed with ATC for clearance.

Dispatchers use flight-planning tools to optimize route, altitude, and speed based on airline priorities (e.g., time, fuel, and ride comfort). Once airborne, each flight is followed by the dispatcher and tracked via a TSD (or position reports for international trips). Flight following provides support to the flight crew with real-time information (e.g., approaching weather hazards and traffic delays). Replanning may be warranted by factors such as errors in forecasted winds and weather along the route, equipment malfunctions onboard the aircraft, ATC delays, or changing weather conditions at the destination. An updated flight plan is then generated with pilot concurrence and coordinated with ATC.

12.2.5.2 A nontypical day in an AOC

During periods of rare-normal operations, the AOC becomes a center of intense activity. Maintaining airline schedule integrity can become quite challenging when weather disruptions impact a hub airport and delays ripple to dependent flights. Airline strategies may vary greatly, depending on airline and circumstances. AOCs can respond by canceling flights, delaying departures, or diverting airborne flights to alternative airports. During these periods, the ATC coordinators work with ATC to understand and mitigate the traffic problems. Meteorology can help plan for breaks in the weather. Crew scheduling and aircraft maintenance are needed to work around logistical bottlenecks created by the delays. The AOC must quickly formulate strategy and coordinate airline assets to recover the schedule. However, due to time constraints, problem complexity, and limited automation tools, recovery strategies are often a best guess rather than a formal system optimization. As the environment changes and new AOC decision support tools are developed, there will be opportunities for the AOC to increase flexibility and improve airline performance.

12.2.6 Flight deck operations

The flight crew of a commercial transport has final responsibility for the success and safety of flight operations (Billings, 1996). Flight crews must have full knowledge of the factors related to the flight, such as weather, ATC constraints, and airline goals. Advances in technology have enabled increasingly sophisticated automation to be introduced into the flight deck. For example, the avionics for a Boeing B777 host more than one million lines of certificated software. This automation supports the flight crews by automating repetitive functions (e.g., attitude control and systems monitoring/management), providing a comprehensive database to plan routes according to international standards, performing trajectory optimization, and providing automatic flight guidance. Coupled with enhanced pilot procedures, this automation has reduced pilot workload, enhanced safety, and improved the economy of operations.

A pilot's basic tasks, in order of priority, are to aviate, navigate, and communicate (Jeppeson Sanderson Inc., 1991). Figure 12.4 illustrates these tasks in terms of categories involving Communication, Flight, Systems, and Task Management (Abbott, 1993). For the purpose of this discussion, and the subsequent focus on ATC interactions, it is most relevant to focus on the Flight Management and Communication Management categories.

Figure 12.4 Interaction among functional categories

Flight Management may be decomposed into subtasks, based on supporting information requirements and the rate at which the subtasks are performed. Figure 12.5 illustrates five key subtasks: flight planning, guidance, navigation, autoflight, and stability augmentation. These tasks are described in the context of a modern digital flight deck.

The flight planning subtask is strategic in nature. Flight crews typically 'download' an AOC-created flight plan to the onboard Flight Management System (FMS) and modify it as necessary to conform to ATC clearance updates and pilot judgment. The FMS provides the pilot with access to a navigation database defining airports, runways, departure/arrival routes, *en route* airways, and published waypoints. The pilot may review and modify the plan, displayed on a 'maplike' Navigation Display, via a Multifunction Control and Display Unit 'keyboard'. The FMS includes an expert system that generates a four-dimensional (4-D) trajectory plan based on knowledge of navigational procedures, atmospheric inputs (wind and temperature), aircraft performance, and strategies to minimize cost (fuel-burn and time). This trajectory provides the guidance subtask with targets for altitude, speed, and direction (track/heading). FMS computations also support pilot decisions regarding altitude and speed (optimum/maximum/minimum), fuel state, and speed schedules for slat/flap extension/ retraction.

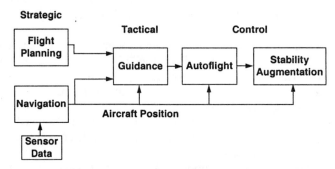

Figure 12.5 Functional representation of flight crew tasks

The guidance subtask is tactical in nature, in that 'real-time' flight-path corrections are made to adhere to the 'active' flight plan. FMS guidance algorithms compare the current aircraft position (from the navigation subtask) to the current leg of the 'flight-planned' trajectory, and determine pitch, thrust, and roll commands to achieve the flight-planned targets (altitude, speed, and direction). Pilots may also perform this task manually by entering pilot-selected altitude/speed/direction commands into the autoflight system via a Mode Control Panel. Pilot selection of targets and control modes is supported by an FMS flight-planning function that presents speed/performance-envelope limits on the Primary Flight Display.

The navigation subtask determines the airplane's 4-D position based on information from air data sensors, inertial sensors, and radio navigation or Global Positioning Satellite (GPS) equipment. The FMS selects the appropriate sensors and combines the data for maximum accuracy. This dynamic subtask must be performed instantaneously for the autoflight and stability augmentation subtasks, fairly frequently for the tactical guidance subtask, and occasionally for the strategic flight-planning subtask. Position information is displayed on the pilot's Primary Flight and Navigational Displays.

The autoflight subtask continuously manipulates the aircraft control surfaces and propulsion system to implement the guidance commands. This subtask is performed based on knowledge of the vehicle and actuator dynamics and the closed-loop properties of the stability augmentation system. Pilots may engage the avionics to perform this subtask, via the Mode Control Panel, or choose to control thrust, pitch, roll, and yaw manually through the throttle levers and flight controls (column/yoke/sidestick and rudder). The stability augmentation subtask supports the autoflight subtask by enhancing aircraft stability accounting for nonlinearities in the control surface and propulsion system actuators.

As for communication tasks, flight crews interact with ATC and AOC via very high frequency (VHF) radio (primarily for ATC communications), the Aircraft Communication and Reporting System (ACARS), the Satellite Communication Systems (SATCOM), and high frequency (HF) radios. ACARS supports two-way data link with AOC as well as limited ATC communications, including Automatic Dependent Surveillance (ADS) reports of airplane position and trajectory intent. ADS, SATCOM, and HF radios primarily support communications in remote airspace (e.g., oceanic).

The state of the art in avionics flight management and communications capability is provided in the 'Future Air Navigation' or FANS-1/A systems offered on Boeing and Airbus aircraft (Bang, 1995). These systems use satellite navigation, ADS position reporting, and Controller/Pilot Data Link Communications (CPDLC) to support flexible, user-preferred (time/fuel-saving) routing in oceanic airspace.

12.2.7   ATC considerations

Mixed fleet capabilities and airspace sectorization are two considerations that significantly influence the design and operation of the NAS.

Airspace is shared by a broad spectrum of aircraft of various sizes, performance, and navigation/communication capabilities. Aircraft, from small commuters to large transports, can vary in performance by 100% or more in terms of speed and rate of climb/descent. Performance differences can pose difficult merge and overtake challenges to controllers. Aircraft weight, which can vary by one to two orders of magnitude, has a direct impact on final approach spacing requirements (for wake vortex avoidance). Rapid advances in the computer and telecommunication industries, coupled with a typical airframe service life of 20–30 years, have led to a wide diversity in flight deck capability (e.g., navigation accuracy, routing options, and approach capability). This non-homogeneous traffic mix increases the complexity for controllers.

Sectorization evolved from the fixed routings associated with yesterday's ground-based navaids. It divides the airspace to facilitate the delegation of authority/responsibility to individual controllers and the distribution of workload. Under low workload, one controller may manage a sector alone. This primary 'tactical' controller separates traffic and communicates with all aircraft in that sector. If workload increases, additional controllers may assist with strategic planning of incoming flights (via analysis of flight-plan strips), direct coordination with neighboring sectors/facilities, and a second set of eyes to monitor the radar tracks. The responsibility of the primary controller centers around the flights that are 'owned' by that sector (generally within the airspace boundaries). This responsibility is transferred by the hand-off procedure. Often, under high-workload conditions, a controller's main goal is to expedite each flight's hand-off to the next sector. This short-term vision, critical to the safety of the system, is not always consistent with efficient trajectory planning, as discussed in section 12.3 below. Although sectorization divides the airspace into manageable pieces, the 'free routing' (area navigation) capability of modern avionics leads to variations in traffic flows that make static sector boundaries suboptimal.

12.2.8   Relative roles and decisions

The purpose of this section is to integrate the operational aspects described above in terms of the relative roles and key decisions surrounding the conduct of a typical flight. The classic roles include the flight crew, dispatcher, and air traffic controller. Although each role has a primary responsibility, their domains of responsibility often overlap, albeit within a defined hierarchy. For example, pilots are responsible for flight planning and the safe conduct of the flight. Dispatchers (required for air carrier operations) are also responsible for flight planning. Under the Federal Aviation Regulations, both roles (pilot and dispatcher) must conclude that a flight is safe before release, and either may decline the flight as unsafe. As a second example, pilots and controllers are responsible for separation. A controller's primary responsibility is to ensure aircraft separation under instrument flight rules (IFR). Pilots must also attempt to see and avoid other traffic either visually, or with aids such as the Traffic

Collision Avoidance System (TCAS). In general, ATC separates flights before the pilots see a problem to avoid. However, if pilots detect a collision risk, it is incumbent on them to avoid it. In the end, the pilot has the final authority for the safe conduct of the flight, and may deviate from ATC direction to meet the demands of an unsafe situation.

In addition to the classic roles described above, the TMC is also a primary decision-maker. Although TFM decisions are relatively strategic in nature compared to controller decisions, they directly impact individual flights. TMCs complement controllers in a manner that parallels the way dispatchers complement pilots. Other supporting roles, such as meteorology, airline facility management, and ATC-airways/facilities management, support pilots, dispatchers, controllers, and TMCs when making decisions that affect the planning and execution of a flight.

Airlines and ATC tend to look at schedule and routing/trajectory decisions from two different points of view. Airlines tend to manage flights from a Lagrangian point of view (i.e., following individual flights within the context of fleet/schedule goals), whereas ATC manages flights from an Eulerian point of view (i.e., analyzing airspace as flights flow through). This difference in point of view makes it challenging to follow the complex airline-ATC interactions, let alone optimize the productivity of the NAS. Figures 12.6 and 12.7 attempt to clarify this interaction by illustrating key decision timelines from the airline and ATC points of view, respectively. Corresponding events will be described from both perspectives to facilitate comparison.

Figure 12.6 presents a model of an airline decision timeline in terms of flight phase (for an individual flight). The horizontal axis represents key phases from predeparture planning to arrival. The vertical axis represents the relative control and flexibility the airline has in managing these phases. The figure presents, for each phase, the primary decision/action (horizontally) and the data/environment affecting that action (vertically). Starting from the left, the airline generates a system schedule (on approximately a monthly basis) based on economic decisions (e.g., markets, equipment, crews). The schedule is optimized based on resource availability and historical averages (e.g., winds and delays) and then published (tickets sold against available seats). As the day and time (of a particular flight) approaches, equipment and crew schedules are updated and flight planning begins. The AOC considers its best knowledge of weather and the state of the NAS (e.g., ground delay programs in effect) to plan routes, delays, and cancellations. Just prior to departure, and with pilot concurrence, the latest weight-and-balance data are uploaded to the airplane. After departure, the flight crew manages the vertical profile and routing for comfort, safety, and economy with assistance from the AOC for re-planning (e.g., around weather). As ATC adjusts the flights for delay or separation, the flight crew complies and/or attempts to negotiate a better clearance. The AOC monitors flight progress and, if delays/deviations are excessive, considers rerouting or diverting (to another airport) as necessary to complete the flight safely with minimal impact on airline schedule. The negative slope of the graph illustrates the reduction in airline decision flexibility as a flight nears completion.

In contrast to the airline decision timeline given in Figure 12.6, the ATC-decision timeline model (Figure 12.7) is presented in terms of airspace 'look-ahead' time. The horizontal axis represents the look-ahead time for a particular segment of airspace (e.g., an *en route* sector that feeds arrivals into a high-density terminal area). Whereas the airline decisions are presented in an absolute time scale, the ATC decisions are presented along a relative time scale that continuously updates in absolute time. The vertical axis represents the degree of 'tactical criticality'. The figure presents a set of key decision event horizons that may impact the planning and execution of a flight. The scale ranges from tactical (safety) events on the left to strategic (planning) actions on the right.

National Airspace System

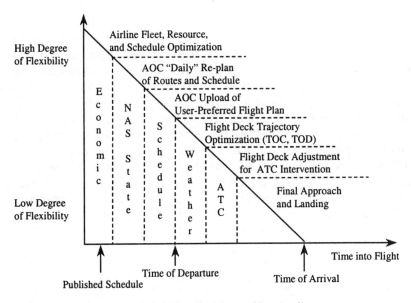

Figure 12.6 Airline decision-making timeline

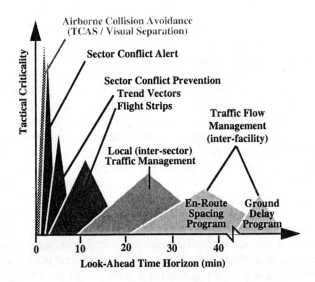

Figure 12.7 ATC decision-making timeline

Key decision events begin on the left with tactical separation. For perspective, the far left illustrates the time horizon for airborne collision avoidance (based on visual or TCAS separation). The tactical side of ATC services is bounded by the conflict alert horizon (approximately 2 minutes prior to conflict). These ATC actions (typically heading vectors or altitude changes) tend to be short-term avoidance maneuvers. The next horizon to the right involves trajectory deviations for conflict prevention as well as conformance with TFM initiatives. These deviations, based on radar-track information (e.g., trend vectors), tend to involve subtler changes in heading/altitude (compared to conflict avoidance) as well as speed changes for sequencing and merging. Within this time horizon, problems often occur that interrupt climb/descent profiles and prevent top-of-descent optimization. Deviations for weather (both pilot-requested and controller-suggested) may also occur within this horizon. The next horizon is based on flight strips that provide controllers with a greater look ahead to plan for incoming flights about to enter the sector. Moving farther out, local traffic management begins looking at problems 10–40 minutes or so in advance. TMCs may initiate route changes at upstream sectors to relieve downstream sectors and airspace impacted by traffic load or weather. TMCs may also invoke local metering of arrivals (à la ASP) within this horizon. The strategic side of the scale includes TFM initiatives that are needed to relieve larger scale capacity problems requiring coordination across multiple ATC facilities. These initiatives may lead to route changes (while airborne) and departure delays while still on the ground.

12.2.9 ATC challenges

Although the current NAS has supported significant traffic growth to date, there are shortfalls that limit user efficiency and future growth in capacity. Airspace user concerns go far beyond the immediate problems associated with an aging NAS infrastructure and focus on the cost of operations. Inefficiencies in the current system have been estimated to cost the U.S. airline industry $4 billion annually (Haraldsdottir, 1997). Given a predicted 2.3% annual growth rate in air traffic (Haraldsdottir, 1997), the relief of system capacity bottlenecks will become an increasingly more important, if not the dominant, economic factor in the success of the air transport industry. Several problems that need to be addressed include accurate real-time analysis of traffic problems, efficient problem resolution across airspace boundaries, and the accommodation of user preferences and flexibility.

There is a need for improved real-time prediction of traffic demand and system capacity. These predictions, which are performed by relatively manual means today, are critical to the accurate analysis of conflicts and delays. System uncertainties mentioned earlier lead to overly conservative controller actions and procedures to maintain safety. If these uncertainties can be reduced, or even removed, airlines and ATC can be more selective in the problems they attempt to solve. The goal would be to exercise control by exception and only for problems of high probability.

Efficient problem resolution requires a 'big-picture' perspective. When problems are divided across sectors, there is no longer a clear and continuous path for resolution. Problem solutions (e.g., delay absorption or conflict resolution) tend to be segmented and inefficient. Somehow, the impact of airspace boundaries must be removed from the character of the problem solutions. Once a problem is identified with a high probability, procedures and tools are needed to enable an efficient resolution (i.e., transparent sector boundaries).

Finally, greater accommodation of user preferences and flexibility are needed, either through clearance negotiation (collaborative decisions) and/or greater maneuvering freedom. Airlines use the NAS to achieve an economic goal. Although ATC carries a heavy responsibility for safety (e.g., separation), it does not have knowledge of, or responsibility for, the economic factors that are critical to the success of the civil aviation industry. Furthermore, airline operational priorities vary over time, depending on weather and delays, and may differ greatly from the priorities of other airspace users. Only individual airspace users can judge the economic performance of their flight operations. This greater variability in traffic flows presents a significant challenge to the human-centered nature of ATC.

12.3 'Free Flight', the Future of ATM

The future of ATM is based on the vision of 'Free Flight'. Free Flight is defined as 'a safe and efficient flight operating capability under IFR in which the operators have the freedom to select their path and speed in real time. Air traffic restrictions are only imposed to ensure separation, to preclude exceeding airport capacity, to prevent unauthorized flight through Special Use Airspace (SUA), and to ensure safety of flight. Restrictions are limited in extent and duration to correct the identified problem. Any activity which removes restrictions represents a move toward Free Flight' (RTCA, 1995). The concept calls for new operational procedures, supported by technology (airborne and ground based), to increase user flexibility to safely optimize operations. A 'mature' Free Flight system would be too complex, risky, and capital intensive to implement in a single step. The economic aspects are further complicated by the differences in FAA and user investment practices. Aircraft capabilities will probably remain mixed across the national fleet as each user strives to maximize its profitability. On the other hand, the FAA must (within the limits of congressional budget practice) continue to provide ATC services while modernizing. A more realistic approach is to implement Free Flight enabling technologies and procedures in incremental, self-justified phases based on the stakeholder's return on investment.

In the near term, the NAS must be modernized to replace aging equipment (e.g., controller displays and computer resources), provide infrastructure for increased information sharing, and deploy the next generation of decision support tools. Although not currently the weakest link, users must anticipate the modernization of flight deck (communication, navigation, and surveillance (CNS)) and AOC (flight planning) capabilities. Near-term benefits would include improved system reliability as well as early capacity, efficiency, and controller productivity improvements that can be achieved with automation support of current procedures.

The longer term role of technology, however, is to enable new procedures to greatly increase user flexibility and flight efficiency. Many of today's inefficiencies are related to procedures designed for fault tolerance (e.g., robustness to lost communication with a flight or a pilot/controller blunder). Maintaining fault tolerance in the more distributed and flexible environment will pose additional challenges. Human factors engineering is absolutely critical to the identification of key problems that inhibit human operators (ATC, flight deck, and AOC) from transitioning to the new procedures.

NAS modernization is too broad in scope to provide a comprehensive description here. The reader is referred to the FAA-NASA interagency research plan (FAA/NASA, 1999), the FAA vision for the NAS architecture (FAA, 1999a), and the NASA concept for Distributed Air-Ground Traffic Management (NASA, 1999). Instead, the following presents a brief overview of several key technologies and systems planned for deployment in the near term and several advanced tools and concepts envisioned for the future.

12.3.1 Evolution of CNS/ATM modernization

A joint government/industry consensus (RTCA, 1997) outlined a path for the FAA and the user community to plan procedural, investment, and architectural decisions that will evolve

Free Flight operational capabilities. This path transitions from the current NAS (Haraldsdottir, 1997) through three distinct time frames: the years 2000–2002 (Free Flight Phase 1 (FFP1)), 2003–2005, and beyond 2005 ('mature-state' Free Flight). This section will describe several key elements in terms of CNS functions and ATC decision support tools.

12.3.1.1 Communications

Next steps will extend Controller-Pilot Data Link Communications (CPDLC) to domestic airspace and introduce Automatic Dependent Surveillance Broadcast (ADS-B). CPDLC will offload congested voice frequencies and reduce pilot/controller workload associated with routine communications and voice errors. CPDLC will also provide direct connectivity between the FMS and ATC automation to enhance the performance (accuracy) and interoperability of airborne and ground-based tools. CPDLC Build-1(A) will introduce a few domestic ATC messages in *en route* airspace in the 2000–2002 time frame. Build-2 will extend the message set to cover most operational ATC communications and facilitate integration of FMS and ATC automation in the 2005 (and beyond) time frame. CPDLC will be facilitated by international development of the Aeronautical Telecommunication Network (ATN), which will integrate communication 'subnetworks' (e.g., VHF, SATCOM, and Mode S) with protocols to increase flexibility, reliability, bandwidth, and airspace-independent connectivity. ADS-B will broadcast aircraft self-surveillance data to increase the flight deck's role in separation assurance.

12.3.1.2 Navigation

Growth and enhancement of satellite navigation will reduce dependence on ground-based navaids and increase the navigational capability/flexibility of the average aircraft (for point-to-point area navigation). Anticipated benefits include reduced separation for *en route* and terminal operations, flexible (curved) approach capability, surface movement guidance, reduced minima for takeoff and approach operations, and improved terrain avoidance.

12.3.1.3 Surveillance

Automatic Dependent Surveillance (ADS) will provide position and intent data from aircraft navigation systems to ATC via data link. ADS data will improve tracking accuracy and/or supplement ground-based surveillance (radar) coverage. The ADS-B application enables participating aircraft to transmit and receive separation assurance data for use on a cockpit display of traffic information. To complement and supplement ground-based surveillance (radar) systems, enhanced traffic displays (using ADS-B) will improve the flight crew's traffic situation awareness beyond TCAS and support a greater flight deck role in separation assurance for key operations. New terrain data bases and FMS functions will enhance flight deck situation awareness to prevent events of controlled flight into terrain. Operational evaluations of these new surveillance applications are under way within the FAA's SafeFlight-21 program in partnership with key user groups.

12.3.1.4 ATC

The FAA's FFP1 program is focused on achieving early capacity and efficiency benefits by introducing automation tools that enhance current practices. FFP1 tools include the Surface Movement Advisor (SMA), Passive Final Approach Spacing Tool (P-FAST), Traffic Management Advisor (TMA), User Request Evaluation Tool (URET), and Collaborative Decision Making (CDM). SMA supports airport surface operations to improve taxi cueing by fusing Tower, airline, and ramp information to improve user and ATC decisions (Glass, 1997). P-FAST (Figure 12.8) improves TRACON arrival throughput by providing

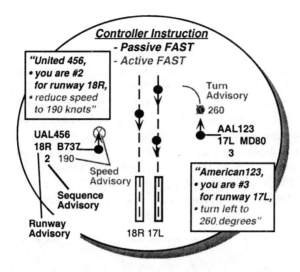

Figure 12.8 Illustration of Passive/Active-FAST capabilities

controllers with optimal runway assignment and sequence advisories directly on their radar displays (Davis *et al.*, 1997). TMA improves the *en route* (Center) arrival flow into the TRACON by supporting TMC flow rate decisions with automated flow rate analysis and more accurate/efficient (than ASP) arrival-metering advisories (sequence and schedule) to the sector controller (Swenson *et al.*, 1997). P-FAST and TMA are an early implementation subset of the Center-TRACON Automation System (CTAS), which includes integrated decision support tool capabilities for efficient arrival/departure management of *en route* and terminal traffic (Erzberger, 1995). URET is an initial conflict probe tool (FAA, 1997a; The MITRE Corporation, 1997) that evolved from the FAA's Automated En Route ATC (AERA) research of the 1980s (The MITRE Corporation, 1993). Figure 12.9 presents a simplified illustration of the capabilities of URET, which alerts controllers of computer-predicted conflicts (up to 20 minutes in advance) and allows controllers to 'trial plan' resolutions before implementing clearances. CDM is a set of technology and procedures designed to support real-time FAA/user collaboration on TFM initiatives with an initial focus on national ground delays. CDM allows participating users to optimize their own dynamic schedule changes in response to potential TFM initiatives without losing competitive position due to unilateral decisions to delay/cancel impacted flights.

In addition to near-term benefits, FFP1 provides the foundation for advanced automation needed to facilitate more substantial changes in operational procedures. Examples that are currently in 'concept development' include the Surface Management System (SMS) for Tower Controllers, the Active FAST (AFAST) and Expedite Departure (EDP) tools for TRACON controllers, and the En route Descent Advisor (EDA) for *en route* (Center) controllers. SMS will extend SMA to provide controllers with improved situational awareness and active guidance to increase the efficiency and safety of surface operations. AFAST, EDP, and EDA are CTAS tools designed to increase the throughput and efficiency of terminal arrival/departure operations as well as the transition to/from unconstrained *en route* airspace. Building on PFAST, AFAST (Figure 12.8) provides terminal controllers with 'active' clearance advisories (vector headings and speeds) leading to accurate, conflict-free spacing on final approach (Davis *et al.*, 1991). EDP represents the departure counterpart to

FAST. For *en route* airspace, EDA introduces new capabilities beyond initial conflict probe to greatly improve the efficiency of traffic flows (Green *et al.*, 1998). Designed to integrate with TMA, EDA provides controllers with active clearance advisories for fuel-efficient, conflict-free trajectories to conform to flow-rate (TFM) restrictions. EDA may be complemented by the Problem Analysis Resolution and Ranking (PARR) capability (i.e., automated conflict resolution) that is being developed from earlier AERA research. As steps toward full EDA implementation, two simpler 'spin-off' capabilities (Spacing Tool and Direct-To) may be deployed earlier as extensions to FFP1 tools. The Spacing Tool applies conflict probe and trial-planning capabilities to help controllers conform to TFM 'miles-in-trail' restrictions without forcing flights onto inefficient 'in-trail' paths (Green and Grace, 1999). When TFM restrictions are not in effect, the CTAS Direct-To capability provides controllers with active advisories for time-saving, conflict-probed short cuts to the active flight plan.

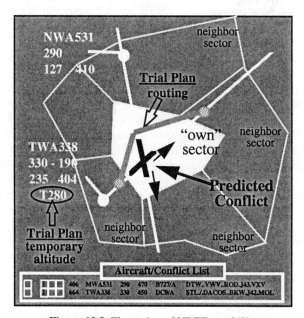

Figure 12.9 Illustration of URET capability

TFM research will include the development of tools to assist in the orchestration of multisector/facility solutions to dynamic *en route* problems. The CDM concept will be extended to the rerouting of airborne flights to mitigate the impact of thunderstorm deviations at the local and regional level.

12.4 Long-Term Challenges in ATM

The designers of future systems and procedures will face many long-term challenges (Jackson and Green, 1998). Five such challenges are presented here to provide the reader with an appreciation for the complexity involved in achieving Free Flight.

12.4.1 Integrated arrival/surface/departure management

Although FFP1 (and its successor) will introduce automation tools to improve terminal operations, these tools are single-focus capabilities. Terminal airspace represents a complicated network of airspace capacity resources. Arrival management must be balanced with surface and departure operations, a task performed manually today by traffic managers. Following the near-term development of arrival/surface/departure tools, automation will be needed to assist traffic managers with the dynamic determination of optimal balances in the allocation of arrival/surface/departure resources. More than just balancing the resources for any one airport, interactions between airports will have an influence on the performance of the NAS, particularly in congested regions such as the Northeast. In other words, the departure flow of one airport may be highly related to the arrival flow of another. These complex capacity-resource interactions require further study to truly understand the system interactions.

12.4.2 Reduced separation standards

Far-term research is also called for to explore avenues for increasing airspace capacity through reductions in terminal and *en route* separation criteria. In the terminal area, this includes the final-approach spacings required for wake vortex avoidance (Hinton *et al.*, 1999). Dynamic determination of minimum wake-vortex separations and their integration into tools such as FAST show great promise for significant increases to airport throughput. Regarding conflicts in both *en route* and terminal airspace, it may be possible to reduce separation criteria through emerging technologies that, in addition to conflict probe, improve aircraft tracking and total system robustness to collisions. There exists a system engineering challenge to combine the best available ground-based and airborne (ADS) surveillance to improve tracking accuracy and reliability. It may also be possible, through increased equipage and improvements to cockpit traffic displays, to raise collision-safety redundancy enough to justify a reduction in separation criteria. This would increase airspace capacity as well as airport arrival throughput.

12.4.3 Increasing user flexibility

In a truly Free Flight environment, ATC would accommodate user preferences as a rule, with deviations only when necessary for system safety and efficiency. User-Preferred Trajectories (UPTs) may be enabled by modifying the current system (ground-based responsibility for separation) to minimize deviations from UPTs (Williams *et al.*, 1993; Green *et al.*, 1997). An alternative approach is to shift separation responsibility to the aircraft along with greater authority and flexibility in flight planning and maneuvering. Although too risky (both technically and economically) for near-term applications, this 'unconventional' approach to separation may be the only viable means of achieving long-term 'stretch' goals for capacity and efficiency improvements. This approach may also provide a foundation to support air traffic growth in less developed nations. Such nations may not be able to afford the ATC infrastructure investment necessary to support growth with ground-based separation services. Aircraft may bring the capabilities necessary to operate in remote airspace.

The feasibility and benefits of the Distributed Air-Ground Traffic Management (DAG-TM) concept is under investigation (NASA, 1999). Although the DAG-TM concept offers a broad range of operational applications that leverage distributed decision making across elements of ATC, the flight deck, and the AOC (e.g., collaborative TFM and clearance decisions), much of the technical work is focused on the distribution of separation responsibility.

Several key challenges must be overcome to successfully develop operationally acceptable procedures for distributing responsibility for separation assurance. One challenge is the creation of a common situational awareness between flight crews and ATC with regard to aircraft states (position and velocity), intent, and conflict detection. An even greater challenge is the need for clear roles and responsibilities of controllers and flight crews under normal and abnormal operating conditions. Conflict resolutions must be coordinated to the level necessary to prevent flexibility from creating unacceptable levels of chaos. Assuming mixed levels of aircraft capabilities, this coordination must encompass both air-to-air (flight-crew-to-flight-crew) and air-to-ground (flight-crew-to-controller) aspects as controllers provide separation assurance services for aircraft that are not equipped for self-separation.

Airborne separation assurance introduces new flight crew issues. First, it is necessary to determine the information a flight crew needs to support the shared separation procedure. A significant human factors challenge is to determine how this information needs to be presented, and what flight crew actions are required to support conflict detection and resolution procedures.

Independent of where separation responsibility resides, ATC must fulfil a critical TFM role to support user preferences. User-preferred speeds and routes depend on, among other things, the user's knowledge/estimation of the state of the NAS (e.g., weather and delays). The value of user flexibility depends upon the 'quality' of the TFM constraints against which users will plan their operations. It is not efficient for users to fly fast (to maintain schedule) into airspace with predictable delays that could have been absorbed more efficiently earlier. On the other hand, it may not be operationally wise for a user to slow for a delay that is uncertain or unfairly distributed by ATC. It is incumbent on ATC to provide users with accurate and timely updates of NAS status (from which intelligent preferences may be selected) and stable, equitable, and minimal TFM constraints within which user preferences may be planned. Although not presented here, a complimentary (and no less critical) element of user flexibility is related to user preferences for TFM constraints (NASA, 1999).

12.4.4 Executing the 'plan'

In many situations, trajectory uncertainties (due to factors such as delays in clearance execution and navigational/tracking accuracy) lead controllers to conservative action, including the introduction of extra separation buffers. This is particularly true within the extended terminal area (i.e., within 200 miles of a major airport), where arrivals converge into smaller regions of airspace. Even though modern aircraft are equipped with sophisticated navigation, guidance, and control systems, the flight deck automation is often turned off during operations in congested terminal airspace. Controller procedures force pilots to revert to the use of simple autoflight modes for heading, altitude, and speed changes.

To enable more efficient use of resources in a Free Flight environment, a more cooperative approach is needed between the evolving flight deck and ATC automation. Air and ground systems may be integrated via two-way data link communications (Green et al., 1997). Exchange of data (e.g., winds, temperature, and aircraft state and intent) will improve the accuracy of both systems. Uplinking clearances directly into the FMS (particularly trajectory-based clearances) will allow aircraft to execute the ATC plan with greater accuracy. The result would be a reduction in the necessity for conservative actions and excess spacing buffers.

Research into closely spaced parallel approach procedures (Ashford, 1998) shows promise for maintaining 'clear weather' airport capacity under instrument meteorological conditions at airports with parallel runways spaced less than 3400 feet apart. This procedure requires precision navigation, aircraft-to-aircraft communications (ADS-B), display of the traffic position on parallel runways, and escape maneuver alerting should aircraft position safety tolerances be exceeded. Although commercial aircraft have been certificated with automatic approach and landing systems that are robust to gusts, turbulence, and other anomalies in the final approach phase of flight, this closely spaced parallel approach procedure may place new control demands on precision approach systems in order to maintain adequate safety margins.

12.4.5 Sector complexity

Maintaining separation between flights within the NAS requires more than just tactical separation at the sector level; it also requires management of airspace resources to prevent any one sector from overloading. Flexible staffing practices go a long way, but even this method has limits. Given the anticipated traffic growth and desired increases in user flexibility (flight planning independent of airway/airspace structure), TFM must be able to dynamically predict and control sector workload levels with minimum user impact. Workload, however, is difficult to quantify and measure, let alone predict and control. The current TFM approach is to monitor an objective metric (that is easily predicted) while subjectively estimating the acceptable metric level for the operational conditions. This subjective estimation serves to account for factors such as weather (e.g., storm cells that cause deviations) and rare-normal events (e.g., lost communications with a flight or a blocked frequency).

The metric used by TFM today is sector traffic count. Within each facility, the ETMS monitor-alert function advises TMCs when a sector's traffic count is predicted to exceed a nominal level (based on track and flight plan data). TMCs then use their judgment to plan and implement flow controls as needed. Enhancements to this process are needed in at least two areas – the quality of the metric and the selection of flow controls.

Although workload is sensitive to sector count (additional flights require additional monitoring and communication), there are other critical factors. Controller workload, including cognitive (e.g., problem solving) and physical tasking (e.g., communications, interacting with automation), is thought to correlate with the complexity of the airspace (Heimerman, 1998). Airspace complexity depends on the static characteristics of the airspace (e.g., sector size, shape, location, flight restrictions, and routes/corridors) and the 'dynamic density' (i.e., complexity) of the traffic within it.

Conceptually, dynamic density is a measure of traffic characteristics that correlate with workload. In addition to the traffic count, dynamic density strives to account for trajectory characteristics (e.g., phase of flight) and interactions between trajectories. Climbs, descents, and turns require more attention and interaction than straight and level flight. Trajectory interactions include merging and, more important, close-proximity encounters (i.e., predicted conflicts and near conflicts). Figure 12.10 illustrates the concept by comparing two cases of equal sector count (four aircraft). Each case shows the predicted trajectories of all aircraft 20–25 minutes into the future. Case A represents a relatively benign situation with all four flights exiting the sector, conflict free, in straight and level flight. Case B, on the other hand, represents a case involving a climb, descent, change in course, and two potential conflicts. Although both cases result in an equal traffic count, the situations differ significantly in complexity. Although real progress has been made (Heimerman, 1998; Laudeman *et al.*, 1998; Sridhar *et al.*, 1998), much research still remains to define and validate an operationally useful dynamic density metric.

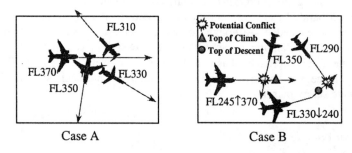

Figure 12.10 Sector complexity: Case A (lower); Case B (higher)

The second enhancement needed for TFM management of airspace complexity is the development of decision support tools for TFM flow control decisions. Simply put, there is a need for tools to advise TMCs on the traffic impact of flow-control strategy selections. A prototype tool, based on the sector count metric, was recently developed to evaluate the impact of TFM rerouting strategies (Carlson and Rhodes, 1998). Once validated for operational TFM use, such a tool may increase user flexibility by facilitating AOC-TFM collaboration on flight/route selection. At the very least, it is necessary to understand the dynamic impact of flow-control strategies on traffic complexity to provide maximum throughput and flexibility to airspace users, while preventing controller workload from exceeding acceptable levels.

Airspace complexity metrics will be critical to other Free Flight enabling applications. Two such applications include dynamic resectorization and management of free-maneuvering airspace. Dynamic resectorization complements flexible sector staffing. Instead of adding controllers to assist a sector, the sector boundaries are adapted to the traffic flow. The ability to predict and control airspace complexity may also lend itself to the determination of regions of airspace and time for which it is appropriate to delegate separation responsibility to the flight deck.

12.5 Conclusions

Although much progress has been made toward developing a strong and vibrant ATC system, much remains to be done to master its complex nature and harness its economic potential. Factors that contribute to its complexity include the distributed nature of the system, the range of user needs and capabilities, human factors, and the sheer scope and breadth of ATC operations. Efforts to modernize the NAS are further challenged by the diversity of its stakeholders and the need to continuously provide ATC services with minimal interruptions. Furthermore, due to its diversity and complexity, the ATM system does not lend itself to the classic 'optimum' operating point. Instead, airspace users desire an open and flexible environment whereby users may operate according to their real-time needs. Near-term enhancements will provide some measure of benefit by introducing automation to improve the efficiency of current procedures. However, the far-term 'stretch' goals will require the air transportation community to develop a far deeper understanding of NAS dynamics to realize the full potential of the longer term concepts. This level of complexity is sure to challenge engineers and human factors specialists well into the future.

References

Abbott, T. (1993) Functional Categories for Future Flight Deck Designs. NASA Technical Memorandum 109005, National Aeronautics and Space Administration.

Ashford, R. (1998) Technological developments in airspace optimization – a summary of NASA research. *Developing Strategies for Effective Airport Capacity Management Conference*, London, February.

Bang, E.S. (1995) 777 FANS. *Journal of Air Traffic Control*, Jan.-Mar.

Billings, C.E. (1996) Human-Centered Aircraft Automation: A Concept and Guidelines. NASA Technical Memorandum 103885, National Aeronautics and Space Administration.

Carlson, L. and Rhodes, L. (1998) Operational Concept for Traffic Management Collaborative Routing Coordination Tools. Technical Report MP 98W0000106, The MITRE Corporation, Center for Advanced Aviation System Development, McLean, VA.

Chew, R.G. (1997) Free Flight: preserving airline opportunity. *Proceedings of the RTCA General Symposium*, September.

Davis, T.J., Erzberger, H., Green, S.M. and Nedell, W. (1991) Design and evaluation of an air traffic control final approach spacing tool. *Journal of Guidance, Control and Dynamics*, **14**(4), July/August.

Davis, T.J. et al. (1997) Operational field test results of the Passive Final Approach Spacing Tool. *Proceedings of the 8th IFAC Symposium on Transport Systems*, Chania, Greece, June.

Erzberger, H. (1995) Design Principles and Algorithms for Automated Air Traffic Management. AGARD Lecture Series No. 200, AGARD-LS-200, November.

FAA (1993) *Airman's information manual/Federal Aviation Regulations*. Tab Books, Blue Ridge Summit, PA.

FAA (1997a) Conflict Probe Operational Concept, Narrative for Initial and Future Capabilities. ATO, Federal Aviation Administration, March.

FAA (1997b) FMS/ATM Next Generation (FANG) team document. *Airline Operational Control Overview*, Federal Aviation Administration, July.

FAA (1998) Air Traffic Control (Air Traffic Controller Handbook). FAA Order 7110.65L, Federal Aviation Administration, Washington, D.C.

FAA (1999a) NAS Architecture Version 4.0. Briefing to the Joint Resources Council, Federal Aviation Administration, ASD-100, January.

FAA (1999b) National Airspace System en route configuration management documents: MD-312 (Route Conversion and Posting), MD-313 (Flight Plan Position Processing and Beacon Code Assignment), MD-320 (Multiple Radar Data Processing), and MD-321 (Automatic Tracking). Federal Aviation Administration Technical Center, Atlantic City International Airport, NJ, June.

FAA/NASA (1999) Integrated Plan for Air Traffic Management Research and Technology Development, Version 3.0. FAA/NASA Interagency Air Traffic Management (ATM) Integrated Product Team (IPT), January.

Glass, B.J. (1997) Automated data exchange and fusion for airport surface traffic management. *Proceedings of the AIAA Guidance, Navigation, and Control Conference*, New Orleans, August.

Green, S.M. and Grace, M.P. (1999) Conflict-free planning for en route spacing: a concept for integrating conflict probe and miles-in-trail. *Proceedings of the AIAA Guidance, Navigation, and Control Conference*, Portland, OR, August.

Green, S.M., Goka, T. and Williams, D.H., (1997) Enabling user preferences through data exchange. *Proceedings of the AIAA Guidance, Navigation, and Control Conference*, New Orleans, August.

Green, S.M., Vivona, R.A., Grace, M.P. and Fang, T.C. (1998) Field evaluation of Descent Advisor Trajectory Prediction accuracy for en-route clearance advisories. *Proceedings of the AIAA Guidance, Navigation, and Control Conference*, Boston, MA, August.

Haraldsdottir, A. et al. (1997) Air Traffic Management Concept Baseline Definition. Technical Report RR-97-3, NEXTOR, University of California–Berkeley.

Heimerman, K.T. (1998) Dynamic density/air traffic control complexity. *Proceedings of the 1st FAA Dynamic Density/Air Traffic Control Complexity Technical Exchange Meeting*, McLean, VA, January.

Hinton, D., Charnock, J., Bagwell, D. and Grigsby, D. (1999) NASA aircraft vortex spacing system development status. *AIAA 37th Aerospace Sciences Meeting*, Reno, NV, January.

IEEE (1993) *IEEE Transactions on Control Systems Technology*, Special Issue on Air Traffic Control, **1**(3), September.

Jackson, J.W. and Green, S.M. (1998) Control applications and challenges in air traffic management. *Proceedings of the American Control Conference*, Philadelphia, PA, June.

Jeppeson Sanderson Inc. (1991) *Private pilot manual*. Jeppeson Sanderson Inc., Englewood, CO.

Kahne, S. and Frolow, I. (1996) Air Traffic Management: evolution with technology. *IEEE Control Systems Magazine*, **16**(4), August.

Laudeman, I.V. et al. (1998) Dynamic Density: An Air Traffic Management Metric. NASA Technical Memorandum 112226, National Aeronautics and Space Administration.

NASA (1999) Distributed Air Ground Traffic Management Concept Definition, Version 1.0. NASA Advanced Air Transportation Technology Project Office, Moffett Field, CA, September.

Nolan, M.S. (1990) *Fundamentals of air traffic control*. Wadsworth Publishing, Belmont, CA.

Perry, T.S. (1997) In search of the future of air traffic control. *IEEE Spectrum*, **34**(8), August.

RTCA (1995) Free Flight Implementation. RTCA Task Force 3, October.

RTCA (1997) A Joint Government/Industry Operational Concept for the Evolution of Free Flight. RTCA Select Committee on Free Flight Implementation, August.

Sridhar, B., Sheth, K.S. and Grabbe, S. (1998) Airspace complexity and its application in Air Traffic Management. *Proceedings of the 2nd USA/Europe Air Traffic Management R&D Seminar*, Orlando, FL, December.

Swenson, H.N. et al. (1997) Design and operational evaluation of the Traffic Management Advisor at the Fort Worth Air Route Traffic Control Center. *Proceedings of the 1st USA/Europe Air Traffic Management R&D Seminar*, Saclay, France, June.

The MITRE Corporation (1993) Full AERA Services, Operational Description. Technical Report MTR93-W0000061, The MITRE Corporation, McLean, VA.

The MITRE Corporation (1997) URET Delivery 3.0 System Level Requirements. Technical Report MWR-97W0000078R4, The MITRE Corporation, McLean, VA.

Williams, D.W., Arbuckle, P.D., Green, S.M. and den Braven, W. (1993) Profile negotiation: an air/ground automation integration concept for managing arrival traffic. *Proceedings of the AGARD 56th Symposium on Machine Intelligence in Air Traffic Management*, Berlin, May.

13

Complex Adaptive Systems: Concepts and Power Industry Applications

A. Martin Wildberger
Electric Power Research Institute

13.1 Introduction

To obtain the benefits of free market competition in energy production and distribution, there is a worldwide trend toward privatization, deregulation, and unbundling of the power industry, along with increased interconnection and increased trade over existing power lines. The North American power network may realistically be considered to be the largest machine in the world, since its transmission lines connect all the electric generation and distribution on the continent. Through this network, every user, producer, distributor, and broker of electricity on the continent has the potential to buy and sell, compete, and cooperate in an 'electric enterprise'. Every industry, every business, every store, and every home is a participant, active or passive, in this continental-scale conglomerate.

The real issue, not yet being faced in the United States (or in many other nations that are moving toward greater competition in electric power), is whether such an open, competitive market can be fair and profitable to all participants while continuing to guarantee to the ultimate consumer of power, at the best possible price, secure, reliable electric service of whatever quality that consumer requires. How to control a heterogeneous, widely dispersed, yet globally interconnected system is a serious technological problem in any case. It is even more complex and difficult to control it for optimal efficiency and maximum benefit to the ultimate consumers, while still allowing all its business components to compete fairly and freely.

The deregulated electric enterprise can be viewed as a complex adaptive system (CAS). Studies of such systems suggest that their behavior is best explained by treating their components as independent, intelligent 'agents' or 'actors' and that their management is best achieved by distributing control to all the agents, while providing them with appropriate goals, learning algorithms, and information.

The Electric Power Research Institute (EPRI) has been developing a generic model of a complete electric power grid based on multiple adaptive, intelligent agents. This experiment in CAS modeling is intended to demonstrate the feasibility of distributed control and to evaluate how optimal and how robust such a control system can be made; the broader objective is to assess whether this new technology can provide a solution to the challenging problem of controlling power grids in a deregulated, competitive environment.

13.2 The Electric Enterprise: Today and Tomorrow

Throughout most of the history of electric power, the institutions that furnished it have tended to be vertically integrated monopolies, each within its own geographic area. They have taken the form of government departments, quasi-government corporations or privately owned companies, subjected to detailed government regulation in exchange for their monopoly status. Selling or borrowing electric power among these entities has been carried out through bilateral agreements between two utilities (most often neighbors). Such agreements have been used both for economy and for emergency backup. The gradual growth of these agreements has had the effect that larger areas made up of many independent organizations have become physically connected for their own mutual support.

In recent years, some of the local monopolies have found it beneficial to be net buyers of power from less costly producers, and the latter have found this to be a profitable addition to their operations. For instance, it is typical in the western United States and Canada for surplus hydroelectric power to be transmitted south for air conditioning in the summer, whereas less expensive nuclear power is transmitted northward in the winter when the reservoirs are low or frozen and only night-time heating is needed in the south. These wide-area sales and the 'wheeling' of power through nonparticipant transmission systems are international in extent, especially in Europe and the Americas.

There is evidence of a worldwide drive to use these interconnections intentionally:

- to create competition and choice, with the hope of decreasing prices;
- to get governments out of operating, subsidizing, or setting the price of electric power; and
- to create market-oriented solutions to deliver increases in efficiency and reductions in prices.

Over the next few years, the electric enterprise will undergo dramatic transformation as its key participants – the traditional electric utilities – respond to deregulation, competition, tightening environmental/land-use restrictions, and other global trends. Although other, more populous, countries such as China and India have greater potential markets, the United States is presently the largest national market for electric power. Its electric utilities have been mostly privately owned, vertically integrated, and locally regulated. National regulations in areas of safety, pollution, and network reliability also constrain their operations to a degree, but local regulatory bodies, mostly at the state level, have set their prices and their return on investment and have controlled their investment decisions while protecting them from outside competition. That situation is now rapidly changing. State regulators are moving toward permitting and encouraging a competitive market in electric power.

13.2.1 Deregulation and competition

In 1978, the United States Federal Government began the movement toward deregulation by allowing competition in several strategic sectors of the economy, starting with the airlines and followed by railroads, trucking, shipping, telecommunications, natural gas, and banking. Adam Smith succinctly stated the philosophy behind this movement in 1776: 'Market competition is the only form of organization, which can afford a large measure of freedom to the individual. By pursuing his own interest, he frequently promotes that of society more effectively than when he really intends to promote it'. More recently, Prof. Alfred Kahn of Cornell University, who guided the airline deregulation as head of the Civil Aeronautics Board, expressed it in a different way: 'Deregulation is an admission that no one is smart enough to create systems that can substitute for markets'.

To unbundle the monopoly structure of electric power generation in the United States, Congress passed the National Energy Policy Act of 1992. National monopolies in the United Kingdom, Norway, and Sweden have been denationalized and unbundled into separate generation, transmission, and distribution/delivery companies. In most approaches to deregulation, transmission is kept as a centrally managed entity, but generation is split into multiple independent power producers (IPPs), and delivery is left to local option. New IPPs are encouraged or, at least, permitted, as are load aggregators and electric power brokers, both of whom own no equipment but are deal-makers who operate on commissions paid by the actual producers and users.

The concept behind this arrangement is that electricity, much like oil and natural gas, is a *commodity* that can be sold in the cash or spot market. As a commodity, it is possible to buy and sell future options and more complex derivative contracts based on electricity prices. The New York Mercantile Exchange (NYMEX) launched its Palo Verde and California-Oregon Border Electric Futures contracts on March 29, 1996, and options trading followed on April 26, 1996. NYMEX President R. Patrick Thompson stated: 'We have not seen this level of interest in a new product since we launched natural gas futures, which went on to become the fastest growing contract in Exchange history'. NYMEX had offered the first natural gas futures contracts in April 1990 and followed up in October 1992 by launching its options on natural gas futures.

It is not clear that electricity meets all the necessary criteria for commodity trading, however. The original assumptions of NYMEX and its traders were based on the model of natural gas, which, unlike electricity, can be stored economically. Once a unit of electricity is produced, it must be consumed almost immediately; in contrast, a true commodity can be stored for some length of time and consumed when and how desired. Electricity storage devices are capable of handling only a small percentage of an area's electricity requirements. Storage limitations and capacity constraints on interregional transfer prevent all available suppliers across the continent from head-to-head competition.

An alternative, and more entrepreneurial, view is that furnishing electricity is a *service* to the end user. Electric service may be segmented into more specific markets such as heating, cooling, lighting, building security, etc., or combined with other consumer services such as telephone, cable TV, and Internet connections. Both views may be reconcilable by separating the *product*, handled by generation and transmission companies, from the *service*, performed by distribution companies.

13.2.2 Re-regulation and new institutions

In its most basic form, an electric power system starts with a generator and ends at a house. In the United States, it is generally assumed that only the two ends, generation and retail sales, will be deregulated. Even distribution (the delivery of power to houses and businesses) is expected to remain a monopoly operated by the locally franchised utility and supervised by regulators. Despite the nominal 'deregulation', most plans for control of the electric power network call for some form of centralized administrative agency. This entity, often called an Independent System Operator (ISO), will have the responsibility and authority to ensure the security of the system and to administer or even set prices for all the ancillary services required for secure operation. California, for example, has established both an ISO and a separate Power Exchange. The transmission (the long-distance movement of electricity) is managed by this regulated ISO, whose role and responsibilities include:

1. To manage transmission access, including the dispatch of the combined transmission facilities of participating utilities, to minimize networkwide transmission congestion and constraints while ensuring that all users are subject to the same access protocols and prices.

2. To coordinate the day-ahead scheduling and real-time balancing for all users of the grid.

3. To comply with all North American Electric Reliability Council (NERC) and Western Systems Coordinating Council (WSCC) operating and reliability standards.

4. To provide non-discriminatory open access to the grid and ancillary services to all users.

5. To provide transparent information flow.

The California Power Exchange operates in a manner similar to a financial exchange. It opens the market to all sources of generation with openly posted prices for all customers to see. It has five major responsibilities:

1. To provide a competitive spot market for power with published hourly prices.

2. To allow power producers to compete using non-discriminatory and transparent rules for bidding into the exchange.

3. To take supply bids from generators and demand bids from utilities, retailers, and power marketers.

4. To rank bids and submit to the ISO a preferred least-cost dispatch schedule for delivery of power.

5. To ensure a visible market-clearing price that will allow customers to make efficient purchasing decisions and to adjust their consumption.

The NERC was formed in 1968 in the aftermath of the November 9, 1965, blackout that affected the northeastern United States and Ontario, Canada. NERC's mission is to promote the reliability of the electricity supply for North America. NERC has stepped into the new regulatory vacuum by defining and proposing standards for Interconnected Operations Services (IOS) (NERC, 1998).

The United States Federal Government is also becoming more active in this area. The Federal Energy Regulatory Commission (FERC) is an independent regulatory agency within the Department of Energy that, among other activities, regulates the transmission and wholesale sales of electricity in interstate commerce. It has issued rules intended to guarantee open transmission access to all parties wishing to 'wheel' electric power through any intermediate network between generation and load (FERC, 1996a). FERC's rules require each public utility to implement standards of conduct that separate transmission and wholesale power merchant functions. FERC is also attempting to establish a *post hoc* price resolution process for unplanned power exchanges that were required to maintain secure operations. Litigation in some of these areas has already begun and may persist.

Pursuant to Congress' National Energy Policy Act of 1992, FERC issued rules 888 and 889 (released April 24, 1996) establishing and governing an Open Access Same-time Information System (OASIS) and prescribing standards of conduct (FERC, 1996b). Under these rules, each public utility (or its agent) that owns, controls, or operates facilities used for the transmission of electric energy in interstate commerce will be required to participate in an OASIS intended to provide information and processes necessary for non-discriminatory access to electric transmission systems. The present version supports posting of available transfer capabilities, as well as the offering of transmission and ancillary services with their associated prices and terms. OASIS allows users to reserve capacity on the transmission system, purchase ancillary services, and resell transmission services to others. In March 1997, the Commission issued *Rehearing Order 889-A* requiring implementation in OASIS of online price negotiation for transmission and ancillary services, the unmasking of identities and prices associated with OASIS reservations, disclosure of discounts given, and other changes (FERC, 1998).

13.2.3 Control and automation issues

Centralized control of electric power within limited geographic regions has been practical because of the single ownership of all the facilities in each region. Even then, control has emphasized security rather than economy because regulators and the public preferred that emphasis. However, competition changes the emphasis to cost cutting and efficiency, while many of the new entrants into the electric enterprise are used to taking risks to gain market share. Furthermore, the control region is now continent-wide, and ownership of the facilities is diverse and subject to rapid change. Owners are unwilling to turn over control of their facilities to a central manager unless they can be compensated for their perceived losses compared to the more risky decisions they might make independently. They want to be free to make the best 'deals' they can with as few constraints as possible, and to change their behavior rapidly for the sake of even a temporary advantage. They are even willing to make mistakes in the process of learning how to operate their own businesses more efficiently and effectively. Still, reasonably secure and stable operation of the entire system is needed if any of the participants is to make a profit.

Fortunately, some recent developments and current research activities improve the outlook for economical, real-time control of electric power systems. The use of Flexible AC Transmission Systems (FACTS) and other high-power, active control devices throughout power delivery systems promises significant increases in the performance, efficiency, and capacity of existing networks, improving their economic operation and making them flexible enough to handle the pressures of deregulation and competition. These solid-state electronic devices, analogous to integrated circuits but operational at multi-megawatt power levels, can convert electricity to a wide range of voltages, numbers of phases, and frequencies with minimal electrical loss and component wear. They enable control and tuning of all power circuits for maximum performance and cost-effectiveness, promising enhanced energy efficiency for the U.S. economy and flexible value-added service offerings from electricity providers (EPRI, 1996). By replacing the slow mechanical switches now used to manage system operations, these controllers offer for the first time the potential to fine-tune transmission dynamically, so that power delivery can respond instantly to changes in demand without the burden of maintaining a large amount of 'spinning reserve' generation (Jalali et al., 1994). A fully coordinated, network-wide system of active controllers could provide capacity increases of 50% or more without a comparable security margin reduction and with improved quality in the power delivered to the customer. Potential savings associated with avoidance of new infrastructure construction are of the order of billions of dollars.

However, these active control devices are a two-edged sword. Increased use of high-speed electronic controllers like FACTS presents more opportunities for large disturbances to occur. The failure of any of these active control devices can shift the power system into an unstable condition in less than half a cycle. Real-time response by a human operator under such conditions is impossible (Wildberger, 1994a,b). These devices, although called 'controllers', are really only actuators or, at most, regulators. They require some combination of local intelligence and coordination with the rest of the network to gain the full benefit of their physical capabilities. Meanwhile, the large-capacity HVDC (High Voltage Direct Current) interconnections, which are also now practical and being used for long-distance power transmission, encourage dependence on geographically remote power sources, making the new control devices associated with them essential for secure and stable operation.

EPRI and the U.S. National Science Foundation (NSF) have been pursuing a joint exploratory research initiative intended to improve utility sensing capabilities along with other parts of the national infrastructure. (Another research initiative that EPRI is involved with that concerns the national infrastructure is discussed in Chapter 14.) This effort, and other ongoing work to combine artificial intelligence technology with innovative sensing methods such as fiber optics, micromechanical devices, and acoustical techniques, is expected to lead in a few years to improvements such as in-service self-calibration; direct local measurement of quantities such as power and imbalance, which must presently be derived centrally from multiple local measurements; and methods for local prediction of problems at the bulk system level, such as instability and voltage collapse. Required attributes for these advanced sensors include a span of observability sufficiently large to encompass local operations over all possible conditions, an ability to monitor and account for power quality and burdens introduced to the system by customers, and functions for monitoring the integrity of components and protecting them from excessive stress so as to allow operation closer to design performance. In another 5 to 10 years, a Wide Area Measurement System (WAMS) that includes an integrated suite of self-adjusting instruments is expected to be available for continuous measurement of all system parameters and monitoring of component status at a high enough sampling rate to be used for practical real-time control.

Ultimately, these sensors must be integrated with locally positioned intelligent controllers because the time between identification of a potential failure or problem, and its occurrence can be too short for effective intervention from a centralized control room. At least enough embedded intelligence is required to allow each combined sensor-controller to communicate only exceptions, based on a context derived from similar exception-based communications from the other intelligent sensor-controllers.

This new configuration of the electric enterprise fits the working definition of a complex adaptive system. It is:

- large and intricate,
- composed of many heterogeneous, interactive components,
- behaving according to each one's individual rules,
- continually adapting to each other's behavior,
- resulting in emergent, global behavior that also changes over time.

Studies of such systems suggest that their behavior is best explained by treating their components as independent, intelligent 'agents' or 'actors', and that their management is best achieved by distributing control to all the agents while providing them with appropriate goals, learning algorithms, and information.

13.2.4 *Distributed versus centralized control*

Distributed control of any relatively complex system has many advantages. This is especially true when the major components are geographically dispersed, as in a large telecommunications, transportation, or computer network. However, even in an individual manufacturing or processing plant, in a military weapon system, or for that matter, in a modern automobile, sensors and actuators are scattered throughout the system. Completely centralized control of the system, plant, or vehicle would require that all the sensor or instrument data be transmitted to a central location, that all decisions be made there, and that detailed commands be sent back to each actuator (see Figure 13.1). In any situation subject to rapid changes, completely centralized control requires a high data-rate, two-way communication, which is susceptible to disruption at the very time when it is most needed: for instance, when the weapon system is under external attack, or the plant or the vehicle is under internal stress from the demand for rapid maneuvering. It is almost always preferable to delegate to the local level as much of the control as practical.

The simplest kind of distributed control would combine remote sensors and actuators to form regulators and adjust their set points or biases with signals from a central location. Replacing the regulators with conventional feedback or PID controllers can improve their response to changing local conditions, but as these distributed controllers increase in number or the interaction among their functions intensifies, it becomes extremely difficult to supervise and coordinate their performance by adjusting their gains and reprogramming their switching points. As complexity increases and decision making approaches real time, the computational burden on the central computer builds until massively parallel processing is required. But even before this point is reached, there can be significant benefit in distributing both the computational load and the detailed decision making among the local sensors and actuators, which have both the data and the ability to act on it (see Figure 13.2).

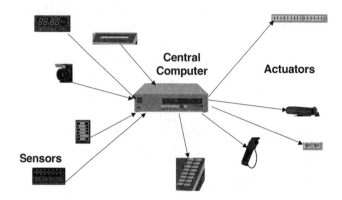

Figure 13.1 Centralized control system architecture

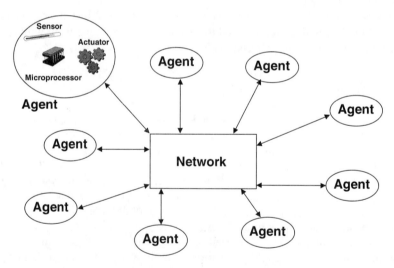

Figure 13.2 Distributed control by autonomous agents

Such an approach requires a different way of modeling – of thinking about, organizing, and designing – the control of a complex, distributed system. Recent research results from a variety of fields, including nonlinear dynamical systems, artificial intelligence, game theory, and software engineering, have led to a general theory of CASs. Mathematical and computational techniques originally developed and enhanced for the scientific study of CASs provide new tools for the engineering design of distributed control, so that both centralized decision making and the communication burden it creates can be minimized. The basic approach to analyzing a CAS is to model its components as independent adaptive 'agents' – partly cooperating and partly competing with each other in their local operations while pursuing global goals set by a minimal supervisory function.

From a computer programming point of view, agent-based modeling and simulation is a natural extension of the prevailing object-oriented paradigm. Agents are simply *active objects* that have been defined to simulate parts of the model. Discrete event simulations with multiple quasi-autonomous agents (usually called actors or demons) have been used for at least 25 years to assist human decision making in areas such as batch manufacturing, transportation, and logistics. The revolutionary new idea that comes from the computer experiments of CAS research is to let the agents evolve – changing to adapt to their environment as it is modified by external forces and by the evolutionary changes in the other agents. This research has found that, given the capability for merely rudimentary intelligent behavior, the agents will self-organize in a way that benefits the CAS, even as it benefits them individually.

To use this method for distributed control of a complex 'plant', the designer

1. creates a 'fitness function' for the agents that represents required goals and constraints for the plant being controlled, and
2. includes in the design of the agents themselves a sufficient variety of resources and strategies from which they can select combinations to use that improve their joint 'fitness' for controlling the plant, even as the uncontrollable parameters affecting the plant change.

The rest of this chapter has two goals:

1. To explain further the concept of a CAS and to show in more detail how it can provide a useful model for the design and implementation of distributed control systems.
2. As an example of this control concept, to describe its application to the distributed control of the electric power network and the evolution of the electric enterprise as it undergoes deregulation and competition.

13.3 Complex Adaptive Systems in General

For the last 10 years, a number of scientists have been studying various forms of complexity, both in the abstract and as it is found in nature (Gell-Mann, 1994). 'Complex' is understood here in the most general sense: that is, large and intricate. These researchers have been attempting to find common characteristics and/or formal distinctions among complex systems that might lead to a better understanding of how complexity occurs, whether it follows any general scientific laws of nature, and how it might be related to simplicity. At the same time, they have been attempting to better understand specific complex systems that are observed in the real world. Much of this work has been conducted by researchers associated with the Santa Fe Institute (Cowan *et al.*, 1994; Cladis and Palffy-Muhoray, 1995).

Most scientific understanding of nature has been gained by the reduction of complex phenomena at a higher level to simpler phenomena at a lower one. Scientific reductionism has been very successful in explaining, for instance, chemical phenomena at the molecular level by using the atomic model from physics. However, reductionism has not worked as well in the life and social sciences. Science has not been very successful in explaining or reproducing the characteristics and behavior of biological organisms – or life itself – from chemistry and physics alone. The kind of complexity exhibited by organisms and societies has not been able to be explained, nor their behavior predicted and controlled, by reduction to the characteristics of their component parts.

Seeking answers to these questions has required multidisciplinary research combining fields as diverse as biology, physics, computer science, and economics, and has developed over the last decade into CASs as a new field of study. This is because one characteristic that these kinds of complex systems all seem to have is adaptability. All these systems can be seen to change, evolve, or 'learn' as their environment changes and as they come into contact with other complex adaptive systems.

Scientists working in this field are trying to develop and test theories by working 'bottom up' rather than 'top down'. They start with the known behavior of component parts, hypothesize additional simple rules or relationships, and attempt to reproduce the complex phenomena they have observed in the real-world entities made up of these components. Top-down modeling requires all rules and relationships, qualitative or quantitative, internal and external, of the large system being studied to be specified explicitly. This enormous task is not possible in any practical case, so top-down models are often oversimplified to the point where they do not adequately reflect any actual situation. Bottom-up modeling with agents focuses on the behavior of less complex individuals that make up the larger system. These relatively simple agents can be modeled in detail. The autonomous interaction among them is the source of all the complex structure that would otherwise have to be programmed. In the most common approach to the mathematical modeling of complex systems, using sets of differential and algebraic-differential equations, the parameters approximate phenomenological attributes of the aggregate population that makes up the system under consideration. They do not identify specific internal mechanisms nor consider variation among individuals. Agent-based modeling simulates the underlying processes believed responsible for the global pattern and allows us to evaluate what mechanisms are most influential in producing that emergent pattern.

A CAS, real or simulated, exhibits a number of essential properties and mechanisms:

1. *Emergence of complex, global behavior from the aggregate interactions of many individual, relatively simple agents.* Biological examples such as beehives or ant colonies are the most obvious, but the behavior of a market also emerges from the interactions of its investors, a phenomenon Adam Smith called the 'invisible hand'. CASs exhibit varying degrees of randomness, deterministic chaos, nonergodicity, heterogeneity, non-stationarity, and/or extremely high dimensionality. This makes CASs difficult to predict, to control, or even to explain on the basis of their simpler components alone.

2. *Strategic learning and adaptation.* Adaptation may be defined as *the capacity for modification of strategies for goal-oriented, individual or collective behavior in response to changes in the environment.* Several aspects of adaptation are important for the study of CAS. First, a distinction should be made between learning as an experience-based response (for example, sorting through a set of predetermined patterns to find the optimal response in the set) and innovation, which requires the development of new patterns not previously known to that agent. It is also useful to distinguish between *active* and *passive* adaptation. A passive adaptive agent responds to changes in its environment without trying to modify that environment. An active adaptive agent exerts some control or

influence on its environment to improve its adaptive power. In effect, it conducts experiments and learns from them. Last, in modeling a CAS, it is necessary to distinguish between individual and population learning. Individual agents must exhibit enough plasticity to respond to environmental conditions and to other agents in a way that enhances their survival or meets other goals. To learn a strategy that increases its 'fitness', the agent has to gather and store enough information to adequately forecast and deal with changes that occur within a single generation. The population then adapts through the diversity of its individuals. This implies that some individuals always survive and their individual actions benefit the population goals, so that the population evolves over many generations, surviving as a recognizable organization. Individual trading rules and investment rules of thumb can be thought of as adaptive strategies in the capital markets.

3. *Nonlinearity and a potential for chaotic behavior.* In nonlinear systems, the aggregate behavior is more complicated than would be predicted by simply summing the behaviors of each component. Although chaos itself is rarely desirable, its ever-present possibility creates critical junctures in the evolution of the CAS that seem to be essential for the emergence of interesting global behaviors. In a CAS, large changes occur as the result of cumulative small stimuli. This phenomenon, called 'self-organized criticality', is a function of the dynamic interactions among the agents in the system (Bak, 1996). The difficulty in forecasting the behavior of capital and commodity markets may be an instance of this effect.

4. *Feedback, both negative and positive.* The multiplier effect in economics is one example of positive feedback. In capital and commodity markets, investors using security price changes as a buy/sell cue produce a momentum, which may be either positive or negative, but is self-reinforcing. Fabled market guru George Soros (1995) has developed a 'theory of reflexivity' based primarily on feedback effects. Feedback loops in CAS may operate at multiple levels. For instance, feedback of success or failure in operational strategies may lead to modification of goals to either seize opportunities or abandon infeasible objectives.

Figure 13.3 illustrates an abstract agent, showing some functionalities that are often included in conceptualizing agents.

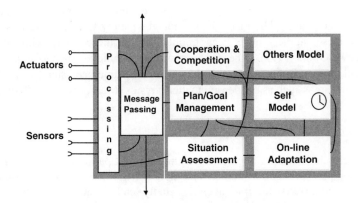

Figure 13.3 Abstract agent architecture

13.3.1 Computer experiments with CAS

Real-world experiments in this area are difficult even to imagine, much less carry out. However, the advent of powerful and inexpensive computers has provided a facility for carrying out computer experiments – simulations – that have given us some insight into the nature of CASs and some hints as to why reduction to their component parts is not enough to explain or control them. In these computer experiments, many simple components engage in local behavior, following a few simple rules and often affecting only their nearest neighbors. But global phenomena emerge that are only recognizable and meaningful in the context of the whole system. Flocking arises in a CAS model of birds, solitons arise in models of gas dynamics, and elasticity in models of polymers. Using computer experiments on a CAS to simulate biological phenomena has been called, somewhat extravagantly, 'artificial life' (Mitchell and Forrest, 1993; Wildberger, 1996).

The attractiveness of these methods for general-purpose modeling, design, and analysis lies in their ability to produce complex emergent phenomena out of a small set of relatively simple rules, constraints, and relationships couched in either quantitative or qualitative terms. Inventing the right set of local rules to achieve the desired global behavior is not always easy – although it often seems obvious afterward. For instance, flocking behavior requires only two basic rules: (1) stay close to the nearest bird and (2) avoid colliding (either with another bird or any obstacle).

Whether or not these computer experiments mirror the natural, the social, or the political world with any great fidelity, the techniques used in them have considerable usefulness for analysis and design of engineered systems, where precise real-world data is often more easily obtained. Using simple rules that govern the actions of individual vehicles ('travel at such-and-such a speed', 'don't hit anything'), Professor Kai Nagel of the University of Cologne has constructed a CAS model for freeway traffic (Nagel and Rasmussen, 1994). The deterministic version produces a high/low-density phase transition at the point of maximum throughput with self-organized criticality driven by the speed of the slowest car. Adding random noise causes spontaneous formation of traffic jams. Their distribution responds to a scaling law near the critical point, and expansion of the scaling region can be characterized by the percentage of drivers using 'cruise control' to reduce their fluctuations at high speed.

Business enterprises, financial markets, and the economy itself can all be viewed as complex adaptive systems, and they give rise to practical problems that are often mathematically intractable. The methods developed to study CASs, as well as the insights derived from these studies, have been applied to all these areas with some success (Friedman, 1991; Gintis, 1997). Other CAS simulation techniques such as spin glass models, sand piles, and random Boolean networks have for some time been standard tools in certain relatively narrow areas such as condensed matter physics (Margolus, 1995; Toffoli, 1995).

13.3.2 Agent-based modeling

The use of agents in modeling engineering systems is not particularly new. For at least 25 years, many computer simulations, much analytical software, and the tools for building such programs have employed multiple agents, usually called 'active objects', 'actors', or 'demons', and represented by separate processes that operate independently and interact by communicating selected information through messages. Until recently, these active objects and their interactions have been specifically modeled to represent relatively unchanging functions, operations, or physical entities. For most of these simulations, consistency and reproducibility were essential requirements. In general, artificial intelligence techniques were used only when necessary to model human decision making, and were couched in the form of policy rules to be followed, for instance, by processes simulating other manned aircraft or additional crew members in a flight crew training simulator.

From a programming point of view, this multiple-actor software was an early example, if not the origin, of the object-oriented programming paradigm. Agents are simply active objects that have been defined to model parts of a real-world system. Other objects in the software provide programming support, record-keeping, visual displays, and so on. All objects/agents have unique ID tags used to address messages to them, internal data storage for their state variables, and rules or functions that handle the messages and provide each object-agent's particular reactive and proactive behaviors.

Agent-based modeling starts with a population of idealized active objects. Parameter values are assigned to each, describing their starting behavior in the context of a computerized environment ('landscape') that includes the initial organizational pattern representing the inter-relationships among all the agents. These parameters depend on the particular system being modeled. They may include:

- Attributes, capabilities, or strategies, usually expressed as references or 'pointers' to algorithms;
- A range of allowable behaviors such as cooperation, competition, or conflict;
- Potential sources of strength and weakness, which may be physical, technological, intellectual, social, or political.

13.3.3 Adaptation in agents

In an adaptive agent-based simulation, the agent community is made to evolve by causing innovative changes in the parameters of individual agents to be generated randomly and/or systematically. These parameter changes, in turn, produce changes in the agents' actions and decisions, so that the agents 'tinker' with the rules and the structure of the system. Agents subjected to increased stress (such as equipment failures, resource shortages, environmental pressures, or financial losses) increase their level of tinkering until some develop strategies that relieve that stress. Some individual agents succeed (meet goals, grow, reproduce, increase their profits) while others fail (collapse, shrink, die, are replaced, bought out). In all CASs that involve humans as participants, and in many automated CASs designed by humans, two different time frames of adaptation must be simulated. In the short term, in 'real time', the agents, whether humans or automated 'softbots', adapt their behavior to the current situation as best they can, based on the strategies they have evolved over the long term. Over that long term, some of them evolve better strategies and solutions than others do. In the real world, more of the successful agents survive, and their successful strategies are imitated by many of the less successful that also survive. In the simulation, these effects can be modeled through evolutionary algorithms, which reproduce more of the successful agents without eliminating all of the less successful ones.

A variety of evolutionary computational methods have been employed to model adaptive agents in software. Most are variations on the genetic algorithms (GAs), originally invented by John Holland (1975) as search and classification techniques by an analogy to biological evolution and the breeding of animals for specialized purposes. The GA structure, based on individual membership in 'species', most easily lends itself to modeling individuals as unique instances of more general (evolving) classes. GAs provide a technique for evolving the global control of complex systems by modeling the local sensors and actuators as multiple, interacting agents, represented simply by bit strings of ones and zeros, that evolve and adapt through competition and cooperation to meet a combination of goals defined by the designer, called their 'fitness functions' by analogy with biological evolution. The bits encode that agent's possession or use of specific characteristics, capabilities, or algorithmic strategies. These can be selected and recombined to form different individuals by 'crossover' (cutting up two strings and patching them together) or by 'mutations' (random changes in bit settings). By comparing randomly selected individuals' relative success with respect to the fitness function and then creating a new 'generation' of individuals from crossover and mutation, biased toward using the most successful individuals, eminently successful individuals are eventually 'evolved'.

Genetic algorithms are examples of a very broad class of systems that can be described as 'adaptive nonlinear networks' (Holland, 1975). Some artificial neural network paradigms also fit into this class. All these systems consist of a large number of units that interact in a nonlinear, competitive manner and are modified by their own interaction or by external operators so that the overall system adapts to its environment, usually defined in terms of the desired output for any given input. The competitive nature of this process does not necessarily imply a zero-sum game in which individual units within the system only benefit at the expense of other units. Both competition and cooperation among units can increase the benefit to the system, and the whole system adapts through change, growth, or even the elimination of individual units.

Mathematically, genetic algorithms are similar to other forms of 'hill climbing' optimization using randomized gradients. However, they allow a convenient representation of a multidimensional problem space when different coordinates of that space are measured by continuous, discrete, and symbolic values. Genetic algorithms also provide for direct parallel implementation, whereas more conventional algorithms require additional effort to convert them from their original sequential form.

13.3.4   Other machine learning methods

Two similar computing approaches have developed in parallel with GAs. One, the 'evolutionary programming' technique of Lawrence J. Fogel (Fogel *et al.*, 1966), is based on evolution at the 'behavioral' (performance) level – where selection for fitness actually takes place – rather than at the genetic level of GA. In this method, evolutionary change is produced solely through random mutation of the most successful candidates; no explicit connection is made between 'behavioral' fitness and 'genetic' change. Despite this difference, evolutionary programming generally produces results similar to GAs. The second approach involves 'evolutionary strategies', a technique introduced by Ingo Rechenberg (1973) and Hans-Paul Schwefel (1981). This method focuses on evolution of machine learning through trial-and-error search procedures with automated reinforcement of successful strategies.

Although evolutionary programming, evolutionary strategies, and GAs developed separately and continue to represent areas of independent research, they are beginning to be viewed as different facets of evolutionary computing rather than as fundamentally different techniques.

Besides evolutionary computing, many other methods of machine learning may potentially be used for agent adaptation. In the simplest approach, the agent is merely rewarded (positively or negatively) by a fixed amount after each trial, based only on whether its performance improved or degraded. Using 'supervised learning', on the other hand, the behavioral algorithm in the agent is adjusted by a variable quantity based on the difference between its performance and the precisely known correct answer. A well known example is the training of artificial neural networks by backpropagation. The weights associated with each neuron are adjusted until the network has learned its training set of input/output data and generalized from those relationships to an abstract mapping that includes any new data statistically 'near' the examples it has experienced. Supervised learning is far more efficient than unsupervised learning, but it requires a large, accurate sample of real-world data.

A popular method of machine learning that represents a compromise between these extremes is called Q-learning (Watkins and Dayan, 1992). An automaton engaged in Q-learning attempts to estimate value functions for selecting its best strategy or control action in any given situation or 'state'. These values reflect its expected reward for taking each of the actions it considers. The rewards can be immediate and/or long term and may be discounted based on time horizon and/or likelihood of occurrence. These 'Q' values are usually stored in a table along with all the possible state/action pairs. The Q-learning algorithm is decentralized and asynchronous. It lends itself to the use of multiple adaptive agents, with each agent developing its own set of values. The agents need not start from the same set of possible states and actions. As they are learning, the agents may compete by hiding their values so that other agents cannot predict their most likely actions, or each may cooperate and benefit from knowing what others are learning.

Q-learning is a 'direct' or model-free method. It allows multiple individual agents to learn how to act optimally by experiencing the results of their actions, without requiring them to build explicit models of their domains. Instead, the Q-learner uses as a model the environment in which it finds itself, usually the real world. At the start, the learner knows essentially nothing about that environment, so some arrangement must be made to ensure that the problem domain is systematically explored, while at the same time successful actions are exploited. One approach is to use a Boltzmann selection process governed by a 'temperature' parameter. At high temperatures (i.e., before the agent has had much experience with the problem domain), each action is selected with approximately the same probability. As experience is gained, the learner 'cools' and action selection approaches the tactic of 'always select the observed best'. A separate temperature can be maintained for each state, so infrequently visited states will cool more slowly.

13.3.5 Complex adaptive systems as models for distributed control

The success of CAS researchers in using intelligent agent simulations with some form of self-modifiability to model natural or societal processes has led engineers to apply these same methods to the control of man-made systems whose complexity would otherwise lead to exponentially increasing computational demands. The chief difference from earlier computational models, using actors or demons, is that these intelligent agents change as time progresses. They evolve by adapting their behavior in both competitive and cooperative ways to meet general goals that are assigned by the designer but whose details may also change through the same 'evolutionary' process.

In the context of a distributed control system, each individual agent represents a local controller, comprising one or more sensors and/or actuators, a microprocessor with some memory, and two-way communications. This configuration could be created by adding some computational power to what would otherwise be a simple regulator or a PID controller with remotely adjustable bias, set point, and/or gain. Conceptually, the added computational power would come from the distribution of the processors and memory in a central computer made up of co-located parallel processors. For a system complex enough to require a powerful, parallel computer for real-time control, such a distributed configuration is only marginally more expensive and has the added potential for reducing the communication burden and providing greater resiliency to equipment failures.

The individual agents, represented by the software in each of the distributed processors, could, of course, be 'hard wired' as a complicated algorithm comprising what its designer believed to be all the possible eventualities that agent might have to handle. But this only increases the difficulty of the design process and leaves no margin for the unexpected, no possibility for improvement from experience under actual operation, and no way to adapt to aging or to temporary component failures in the plant. For any plant or system complex enough to warrant the distribution of computation along with control, it is computationally intractable to handle all possible combinations of capabilities, requirements, errors, and failure modes. It is better to design a control system that can adapt its own behavior reasonably and safely, if not always optimally. Given the time to act, experienced human operators do this remarkably well. Distributed intelligent control with adaptive agents can provide the same robust response to a fast-changing process in real time. It even has the potential for achieving robustness to human error and intentional sabotage, provided it, like the human operator, has had the opportunity to experience these errors and failures either in the real world or in a training simulation.

13.4 Simulator for Electrical Power Industry Agents

Based on CAS concepts, the Electric Power Research Institute has been developing a generic model of a complete electric power grid (including generation, transmission, distribution, and loads). This experiment in CAS modeling is intended to demonstrate the feasibility of distributed control and to evaluate how optimal and how robust such a control system can be made. Intelligent, agent-based, distributed control may be the only practical way to achieve true real-time dynamic control while getting full advantage from FACTS and other high-power electronic devices, making maximum use of the intelligent sensors and wide-area measurement systems now being developed and maintaining overall system security in a completely competitive economic environment. A team including researchers from EPRI, Honeywell Technology Center, and the University of Minnesota have taken the first steps toward realizing this CAS model of the grid in a computer-based simulation (Harp et al., 1999).

Development of the Simulator for Electrical Power Industry Agents (SEPIA) began by modeling relatively large components, such as whole power plants and sections of the grid at the level of abstraction where economic decisions actually take place. By structuring the simulation so that competition is ultimately limited only by the physics of electricity and the topology of the grid, SEPIA allows its users to test whether any central authority is required, or even desirable, and whether free economic cooperation and competition can, by itself, optimize the efficiency and security of network operation for the mutual benefit of all.

SEPIA is an object-oriented, fully integrated Windows application with plug-and-play agent architecture. Simulation time is advanced by events, and synchronous checkpointing is performed at regular intervals. SEPIA agents are autonomous modules that encapsulate specific domain behaviors. They are implemented as independent ActiveX applications, which communicate with each other by messages sent through the SEPIA agent bus and perform a simulation cycle when invoked by the simulation engine. All messages are delivered to mailboxes maintained by SEPIA. Multicasting operations are managed without message replication, and delayed messages add flexibility to the simulation. The user interface (see Figure 13.4), based on the familiar Windows graphical user interface (GUI), allows users to specify agents and agent relationships, permits agent modification, and provides mechanisms for simulation steering and monitoring. Icons represent generation plants (fossil, nuclear, and hydroelectric), generating companies, electric loads, and aggregated electricity consumers. Various attributes of agents, such as total production and quoted price for generating companies, can be viewed graphically as the simulation evolves.

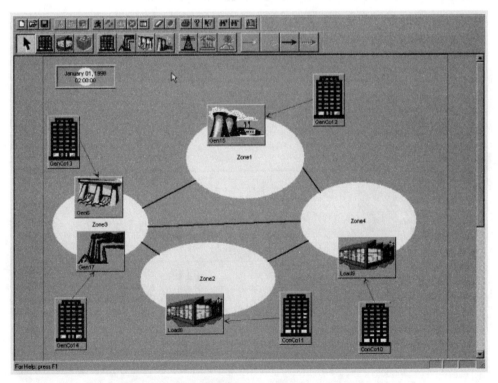

Figure 13.4 SEPIA interface screen

In Phase 1 of SEPIA, the agent model, the simulation engine, and the GUI have been implemented, as well as user-selectable adaptation strategies for the agents. Simulations exploring real-time pricing scenarios under open access transmission can be conducted with agents adapting, for example, pricing structures to achieve maximum profitability given demand profiles and generation and transmission constraints.

The first set of scenarios implemented in SEPIA model the wholesale world of open access electric utility operations. The agents involved in these scenarios include generation company agents ('GenCos'), generator agents, consumer company agents ('LoadCos'), load agents, and a transmission network operator agent. Each of the generation companies controls the set of generator agents it owns and each consumer company agent controls its set of load agents. The transmission network operator agent is an independent and impartial agent that controls the flow of electricity between the other agents. In these scenarios, consumer company agents purchase all the energy they need from generation company agents through direct 'bilateral' contracts. Periodically, each generation company determines the unmet hourly power needs of each of its loads for the next week, and broadcasts a 'request for quotes', or RFQ, to all generation company agents. Generation companies receive such broadcasts and determine whether to submit a quote for some or all of the power requested by the RFQ. Deciding whether to respond to an RFQ and determining the price to charge for the energy are difficult problems that are further complicated by the limits of the transmission network. Figure 13.5 shows the results of a SEPIA simulation, which can be plotted and analyzed with a built-in charting package.

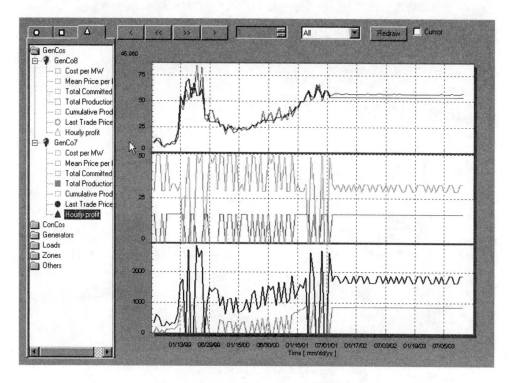

Figure 13.5 SEPIA visualization package

13.4.1 Agent adaptation

Agent adaptation in SEPIA means that the agent's online algorithms modify its internal state based on experience. Agent design determines when and how this occurs. All learned knowledge is stored in the internal states of agents, but it is also possible to have adaptation at multiple levels of organization (i.e., distributed over a population of agents or within a cohort of related agents as well as internal to a single agent). The current version of SEPIA offers two reusable adaptation algorithms:

1. Learning Classifier System (LCS) using rule representation, with discovery via a genetic algorithm and blackboard architecture for reinforcement learning. The LCS is implemented as a generic modular LCS C++ class template. The rules, conditions, and actions are separate classes of objects and may be reused for different kinds of agents, instantiated with different conditions and actions. The incremental genetic algorithm, triggered periodically, uses crossover, mutation, tournament selection, and a simplified bucket brigade algorithm.

2. Q-learning with tabular representation. Agents learn value functions to use for selecting the best control action in any given situation or state. The values reflect the immediate and (discounted) long-term expected reward for taking a given action. These 'Q' values are stored in a table that enumerates all possible state/action pairs.

13.4.2 Enhancements to SEPIA

Several enhancements to the first version of SEPIA are envisioned, emphasizing improvements to both physical and market realism. The physical realism can be enhanced with models of Flexible AC Transmission devices, superconducting cables, and storage. These extensions will allow users to evaluate potential technological investments. Improvements to market realism can include a futures market, options and various derivatives, exchange and bilateral contracts, and exogenous inputs. This will permit the development of (1) scenarios involving the revenue impact of load forecasting and (2) various control algorithms. Further enhancements can emphasize greater fidelity in modeling the implication for each transaction of the resulting power flow (stability, security, etc.) on the existing network. Parallel processing, agent template libraries, and more readily customizable agents will enhance performance and flexibility of the tool itself.

13.5 Conclusion

By modeling the newly deregulated and competitive electric enterprise as a CAS and testing this conceptual model through computer simulation, the first steps are being taken toward the design for distributed control of the power system of the 21st century by intelligent agents operating locally with minimal supervisory control. The design integrates computation, sensing, and control to meet the goals of efficiency and security in a geographically distributed system, subject to unavoidable natural disasters and operated by partly competing and partly cooperating business organizations.

This vision for a revolutionary power grid control system, based on complex adaptive system principles, has far-reaching implications for autonomous, safe, and optimized operation:

- Artificial-intelligence-based, self-calibrating, self-diagnostic sensors will measure all key network parameters and assess the failure potential of all components.
- Locally autonomous intelligent controllers will employ sensor data for modeling local conditions, communicate important data to each other, and automatically control local operations via FACTS or other active devices.
- The network of local controllers will act as a parallel, distributed computer, communicating, via microwaves, optical cables, or the power lines themselves, only that information necessary to achieve global optimization and facilitate recovery after failure.
- If a failure occurs within a local area, its intelligent controller will immediately respond to minimize impact on the overall network. If a failure causes system breakup, local controllers will guide isolated areas to operate independently, in a suboptimal manner, while preparing them to rejoin the network without creating unacceptable local conditions either during or after the transition.

The communication of information and decisions over large distances is a key requirement for this ambitious vision to be realized. A few years ago, no standard mechanism for this communication existed, and few if any practical applications of complex adaptive systems were available. Today, with the World Wide Web, this problem has been solved. As a result, there is renewed interest in CAS research, and in taking this research out of the laboratory (see, e.g., Joshi and Singh (1999); Ma (1999)). Exploration and analysis tools such as SEPIA represent the first wave of intelligent multiagent systems. In the future we can anticipate CAS-based systems that do not just share and manipulate information, but that control large-scale physical systems such as national and international power grids as well. Many issues, and not just technological ones, will need to be resolved before such a dramatic transformation of today's control system technology is feasible, but the issues no longer appear insuperable.

References

Bak, P. (1996). *How nature works*. Springer-Verlag, New York.

Cladis, P.E. and Palffy-Muhoray, P. (eds.) (1995) *Spatio-temporal patterns in nonequilibrium complex systems*. Addison-Wesley, Reading, MA.

Cowan, G.A., Pines, D. and Meltzer, D. (eds.) (1994) *Complexity: metaphors, models, and reality*. Addison-Wesley, Reading, MA.

EPRI (1996) High-Power Electronics: Advanced Technology Program. EPRI Brochure BR-106800, Electric Power Research Institute Dist. Ctr., Pleasant Hill, CA.

Friedman, D. (1991) Evolutionary games in economics. *Econometrica*, **59**(3) (May), 637–666.

Fogel, L.J., Owens, A.J. and Walsh, M.J. (1966) *Artificial intelligence through simulated evolution*. Wiley, New York

Gell-Mann, M. (1994). *The quark and the jaguar: adventures in the simple and the complex*. W.H. Freeman, New York.

Gintis, H. (1997) A Markov Model of Production, Trade, and Money: Theory and Artificial Life Simulation. Technical Report 97-01-006, Santa Fe Institute, Santa Fe, NM.

Harp, S.A. *et al.* (1999) Simulation of Complex Systems for the Power Industry with Adaptive Agents. Technical Report TR-112816, EPRI, Palo Alto, CA.

Holland, J.H. (1975) *Adaptation in natural and artificial systems.* University of Michigan Press, Ann Arbor, MI.

Jalali, S.G., Lasseter, R.H. and Dobson, I. (1994) Dynamic response of a thyristor controlled switched capacitor. *IEEE Transactions on Power Delivery,* **9**(3), 1609–1615.

Joshi, A. and Singh, M.P. (eds.) (1999) Multiagent systems on the net. *Communications of the ACM,* **42**(3).

Ma, M. (ed.) (1999) Agents in E-commerce. *Communications of the ACM,* **42**(3).

Margolus, N. (1995) Ultimate computers. *Proceedings of the Seventh SIAM Conference on Parallel Processing for Scientific Computing,* San Francisco, February.

Mitchell, M. and Forrest, S. (1993) Genetic Algorithms and Artificial Life. Technical Report SFI 93-11-072, Santa Fe Institute, Santa Fe, NM.

Nagel, K. and Rasmussen, S. (1994) Traffic at the Edge of Chaos. Technical Report SFI 94-06-032, Santa Fe Institute, Santa Fe, NM.

North American Electric Reliability Council (NERC) (1998) http://www.nerc.com.

Rechenberg, I. (1973) *Evolutionsstrategie: Optimierung technischer Systeme nach Prinzipien der biologischen Evolution.* Frommann-Holzboog, Stuttgart.

Schwefel, H.-P. (1981) *Numerical optimization of computer models.* Wiley, Chichester, UK.

Soros, G. (1995) *Soros on Soros: staying ahead of the curve.* Wiley, New York.

Toffoli, T. (1995) Fine-grained models and massively-parallel architectures: the case for programmable matter. *Proceedings of the Seventh SIAM Conference on Parallel Processing for Scientific Computing,* San Francisco, February.

FERC (1996a) Promoting Wholesale Competition Through Open Access Non-discriminatory Transmission Services by Public Utilities; Recovery of Stranded Costs by Public Utilities and Transmitting Utilities. Order No. 888, United States Federal Energy Regulatory Commission, April 24.

FERC (1996b) Open Access Same-Time Information System (formerly Real-Time Information Networks) and Standards of Conduct. Order No. 889, United States Federal Energy Regulatory Commission, April 24.

United States Federal Energy Regulatory Commission (FERC) (1998) http://www.ferc. fed.us/.

Watkins, C.J.C.H. and Dayan, P. (1992) Q-Learning. *Machine Learning,* **8**, 279–292.

Wildberger, A.M. (1994a) Stability and nonlinear dynamics in power systems. *EPRI Journal,* **19**(4), (June), 36–39.

Wildberger, A.M. (1994b) Automated management for future power networks: a long-term vision. *Public Utilities Fortnightly,* **132**(20) (Nov.), 38–41.

Wildberger, A.M. (1996) Introduction & overview of 'Artificial Life' – evolving intelligent agents for modeling & simulation. *Proceedings of Winter Simulation Conference (WSC'96),* San Diego, CA, Dec. 8–11, 161–168.

14

National Infrastructures as Complex Interactive Networks

Massoud Amin
Electric Power Research Institute

14.1 Introduction: Complex Interactive Networks[1]

The increasing complexity and interconnectedness of energy, telecommunications, transportation, and financial infrastructures pose new challenges for secure, reliable management and operation. No single entity has complete control of these multiscale, distributed, highly interactive networks, or has the ability to evaluate, monitor, and manage them in real time. In addition, the conventional mathematical methodologies that underpin today's modeling, simulation, and control paradigm are unable to handle their complexity and interconnectedness. Complex interactive networks are omnipresent and critical to economic and social well-being. Many national and international critical infrastructures are complex networked systems, including:

- Electric power grid
- Oil and gas pipelines
- Telecommunication and satellite systems
- Computer networks such as the Internet
- Transportation networks
- Banking and finance
- State and local services: water supply and emergency services.

---

[1] This chapter is based primarily on a series of presentations given by the author at several universities and professional events to provide a program framework for the EPRI/DoD's Complex Interactive Networks/Systems Initiative (CIN/SI). More information is available at: http://www.epri.com/targetST.asp?program=83.

Interactions between networks such as these increase the complexity of operations and control. The networks' interconnected nature makes them vulnerable to cascading failures with widespread consequences. Secure and reliable operation of these systems is fundamental to our economy, security, and quality of life, as was noted in 'Critical Foundations – Protecting America's Infrastructures', a report by the U.S. Presidential Commission on Critical Infrastructure Protection published in October 1997 (CIAO, 1997), and the subsequent Presidential Directive 63 on Critical Infrastructure Protection, issued on May 22, 1998.

Management of disturbances in all such networks, along with prevention of cascading effects throughout and between networks, requires a basic understanding of true system dynamics as well as effective distributed control so that, after a disturbance, parts of the networks will remain operational and even automatically reconfigure themselves. Interactive networked systems present unique challenges for robust control and reliable operation, such as:

- the multiscale, multicomponent, heterogeneous, and distributed nature of these large-scale interconnected systems;

- their vulnerability to attacks and local disturbances that can lead to widespread failure almost instantaneously;

- the many points of interaction among a variety of participants – owners, operators, sellers, buyers, customers, data and information providers, data and information users – that characterize these networks;

- the dramatic increase in the number of possible interactions as the number of participants grows, resulting in the complex activity of these networks greatly exceeding the ability of a single centralized entity to evaluate, monitor, and manage them in real time;

- the fact that they are too complex for conventional mathematical theories and control methodologies.

As an example, the U.S. electric power system developed over the last hundred years without a conscious awareness and analysis of the system-wide implications of its current evolution under the forces of deregulation and interaction with other infrastructures. The possibility of power delivery beyond neighboring areas was a distant secondary consideration.

From a broader view, infrastructure networks with several functional, operational, and management layers as well as many independent points of interaction between owners, operators, sellers, and buyers are considered complex because the number of possible interactions rises at a dramatically higher rate than the number of participants. Infrastructures that interact with their users and other networks (e.g., an automatic switching system for telephone calls) create additional complexity, because the interaction of their elements further increases the number of possible outcomes.

The various areas of interactive infrastructure networks present numerous theoretical and practical challenges in modeling, prediction, simulation, analysis, optimization, and control of coupled systems comprising a heterogeneous mixture of dynamic, interactive, and often nonlinear entities, unscheduled discontinuities, and numerous other significant effects. The science of complex adaptive systems (see Chapter 13) is considered particularly relevant to interactive infrastructure dynamics and security. In many complex networks, for instance, in the organization of a corporation, the human participants are both the most susceptible to failure and the most adaptable in the management of recovery. Modeling these networks, especially in the case of economic and financial market simulations, will require modeling the bounded rationality of actual human thinking, unlike that of a hypothetical 'expert' human, as in most applications of artificial intelligence. Furthermore, a pertinent question is, at what resolution should sensing, modeling, and control be started to achieve the overall objectives of efficiency, robustness, and reliability?

In this chapter, many of these challenges are presented; we first present a brief overview of an initiative in this area, followed by the motivation and rationale for considering modeling, simulation, and control methodologies for such large-scale network problems. To this end, an overview of four infrastructures and their interactive vulnerabilities that we consider pertinent is included. This is followed by a discussion of self-healing for infrastructure networks. A more detailed discussion of the objectives and program content for the Complex Interactive Networks/Systems Initiative (CIN/SI) is followed by the conclusion.

14.2 Complex Interactive Networks/Systems Initiative

In a joint initiative with the Deputy Under Secretary of Defense for Science and Technology, through the Army Research Office (ARO), EPRI is working to develop new tools and techniques that enable large national infrastructures to function in ways that are self-stabilizing, self-optimizing, and self-healing.

The Complex Interactive Networks/Systems Initiative is a 5-year, $30 million program of Government Industry Collaborative University Research (GICUR), funded equally by the U.S. Department of Defense (DoD) and EPRI. GICUR research focuses on breakthrough concepts to address major long-term challenges in complex interactive networks. Commonwealth Edison Co. and Tennessee Valley Authority are also participating directly in the program, providing staff expertise, data, and test sites.

CIN/SI was initiated in mid-1998, and work began in spring 1999. Major technical challenges are being addressed in modeling and simulation; measurement, sensing, and visualization; control systems; and operations and management. CIN/SI aims to develop modeling, simulation, analysis, and synthesis tools for robust, adaptive, and reconfigurable control of the electric power grid and infrastructures connected to it. Through this initiative, we are investigating new computation and control methods to enable critical infrastructures to adapt to a broad array of potential disturbances, including attacks, natural disasters, and inadvertent equipment failures.

There are clearly many opportunities for modeling, simulation, optimization, and control in this area. Mathematical models of such interactive networked systems are typically vague (or may not even exist); moreover, existing and classical methods of solution are either unavailable or are not sufficiently powerful. Management of disturbances in all such networks, and prevention of undesirable cascading effects throughout and between networks, requires a basic understanding of true system dynamics, rather than mere sequences of steady-state operations. Effective, intelligent, distributed control is required that would enable parts of the networks to remain operational and automatically reconfigure in the event of local failures or threats of failure. Detailed discussion of these issues as well as the CIN/SI program content is presented later in this chapter, after a discussion of the motivation for this effort, including the societal context, examples of interactive networks, and the nature of vulnerabilities in these large-scale systems.

14.3 Societal Context: Infrastructures and Population Pressure – a 'Trilemma' of Sustainability

These infrastructures, faced with increased density in today's urban population centers, are becoming more and more congested. Human population centers have grown dramatically in the past century, creating a 'trilemma' of sustainability issues: population, poverty, and pollution. In 1900 there were no cities with 10 million or more people. By 1950 London and New York had crossed this threshold; by 2020 there will be more than 30 of these 'megacities', and by 2050 there will be nearly 60 cities of this size. The stress these megacities place on infrastructures will be immense.

What steps can be taken, then, to deal effectively with this trilemma? There are technology megatrends that may enable us to tackle such problems while managing resources more efficiently and improving the quality of life. These trends include:

- *Information revolution*: the ability to disseminate knowledge instantaneously around the globe via the Internet; information technology giving rise to virtual communities and an expanded international economy;

- *Materials advances*: designer alloys, ceramics, polymers, nanotechnology, and biomimetics offering new capabilities (computer memory and speed, sensors, superconductivity, and superstrength);

- *The new genetics*: the Human Genome Project providing the information foundation for medical advances; agricultural biotechnology offering the potential for feeding the world's population using less land.

Such technologies may help us manage international economies more effectively and feed the world's population while using less land. However, application of these advances will require many new developments as well as rethinking the operation of national infrastructures. As an example, complex interactive networks can be viewed as multilayered, multiresolutional intertwined grids; some of these networks have physics (or first principles) superimposed on graphs, and each one of these infrastructures has many different levels. CIN/SI's goal is to develop methodologies, protocols, and controls that can self-heal and stabilize such systems.

14.4 Genesis: President's Critical Infrastructure Protection Report

CIN/SI complements a federal study of national infrastructures and their vulnerability to cyberattack. Key conclusions of the study include the following:

- Our most critical infrastructures are not in jeopardy, but the danger and number of threats to them are increasing.

- Now is a good time to develop ways to protect our infrastructures, since many industries are in a period of transition due to adjustment to the new Information Age, deregulation, or both.

- Protecting infrastructures should be a cooperative effort among government agencies, infrastructure owners and operators, and the research community.

The report by the Presidential Commission on Critical Infrastructure Protection was published on October 13, 1997. Subsequently, several studies and reports were sponsored by other agencies (e.g., Sandia National Laboratory, 1998); the Presidential Decision Directive 63 on Critical Infrastructure Protection was issued on May 22, 1998 (www.info-sec.com/ciao/ 63factsheet.html and www.info-sec.com/ciao/paper598.html).

14.5 Examples of Complex Interactive Networks

In what follows we provide four examples of interactive networks and then discuss their vulnerabilities, beginning with the electric power grid.

14.5.1 Example: The U.S. electric power system

The U.S. electric power system evolved in the first half of the 20th century without a clear awareness and analysis of the system-wide implications of its evolution. This continental-scale electric power grid is a multiscale, multilevel hybrid system (Fouad, 1992; Kundur, 1994; Machowski *et al.*, 1997). This infrastructure underlies every aspect of our economy and society. Possibly the largest machine in the world, its transmission lines connect all generation and distribution on the continent. This system of vertically integrated hierarchical networks consists of the generation layer and the following three basic levels (Figure 14.1):

- *Transmission level*: a meshed network combining extra high voltage (above 300 kV) and high voltage (100–300 kV) connected to large generation units and very large customers; tie lines to transmission networks and to the subtransmission level;

- *Subtransmission level*: a radial or weakly coupled network including some high voltage (100-300 kV but typically only 5–15 kV) connected to large customers and medium-sized generators; and

- *Distribution level*: typically a tree network including low voltage (110–115 or 220–240 V) and medium voltage (1–100 kV) connected to small generators, medium-sized customers, and local low-voltage networks for small customers.

In its adaptation to disturbances, a power system can be characterized as having multiple states, or 'modes', during which specific operational and control actions and reactions are taking place:

- *Normal Mode*: economic dispatch, load frequency control, maintenance, forecasting, etc.

- *Disturbance Mode*: faults, instability, load shedding, etc.

- *Restorative Mode*: rescheduling, resynchronization, load restoration, etc.

Some authors include an Alert Mode before the disturbance actually affects the system (Dy-Liacco, 1967). Others add a System Failure Mode before restoration is attempted (Fink and Carlsen, 1978). Besides these many operational, spatial, and energy levels, power systems are also multiscaled in the time domain, from nanoseconds to decades, as shown in Table 14-1.

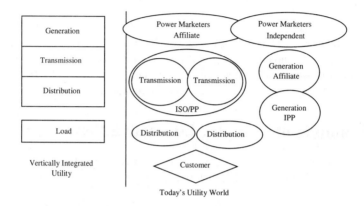

Figure 14.1 Vertically integrated hierarchical networks for traditional power systems and the structural changes now in progress

Table 14.1 Multiscale time hierarchy of power systems

Action/Operation	Time Frame
Wave effects (fast dynamics, lightning-caused overvoltages)	Microseconds to milliseconds
Switching overvoltages	Milliseconds
Fault protection	100 milliseconds or a few cycles
Electromagnetic effects in machine windings	Milliseconds to seconds
Stability	60 cycles or 1 second
Stability augmentation	Seconds
Electromechanical effects of oscillations in motors and generators	Milliseconds to minutes
Tie line load frequency control	1 to 10 seconds; ongoing
Economic load dispatch	10 seconds to 1 hour; ongoing
Thermodynamic changes from boiler control action (slow dynamics)	Seconds to hours
System structure monitoring (what is energized and what is not)	Steady state; ongoing
System state measurement and estimation	Steady state; ongoing
System security monitoring	Steady state; ongoing
Load management, load forecasting, generation scheduling	1 hour to 1 day or longer; ongoing
Maintenance scheduling	Months to 1 year; ongoing.
Expansion planning	Years; ongoing
Power plant site selection, design, construction, environmental impact, etc.	10 years or longer

Making adaptive self-healing practical in agent-based, distributed control of an electric power system will require the development, implementation, and widespread local installation of Intelligent Electronic Devices (IED) combining the functions of sensors, computers, telecommunication units, and actuators. Several current technological advances in power developments can provide these necessary capabilities when combined in an intelligent system design. Among them are:

- Flexible AC Transmission System (FACTS), high-voltage electronic controllers that increase the power-carrying capacity of transmission lines (already fielded by American Electric Power).

- Unified Power Flow Controller (UPFC), a third-generation FACTS device that uses solid-state electronics to direct the flow of power from one line to another to reduce overloads and improve reliability.

- Fault Current Limiters (FCLs) that absorb the shock of short circuits for a few cycles to provide adequate time for a breaker to trip.

- Wide Area Measurement System (WAMS) based on satellite communication and time stamping using GPS, which can detect and report angle swings and other transmission system changes (in limited use within the Western Area Power Administration).

- Several innovations in the areas of materials science and high-temperature superconductors, including the use of ceramic oxides instead of metals, oxide-power-in-tube (OPIT) wire technology, wide-bandgap semiconductors, and superconducting cables.

- Distributed resources such as small combustion turbines, solid-oxide fuel cells (SOFCs), photovoltaics, superconducting magnetic energy storage (SMES), transportable battery energy storage system (TBESS), etc.

- Information systems and online data processing tools such as: Open Access Same-time Information System (OASIS) and Transfer Capability Evaluation (TRACE). The latter software determines the total transfer capability for each transmission path posted on the OASIS network, while taking into account the thermal, voltage, and interface limits. (OASIS, Phase 1, is now in operation over the Internet.)

Increased use of electronic automation raises significant issues regarding the adequacy of operational security: (1) reduced personnel at remote sites makes them more vulnerable to hostile threats; (2) interconnection of automation and control systems with public data networks makes them accessible to individuals and organizations from any worldwide location using an inexpensive computer and a modem; and (3) use of networked electronic systems for metering, scheduling, trading or e-commerce imposes numerous financial risks implied by the use of this technology.

Furthermore, competition and deregulation have created multiple energy producers who must share the same regulated energy distribution network – such that this network now lacks the carrying capacity or safety margin to support anticipated demand (CEC, 1997; CA-ISO, 1998; CA-CTA, 1998). In the U.S. electrical grid, actual demand has increased some 35% while capacity has increased only 18% (Figure 14.2). The complex systems used to relieve bottlenecks and clear disturbances during periods of peak demand are now at greater risk to serious disruption – and technological improvements for these systems are needed.

The electric power grid's emerging issues include creating distributed management through using active-control high-voltage devices; developing new business strategies for a deregulated energy market; and ensuring system stability, reliability, robustness, and efficiency in a competitive marketplace. Power systems are a rich area for research and development of tools, especially now because of increased competition and the embracing of technology for competitive advantage, as well as the challenge of providing the reliability and quality consumers are looking for.

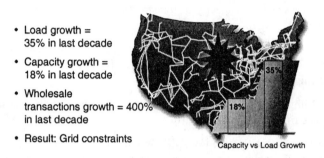

- Load growth = 35% in last decade
- Capacity growth = 18% in last decade
- Wholesale transactions growth = 400% in last decade
- Result: Grid constraints

Capacity vs Load Growth

Figure 14.2 Increased transactions and impact of deregulation on demands for access to the U.S. electricity grid

14.5.2 Example: Telecommunication networks

Emerging issues in telecommunication networks are similar to those found in the energy grid. Parallel algorithms are sought with distributed intelligence that can handle network disruptions and delays, and the telecommunication system's interconnection to the power grid is also the subject of investigations (Figure 14.3). These include determination of the true dynamics of this interaction as well as insights into what are the interactive vulnerabilities and when they potentially may lead to cascading failures. This will require developments in reliability theory, risk assessment, modeling, and mathematical analysis of the underpinnings involved in cascading, along with development of control and management tools for prevention and mitigation of such events. In addition, the following are desirable:

- parallel algorithms with distributed intelligence to handle network delays and disruptions;
- smart onboard orbit management of telecommunication satellites.

Telecommunication networks offer fixed or mobile services, including:

- *Fixed network services*: public switched telephone networks, public switched data networks;
- *Wireless network services*: cellular, PCS, wireless ATM;
- *Computer networks*: the Internet.

Currently in service are 172 nonmilitary satellites, 3625 transponders, and 81 spacecraft (on order). In the next five years there are expected to be 2251 new transponders with 15- to

25-year lifetimes. Present applications include telephone, video, Internet/data, military, GPS, and transponder. Current and future applications to electric power systems applications include (Heydt, 1999):

- surveying overhead transmission circuits and determination of rights of way;
- transmission of system data/SCADA systems (usually via telephone circuits);
- overhead conductor sag measurement;
- phasor measurement (a precise timing signal derived from GPS to time-tag measurements of AC signals);
- fitting of sine waves to signals and determination of magnitudes and phases of voltages and currents in remote locations;
- generation of real-time pictures of system states and real-time power flow;
- fast response: control data from low Earth orbit satellites can be more than 100 times less delayed than high Earth orbit satellites; parallel data stream facilities exist – effectively a high-speed global RS-232 channel.

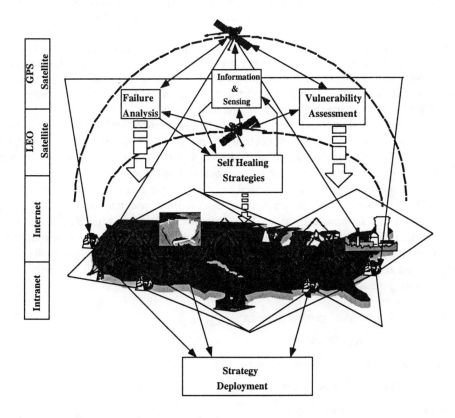

Figure 14.3 An example of complex interactive networks: telecommunication and the electric power grid (courtesy of Prof. C.-C. Liu, University of Washington)

14.5.3 Example: Banking and finance networks

In the realm of banking and finance networks, two primary concerns are banking and commodity transactions over electronic media. How do we perform data mining and knowledge discovery in multiple, mixed-structure financial databases? How do we create secure networks for real-time financial transactions? What new business strategies are needed for internal reorganization, external partnerships, and market penetration? And how can multiple intelligent agent modeling and simulation of corporate entities help?

14.5.4 Example: Transportation and distribution networks

From a broader historical perspective, rapid lines of transportation and communication constitute the foundation of all prospering societies; they are fundamental to national economies and quality of life. The evolution of the United States into a world power is closely related to the development of the railway system that opened up a vast continent with all its natural resources. Today, the backbone of the U.S. transportation system and economy is an interstate highway system that has continually evolved since the 1930s (Figure 14.4). This infrastructure, however, faced with increased density in today's urban population centers, is becoming more and more congested. The United States, along with many other nations, is seeking a solution to this worsening traffic congestion problem. Such solutions have to be viewed in terms of the economic, social, and political environments together with the technological capability of the nation. Furthermore, the costs associated with generating and maintaining the road infrastructure are rising; the impact of inefficiencies can be measured in quantifiable terms of loss of labor hours in the workplace and loss of fuel, as well as intangibly in terms of pollution and the generally increased stress level of a work force using these transportation channels.

Within the United States, the Department of Transportation (USDOT) estimates that the annual cost of congestion in lost productivity alone is more than $100 billion. In addition, more than 40,000 persons are killed and another five million injured each year in traffic accidents. Where feasible, increasing the number of lanes will expand the present system capacity, but such measures are recognized to be expensive and disruptive. The Intelligent Transportation Systems (ITS) program in the United States was launched as part of Public Law 102-240, the Intermodal Surface Transportation Efficiency Act (ISTEA), passed by Congress in 1991. The goals are to improve the safety, capacity, and operational efficiency of the surface transportation system while at the same time reducing the environmental footprint and energy impacts. Along these lines, the U.S. Department of Transportation has identified six generic ITS thrusts:

- Advanced Traffic Management Systems
- Commercial Vehicle Operations
- Advanced Traveler Information Systems
- Advanced Public Transportation Systems
- Advanced Vehicle Control Systems
- Advanced Rural Transportation Systems.

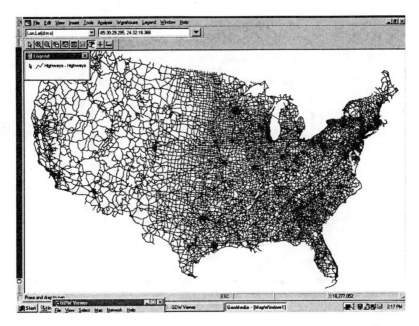

Figure 14.4 U.S. highway system (courtesy of Dr. Joseph Knickmeyer, MTMC)

We believe that the area of Advanced Traffic Management Systems (ATMS), because of its broad geographical coverage and direct impact to regional economies, perhaps offers the greatest analytical challenge and the most substantial payoff. In the case of ATMS, recent travel demand and capacity projections indicate that it will no longer be possible to meet the growing demand for travel through the addition of roads. Hence, the ATMS challenge is to improve travel capacity through management of the existing road infrastructure (Amin *et al.*, 1995; Garcia-Ortiz *et al.*, 1998; Amin, 1999).

ITS will rely on the consolidation of information technology combined with automotive and highway technology to achieve its objective. This will be accomplished by placing sensors and communication devices alongside the road as well as in the vehicle (Figure 14.5). These will allow the road to 'know' its operational status, which will then be 'communicated' to the vehicle, which will in turn help to make informed decisions about which routes to take and daily activity planning. The road data collected will also be used by public and private concerns to plan, implement, and manage their daily operations (i.e., traveler information, traffic management, public and rural transportation management, priority vehicle management, and freight and fleet management).

In the area of transportation and distribution networks (air, land, and sea), emerging issues include: electrification of transportation; links with sensors, telecommunications, and satellites; traffic modeling, prediction, and management; multiresolutional simulations; real-time optimization with provable performance bounds; and how to use the intersection of mathematics, systems and control science, economics, computer science, artificial intelligence, biology, and so on, to tackle these problems.

The realization of all of this will not come from one megasystem, but rather from the development of a wide range of small, complementary systems – from electronic route guidance to preemptive signal control and to automated highways.

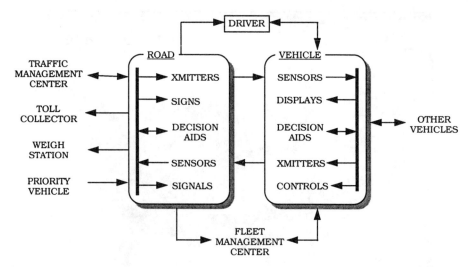

Figure 14.5 Sensors and communication devices allow the road and the vehicle to exchange data that is used to plan, implement, and manage daily activities

14.6 Vulnerabilities

Any of these infrastructures can be vulnerable to both deliberate and accidental disturbances. Their interconnected nature means that single, isolated disturbances can cascade through and between networks with potentially disastrous consequences. Large-scale failures (i.e., failures in areas geographically remote from the original problem) and failures in seemingly unrelated businesses can occur (see Figure 14.6). Since these networks support critical services and supply critical goods, disturbances can have serious economic, health, and security impacts. Therefore, there is a need to develop an ability for these infrastructures to self-heal and self-organize to mitigate the effects of such disturbances.

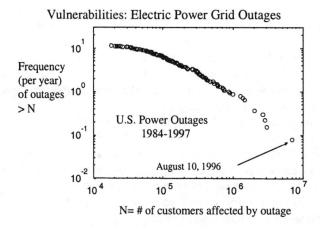

Figure 14.6 Major electric power outages in the U.S., affecting nearly 700,000 customers once a year and 7 million customers once every decade (data from NERC http://www.nerc.com/dawg; log-log plot courtesy of Prof. John Doyle, California Institute of Technology)

The occurrence of several cascading failures in the past has helped focus attention on the need to understand the complex phenomena associated with these interconnected systems. In November 1965, a blackout in the northeastern United States created a power system collapse in 10 states; a blackout in the tri-state area of Pennsylvania, New Jersey, and Maryland occurred in 1967; New York City experienced a blackout in July 1977; and a voltage collapse in France triggered a blackout in December 1978. More recently, the Western U.S. grid experienced outages in July and August of 1996 affecting 11 states and two Canadian provinces. Most recently, on December 8, 1998, a blackout in the San Francisco Bay Area cascaded from San Mateo and affected 2 million people for up to seven hours.

The failure in the northwest United States in August 1996 has several important effects and noteworthy lessons; it shows how cascading can lead to widespread network consequences:

- faults in Oregon at the Keeler-Allston 500 kV line and the Ross-Lexington 230 kV line resulted in excess load . . .

- which led to the tripping of generators at McNary Dam, causing 500 MW oscillations . . .

- which led to the separation of the North-South Pacific Inter-tie near the California-Oregon border . . .

- which led to islanding and blackouts in 11 U.S. states and two Canadian provinces.

The estimated total cost to the people affected reached $1.5 billion, and some experts believe that such cascade effects could have been prevented by shedding, or dropping, approximately 0.4% of total load on the network for 30 minutes. If we can gain such insights afterward, can one develop predictive capabilities that understand the true dynamics of the system and can predict *a priori* what kinds of problems may arise? If not, is there a way to at least mitigate and localize the effects in situ? (It is important to note that load shedding is not typically a first option.)

As an example of interdependencies between the markets and the electric grid, in summer 1998 price spikes showed the energy infrastructure's inadequacy affecting markets (cf. June 22–26, 1998, price spikes, http://www.ferc.fed.us/electric/mastback.pdf). As an example of interdependencies between the electric grid and telecommunication networks, an industry-wide Y2K readiness program identified telecommunications failure as the biggest source of risk of the lights going out on the rollover to 2000.

Another well known failure occurred in May 1998. This failure, dubbed 'The Day the Beepers Died' by *Newsweek*, was triggered when the *Galaxy-IV* satellite was disabled. Because the *Galaxy-IV* was equipped with the latest technology, causing almost all paging companies to use it, it became a critical node in the graph. When it failed, 40 million pagers were affected, National Public Radio went off the air, airline flights were delayed, and data networks had to be manually switched to older satellites. PageNet, the largest pager provider in the United States, has about 3000 satellite dishes. When *Galaxy-IV* was disabled, each of those 3000 dishes had to be manually realigned – a task that required one to two hours per satellite dish.

As another example of network vulnerabilities and interdependencies, computer simulations have indicated that the mobile phone satellites could lead to their own downfall – a large part of the network of satellites could self-destruct if just one of the satellites is hit by space debris (Rossi et al., 1999). Most space debris gradually drifts into the atmosphere and burns up on reentry. According to the researchers who undertook this study, there is a 10% chance that one of the Iridium satellites will be destroyed by debris within a decade, but the probability will increase to 10% within five years if one of the satellites is destroyed. If such a chain reaction starts, it could make the entire low Earth orbit unsuitable for satellites within 100 years – five times faster than current estimates. Space agencies have become increasingly concerned about space debris after both the Russian *Mir* Space Station and the United States Space Shuttle were damaged by flecks of paint traveling at more than 13 km/sec. In 1998, a French spy satellite became the first satellite to be completely knocked out of operation by space debris. There are already 10 million pieces of debris larger than 1 mm surrounding the Earth. The danger of collisions is particularly acute for satellites orbiting at an altitude between 800 and 1400 km. Mobile phone companies are expected to launch hundreds of small satellites into this orbit in the next few years.

14.7 R&D Objectives of CIN/SI

Avoiding failures in our critical infrastructures is a challenge because of their large-scale, nonlinear, and time-dependent behavior; furthermore, mathematical models of such systems are typically vague or nonexistent. In many cases, there are no methodologies for understanding the behavior of these complex systems. Control of these systems is a major challenge, since the number of components and types of possible interactions surpass what a central system could hope to manage.

CIN/SI aims to develop tools/techniques that enable large-scale and interconnected national infrastructures to self-stabilize, self-optimize, and self-heal. To achieve this goal, the following objectives were defined:

- *Modeling – understanding the 'true' dynamics*: to develop techniques and simulation tools that help build a basic understanding of the dynamics of complex infrastructures.
- *Measurement – knowing what is or will be happening*: to develop measurement techniques for visualizing and analyzing large-scale emergent behavior in complex infrastructures.
- *Management – deciding what to do*: to develop distributed systems of management and control to keep infrastructures robust and operational.

14.7.1 Objectives: Modeling

The modeling research objective relates to qualitative and quantitative models of complex interactive systems, including:

- Formal methods for modeling of true dynamics and for real-time computation to cope with system uncertainties; computation of provable performance bounds;
- Multiresolutional simulations, with the ability to go from the macro to the micro level, and *vice versa*;

- 'Artificial life' (cellular automata and multiagent models) for modeling and solving otherwise intractable problems in networked systems;
- Optimization and control theory along with decision analysis to model hybrid (mixed discrete/continuous) systems;
- Techniques for online mathematical modeling and decision support with partial inputs and in the presence of errors.

Currently, no mathematical models exist that can create useful top-down models for these systems – that is, models that start from large-scale graphs, systematically map them into decoupled subsystems and investigate the interactions between them. As there are so many components and potential interactions, deriving all-encompassing rules for complex infrastructures is impractical. Therefore, top-down models offer some insight, but cannot adequately reflect real-world situations for complex infrastructures. Traditional top-down models use algebraic/differential equations to simulate aggregate populations within a complex system. Specific internal mechanisms such as adaptation and learning are ignored, as is variation that might exist among individuals.

An alternative would be developing a bottom-up model (e.g., Wildberger *et al.*, 1999), for example, for power industry applications. By concentrating on smaller parts of the system, deriving rules becomes more practical. Bottom-up models use agents to simulate the processes responsible for global patterns within the system, and agent-based models let us evaluate the local mechanisms that produce emergent patterns. Real-world complexity is modeled by letting the individual 'agents' interact independently – which can provide a better understanding of the local mechanisms that produce emergent behavior, as opposed to the centralized control inherent in top-down models.

Through agent-based models, we can use simulation to better understand the large-scale, nonlinear behavior of our complex infrastructures in the hope of better amelioration of disturbances and prevention of disastrous cascading effects. These complex interactive networks have also had parallels and applications to natural systems:

- The behavior of social insects (bees, ants)
- Ecologies (predator-prey relationships)
- Cellular interaction (immune and nervous systems).

The overall behavior of such systems emerges through the simpler independent behavior of many individual components – a phenomenon known as self-organized complexity. Economic and political movements, albeit quite different than engineered systems, can also exhibit these characteristics.

For engineered systems, such simulations can be used as a tool for helping build agent-like characteristics into infrastructure components so they can actively respond to their real-world environment automatically, independently, and cooperatively with other components, and for developing complex infrastructures that are self-optimizing and self-healing through distributed management and control.

14.7.2 *Objectives: Measurement*

Another objective is the development of analytical and computational tools for measuring large-scale complex networks, including:

- real-time survey and status monitoring of systems;
- real-time processing of large data sets; pattern extraction (data mining and cluster analysis);
- techniques for correlating information from separate data sources/sensors;
- intelligent sensors and actuators;
- tools and techniques for system verification and validation;
- adaptive strategies that help components discern their interactions with the environment;
- methods for providing feedback about key environmental variables; ways to generate appropriate commands using local computational devices.

14.7.3 *Objectives: Management*

Finally, a comprehensive framework for distributed network management is also sought, including:

- real-time system state analysis;
- open architectures, intelligent devices, and distributed multilevel controllers;
- methods for reasoning, planning, negotiation, and optimization; methods for rule generation and modification;
- automatic verification of real-time, adaptive systems using formal proofs from specifications;
- task coordination of multiple intelligent agents (both artificial and human) in uncertain dynamic systems;
- tools for automated negotiation and risk management among self-interested agents (e.g., game theory with computational and resource bounds);
- algorithms for 'optimal' performance by independent agents with independent objectives;
- overall control techniques in environments where intelligent response devices may be acting against each other;
- methods for accommodating structural uncertainty and limiting impacts of system disturbances;
- methods for predicting impending failures: root-cause modeling for real-time diagnosis; early warning and failure forecasting;
- methods for recovering from emergencies.

14.8 Distributed, Multilevel Control

Although past studies in agent-based models of the electric power industry emphasize the solution of economic questions, they are intended as a major step toward multilevel, distributed control of the grid by the agents themselves (Wildberger, 1997; Wildberger *et al.*, 1999; Amin, 1998; Amin *et al.*, 1998). As these simulations become more detailed and physically realistic, intelligent agents will represent all the individual components of the grid. Advanced sensors, actuators, and microprocessors associated with the generators, transformers, buses, etc., will convert those grid components into intelligent robots, fixed in location, that both cooperate to ensure successful overall operation and act independently to ensure adequate individual performance. These agents will evolve, gradually adapting to their changing environment and improving their performance even as conditions change. For instance, a single bus will strive to stay within its voltage and power flow limits, while still operating in the context of the voltages and flows imposed on it by the overall goals of the power system management and by the actions of other agents representing generators, loads, transformers, etc. All lines, and most other components, have safety and capacity restrictions that are relatively 'soft' since they are based chiefly on the estimated life reduction resulting from operation outside their specifications. High and low voltage limits, for instance, may not be exceeded by specified percentages for more than specified time periods. Maximum thermal limits, expressed in megavolt-amperes (MVA), are also set in percentage time, but ultimately, most components would fail catastrophically and, for instance, overhead lines would sag until they caused a short circuit to the Earth.

More complex components, such as a generating plant or a substation, will be analyzed using object-oriented methods to model them as class and object hierarchies of simpler components, thus creating a hierarchy of adaptive agents (EPRI, 1996; Wildberger, 1997; Amin, 1998; Amin *et al.*, 1998). These agents and subagents, represented as autonomous 'active objects', can be made to evolve by using a combination of genetic algorithms and genetic programming. In this context, classes are treated analogously to biological genotypes and objects instantiated from them analogously to their phenotypes. When instantiating objects to form individual agents, their class attributes, which define all the potential characteristics, capabilities, limitations or strategies that these agents might possess, can be selected and recombined by the operations typical of genetic algorithms, such as crossover and mutation. The physics specific to each component will determine the allowable strategies and behaviors of the object-agent representing that component. Existing instrumentation and control capabilities can be augmented and computer experiments run with hypothetical, optional capabilities to evaluate their benefits and the actual ways in which their use might evolve. The operational parts of each class, its services or methods, may also be evolved through genetic programming using similar techniques. Their evolution will be governed by a 'fitness' function embodying a combination of ensuring their own survival and meeting the global security and efficiency goals.

Modeling the electric power industry in a control theory context is especially pertinent since the current movement toward deregulation and competition will ultimately be limited only by the physics of electricity and the topology of the grid. The Complex Adaptive System (CAS) simulation will test whether any central authority is required, or even desirable, and whether free economic cooperation and competition can, by itself, optimize the efficiency and security of network operation for the mutual benefit of all. The topic of complex adaptive systems is discussed in detail in Chapter 13, with specific reference to the networked electric enterprise.

A CAS model is also appropriate for any industry made up of many geographically dispersed components that can exhibit rapid global change as a result of local actions. This is characteristic of other industries that make up a national or international 'infrastructure', including telecommunications, transportation, gas, water and oil pipelines, and even the collection of satellites in Earth orbit. The whole combined infrastructure is itself a CAS, consisting of many individual, and often autonomous, components.

All the infrastructure industries in the United States, and to a significant extent in many other nations, have multiple ownership and management, and they operate with only a minimum of regulation by the national government. The international infrastructure is operated by a combination of national governments, national corporate entities, and increasingly, by international corporations. It is regulated only by treaties whose enforcement depends heavily on the goodwill and cooperation of all parties. *A useful approach to analyzing these national and international infrastructures is to model their components as independent adaptive 'agents'* – partly cooperating and partly competing with each other in their local operations while pursuing global goals set by a minimal supervisory function. Computer simulations can be used to test the feasibility of distributed management and control of these infrastructure industries, their viability under stress, and their potential for self-healing after disturbances.

This approach is not applicable only to infrastructure industries, however. The CAS concept has the potential to provide a new paradigm for the design, control, operation, and maintenance of any complex, interconnected system, especially one that possesses significant technological content.

The CAS paradigm is also useful for the design and analysis of interactive systems in which humans directly participate. In a simulation of the system, adaptive agents may be used to model the humans as well as those parts that are completely automated. Computer experiments can suggest the appropriate tasks and specific goals for both the humans and the robotic parts of the system, as well as the information required by the humans and its most efficient sources. This approach may even be applied to the actual operation of interactive systems or of organizations composed mainly of humans. A recent article (Cebrowski and Garstka, 1998) suggests that a CAS model of military operations may lead to a revolution in military affairs not seen since the Napoleonic era, since only an approach based on bottom-up organization, continuous play, co-evolution and self-synchronization can produce the speed of command required in a complex, fast-changing, and information-rich environment.

14.9 CIN/SI Program Content

The EPRI/DoD CIN/SI goal is to develop:

- methodologies for robust distributed control of heterogeneous, dynamic, widely dispersed, yet interconnected systems;
- techniques for exploring interactive networked systems at the micro and macro levels;
- tools to prevent/ameliorate cascading effects through and between networks;
- tools/techniques that enable large-scale and interconnected infrastructures to self-stabilize, self-optimize, and self-heal.

The plan is to develop many of these new capabilities in the next decade to effectively model, analyze, and operate geographically dispersed but operationally interconnected industries, including energy, telecommunications, transportation and distribution, and banking and finance. All of these are themselves complex nonlinear networks, interacting both among themselves and with their human owners, operators, and users. As the complexity of these intertwined operations has increased, they have become responsible for much of 'the good life' that we, at least in the more developed countries, lead today. However, with those increasing benefits come increasing risks. A common characteristic of all these networks is that local actions have the potential to create global effects by cascading throughout the network and even into other networks. The challenge of developing self-regulating systems includes these areas:

- Measurement, sensing, and visualization
- Modeling and simulation
- Control systems
- Operations and management.

The EPRI/DoD CIN/SI has funded six consortia, consisting of 28 universities, to address these challenges. Funded research includes:

1. Power Laws and the Power Grid: a Mathematical Foundation for Complex Interactive Networks (Consortium: Caltech, MIT, UCLA, UC–Santa Barbara, University of Illinois). Technical areas include: Basic theory of complex interactive systems with application to power and communication networks. Short-term objectives:

    — Power laws and the power grid – analysis of optimization-induced power-law behavior in idealized networks (thousands of nodes);

    — Development of nonlinear model-reduction capability to study dynamic models of the power grid of moderate scale (up to a hundred generators).

2. Context-Dependent Network Agents (Consortium: Carnegie Mellon University, RPI, Texas A&M University, University of Minnesota, University of Illinois). Technical areas include: development of context-dependent network (CDN) agents; agent templates and components and development of models for restructured power system models. Short-term objectives:

    — Power system modeling framework that maps global conditions onto a distributed decision-making architecture;

    — Real-time infrastructure for the agent modules and robust operating system for data communication: Development and demonstration of a context-dependent agent module architecture and preliminary agent coordination rules on two test scenarios.

3. Minimizing Failures While Maintaining Efficiency of Complex Interactive Networked Systems (Consortium: Cornell University, George Washington University, University of California–Berkeley, University of Illinois, Washington State University, University of

Wisconsin). Technical areas include: implications of a wide distribution of independent entities actively interacting through and with complex interactive infrastructure networks. Potential for local actions to create effects that cascade throughout the network and/or into other networks. Stochastic analysis and reliable performance of complex networks. Short-term objectives:

— Definition and formulation of the layered networks and their attributes. Coordinated approach to the several communication systems needed for the emergent power system.

4. Modeling and Diagnosis Methods for Large-Scale Complex Networks (Consortium: Harvard University, Boston University, MIT, University of Massachusetts–Amherst, Washington University–St. Louis). Technical areas include: discrete event dynamical systems (DEDS); simulation modeling of complex networks; modeling and diagnosis methods for large-scale networks. Short-term objectives:

— Development of advanced simulation tools (perturbation analysis and ordinal optimization) to improve robust stability and quick detection of faults in complex systems.

5. Intelligent Management of the Electric Power Grid: Anticipatory Multi-Agent, High Performance Computing Approach (Consortium: Purdue University, University of Tennessee, Fisk University, Commonwealth Edison Co., Tennessee Valley Authority). Technical areas include: energy infrastructure management and anticipatory multiagent, high-performance computing. Short-term objectives:

— Modeling of part of the grid (ComEd/TVA) for use in the development and testing of an anticipatory approach.

6. Innovative Technology for Defense Against Catastrophic Failures of Complex, Interactive Power Networks (Consortium: University of Washington, Arizona State University, Iowa State University, Virginia Tech). Technical areas include: self-healing strategies for competitive environments; interactive power networks; Wide-Area Measurement Systems (WAMS) for infrastructures. Short-term objectives:

— Conceptual design of the Strategic Power Infrastructure Defense (SPID) system: Model of the electricity markets; control and protection strategies against catastrophic failures; sensing, information, computing, and control strategies.

During 1999, these consortia achieved significant progress, as well as a number of key technical milestones. An EPRI report (EPRI, 2000), 'Complex Interactive Networks/Systems Initiative: First Annual Report', describes the initiative and its progress as of January 2000. Six additional annual reports (EPRI, 2000, TP-114661 through TP-114666) provide more details on this program.

Areas of research being investigated by each consortium are depicted in Figure 14.7 as a canvas of research and development.

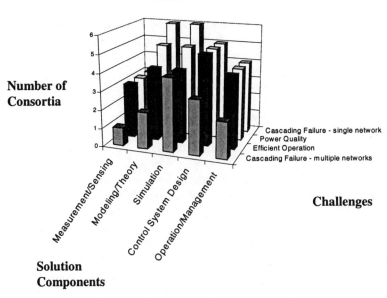

Relative level of effort vs. challenges and solution components

	Measurement and Sensing (including Visualization)	Modeling/ Theory	Simulation	Control Systems Design	Operation and Management
Efficient Operation	Harvard Purdue UWA	Caltech Cornell Harvard Purdue	CMU Harvard Purdue UWA	Caltech CMU Cornell Harvard UWA	CMU Harvard Purdue UWA
Power Quality	Harvard Purdue UWA	Caltech CMU Cornell Harvard Purdue	Caltech CMU Harvard Purdue UWA	Caltech CMU Cornell Harvard UWA	CMU Harvard Purdue UWA
Cascading Failure (single network)	Harvard Purdue UWA	Caltech CMU Cornell Harvard Purdue UWA	Caltech CMU Cornell Harvard Purdue UWA	Caltech CMU Cornell Harvard UWA	CMU Cornell Purdue UWA
Cascading Failure (multiple networks)	Harvard	Caltech Cornell	Caltech Cornell Harvard UWA	Cornell Harvard UWA	Cornell UWA
	Measurement and Sensing (including Visualization)	Modeling/ Theory	Simulation	Control Systems Design	Operation and Management
	Solution Component				

Challenges of Reliable and Robust Operation vs. Solution Components.
Consortia Lead Universities: Caltech; Carnegie Mellon University; Cornell University; Harvard University; Purdue University; University of Washington

Figure 14.7 A canvas of research and development for reliable and robust operation vs. solution components

14.10 Conclusion

Many national critical infrastructures are complex interdependent networked systems; prime examples are the highly interconnected and interactive industries, which make up a national or international infrastructure, including telecommunications, transportation, gas, water and oil pipelines, the electric power grid, and the collection of satellites in Earth orbit.

Interactions between networks such as these increase the complexity of operations and control. Secure and reliable operation of these systems is fundamental to our economy, security, and quality of life. These large-scale networks are characterized by many points of interaction among a variety of participants – owners, operators, sellers, and buyers. The networks' interconnected nature makes them vulnerable to cascading failures with widespread consequences.

The EPRI/DoD Complex Interactive Networks/Systems Initiative, which began in mid-1999, is leading toward a concept of controls that are 'self-healing' in the sense that they make the system automatically reconfigurable in the event of material failures, threats, or other destabilizing disturbances. This work raises the question as to whether there is a unifying paradigm for the modeling, simulation, and optimization of time-critical operations (both financial transactions and actual physical control) in any multiscale, multicomponent and distributed system. These are the characteristics of any industry made up of many geographically dispersed components that can exhibit rapid global change as a result of local actions.

With the advent of deregulation, unbundling, and competition in the electric power industry, new ways are being sought to improve the efficiency of that network without seriously diminishing its reliability. The complexity of the electric power grid combined with deregulation and the ever increasing interaction between interconnected infrastructures offers new and exciting scientific and technological challenges.

From the viewpoint of strategic research and development, many scientific and technological challenges are posed by the lack of a unified mathematical framework with robust tools for modeling, simulation, control, and optimization of time-critical operations in hybrid, complex, distributed, and interactive networks with multiscale and multiple components. The EPRI/DoD CIN/SI emphasizes the mathematical foundations and robustness of complex, networked systems; it aims to develop:

- Methodologies for robust distributed control of heterogeneous, widely dispersed, yet interconnected systems;
- Techniques for exploring interactive networked systems at the micro (individual component) and macro (emergent property) levels;
- Tools to prevent/ameliorate cascading effects through and between networks;
- Tools/techniques that enable large-scale and interconnected national infrastructures to self-stabilize, self-optimize, and self-heal;
- Ways to enable these infrastructures to help address the 'trilemma' of population, poverty, and pollution.

Acknowledgments

The author wishes to express his gratitude to Dr. Martin Wildberger, Dr. Gail Kendall, and Dr. Robert Holmes for many fruitful discussions, collaborations, and insightful suggestions.

References

A presentation package for the EPRI/DoD University Research Initiative on Complex Interactive Networks/Systems is available on EPRI's Web site at http://www.epri.com/targetST.asp?program=83. The CIN/SI research announcement, DAAG55-98-R-RA08, is at http://www.aro.army.mil/research/complex.htm.

Amin, M. (1998) Complexity and the deregulation of the electric power industry. *Proceedings of the 3rd Embracing Complexity (EC3) Conference,* Cambridge, MA, 101–106.

Amin, M., Wildberger, M. and McCarthy, G. (1998) Self-healing electric power grid as a complex adaptive system. *Proceedings of the 4th Joint Conf. on Information Sciences (JCIS'98),* Research Triangle Park, NC, **I**, 295–301.

Amin, M. (ed.) (1999) Operations research methods in intelligent transportation systems. *International Transactions in Operational Research,* **6**(1).

Amin, M., Garcia-Ortiz, A. and Wootton, J. (eds.) (1995) Network, control, communications and computing technologies in intelligent transportation systems. *Mathematical and Computer Modeling,* **22**(4-7), 454 pp.

CA-CTA (1998) California: An Economic Profile. California Trade & Commerce Agency, Office of Economic Research, http://commerce.ca.gov/economy.

CA-ISO (1998) 1998 Transmission Reliability Report. Prepared by the California Independent System Operator in consultation with The AB 1890 Report Steering Committee.

Cebrowski, A.K. and Garstka, J.H. (1998) Network centric warfare – its origin and future. *U.S. Naval Institute Proceedings,* 28–35.

CEC (1997) A Survey of the Implications to California of the August 10, 1996 Western States Power Outage. California Energy Commission, Report P700-97-003, June.

CIAO (1997) Critical Foundations: Protecting America's Infrastructures. *Report of the President's Commission on Critical Infrastructure Protection,* October 1997, Critical Infrastructure Assurance Office, Washington, D.C., http://www.ciao.ncr.gov.

Dy-Liacco, T.E. (1967) The adaptive reliability control system. *IEEE Transactions on Power Apparatus and Systems,* **86**(5), 517–561.

EPRI (1996) Integrated Knowledge Framework (IKF) for Coal-Fired Power Plants. Technical Report TR-106211-V1/2/3, Electric Power Research Institute Dist. Ctr., Pleasant Hill, CA.

EPRI (2000) Complex Interactive Networks/Systems Initiative: Overview and Progress Report for Joint EPRI/Dept. of Defense University Research Initiative. Technical Report TP-114660, Electric Power Research Institute Dist. Ctr., Pleasant Hill, CA.

EPRI (2000) Conceptual Design of a Strategic Power Infrastructure Defense (SPID) System (Authors: PI and Co-PIs at the University of Washington, Arizona State University, Iowa State University, Virginia Tech.). Technical Report TP-114661, Electric Power Research Institute Dist. Ctr., Pleasant Hill, CA, March 31.

EPRI (2000) Intelligent Management of the Power Grid: An Anticipatory, Multi-Agent, High-Performance Computing Approach (Authors: PI and Co-PIs at Purdue University, the University of

Tennessee, and Fisk University). Technical Report TP-114662, Electric Power Research Institute Dist. Ctr., Pleasant Hill, CA, March 31.

EPRI (2000) Modeling and Diagnosis Methods for Large-Scale Complex Networks (Authors: PI and Co-PIs at Harvard University, Boston University, MIT, the University of Massachusetts–Amherst, and Washington University–St. Louis). Technical Report TP-114663, Electric Power Research Institute Dist. Ctr., Pleasant Hill, CA, March 31.

EPRI (2000) Minimizing Failures While Maintaining Efficiency of Complex Interactive Networks and Systems (Authors: PI and Co-PIs at Cornell University, George Washington University, the University of California–Berkeley, the University of Illinois, Washington State University, and the University of Wisconsin). Technical Report TP-114664, Electric Power Research Institute Dist. Ctr., Pleasant Hill, CA, March 31.

EPRI (2000) Context-Dependent Network Agents (Authors: PI and Co-PIs at Carnegie Mellon University, Texas A&M University, the University of Illinois, and the University of Minnesota). Technical Report TP-114665, Electric Power Research Institute Dist. Ctr., Pleasant Hill, CA, March 31.

EPRI (2000) From Power Laws to Power Grids: A Mathematical and Computational Foundation for Complex Interactive Networks (Authors: PI and Co-PIs at CalTech, MIT, UCLA, the University of California–Santa Barbara, and the University of Illinois). Technical Report TP-114666, Electric Power Research Institute Dist. Ctr., Pleasant Hill, CA, March 31.

Fink, L.H. and Carlsen, K. (1978) Operating under stress and strain. *IEEE Spectrum,* March, pp. 48–53.

Fouad, A.A. (1992) Power System Dynamics. Power System Operators' Short Course, Iowa State University, 29 pp.

Garcia-Ortiz, A., Amin, M. and Wootton, J. (eds.) (1998) Intelligent transportation systems – traffic sensing and management. *Mathematical and Computer Modeling,* 27, 9–11, May-June.

Heydt, G.T. (1999) Applications of Satellite Technology in Power Engineering. Arizona State University, presentation material to EPRI, May 16.

Kundur, P. (1994) *Power System Stability and Control,* EPRI Power System Engineering Series. McGraw-Hill, New York.

Machowski, J., Bialek, J.W. and Bumby, J.R. (1997) *Power System Dynamics and Stability.* Wiley, U.K.

Rossi, A., Valsecchi, G.B. and Farinella, P. (1999) Risk of collisions for constellation satellites. *Nature,* 399, 743–744.

Sandia National Laboratory (1998) U.S. infrastructure assurance strategic roadmaps. Publication SAND98-1496, August.

Wildberger, A.M. (1997) Modeling with independent intelligent agents for distributed control of the electric power grid. *Proceedings of the American Power Conference,* 54-I, 361–364.

Wildberger, M., Amin, M., Harp, S. and Morton, B. (1999) Simulating the evolution of the electric enterprise with autonomous adaptive agents. *Proceedings of the 32nd Annual Hawaii International Conference on System Sciences (HICSS),* 13.

15

Multiscale Networking, Robustness, and Rigor

John Doyle

California Institute of Technology

15.1 Introduction

The 20th century may be viewed as bringing near closure to the first scientific 'revolution', which aimed for a simple, certain, reproducible view of nature, in part by a radical denial of the complex and uncertain. Mainstream science has focused overwhelmingly on characterizing the fundamental material and device properties of natural systems, from subatomic particles to the molecular basis for life. In contrast, it has provided few rigorous and predictive tools for dealing with the complexity and uncertainty of the real world outside the laboratory, particularly the complex networks that are certain to be the focus of both engineering and biology. Unfortunately, many mainstream advocates of a 'new science of complexity' have further abandoned rigor and predictability in favor of attractively vague notions of emergence and self-organization.

The next decade and century may be the beginning of a second and far more profound scientific and technological revolution associated with networks and systems, with all their complexity. The management of complex systems to date has depended on taming matter, energy, entropy, and information. Although each has been important throughout human history, substantial progress has resulted when scientific discoveries, based on deep principles, made possible the replacement of *ad hoc* and implicit treatment by systematic and explicit management. The advent of steam engines motivated a deeper understanding of entropy, which in turn led to systematic management of entropy and thus higher efficiency. More recently, the structured and systematic management of information in networks and computers, from VLSI design to coding theory to network protocols to object orientation, has created an astonishing explosion in the complexity of our systems. The post-genomic era also promises to focus attention in biology on networks and systems, from gene regulation to signal transduction to neural systems to ecosystems.

This informal and nontechnical essay is an attempt to outline the principles on which a rigorous new science of complex systems might be based, one that places convergent computation, communication, and control clearly at its center. These principles will be explained in the context of four themes:

1. Convergent, ubiquitous, pervasive networking.
2. Multiscale/resolution, heterogeneity.
3. Robustness, reliability, high confidence.
4. Rigor.

These four themes will be reviewed briefly and then a selection of specific examples will be sketched to illustrate the themes. We are building increasingly integrated, interconnected, and automated networks for information, energy, transportation, and business, yet we are investing relatively little in research directed at understanding how to make these systems robust and predictable. The 20th century's understanding of chaos and undecidability, together with relativity and quantum mechanics, changed completely the 19th century's view of the universe as a clockwork mechanism, but otherwise had surprisingly little impact on the reductionist program. If the 21st century has surprises in store, beyond simply unifying existing theories or cataloging the components of cells, they are likely to be in dealing with the complexity of the networks from which we are built and are now building.

15.2 Themes for a New Science

15.2.1 Convergent, ubiquitous, pervasive networking

Everyone is aware of the Internet's impact on the ongoing convergence of data, voice, and video, as well as the convergence of data, commerce, manufacturing, and transportation in e-commerce. Various deregulations of utility industries have also created growing convergence of information, financial, transportation, and energy networks. These trends are only mild precursors of the future of ubiquitous, pervasive networking where every object in our world will have the equivalent of a telephone number or an IP address. The currently distinct networks associated with communications, computing, transportation, energy, consumer products, utilities, health, finance, and manufacturing will certainly blur into a single integrated network of networks.

The future of biology is poised to parallel that of engineering systems. Just as hardware increasingly becomes a commodity and technological value added occurs at the network level, so too do the component molecules of biology and the experimental techniques to explore them become commodities. The focus is beginning to shift to understanding the vast networks that biological molecules create which regulate and control life. We can reasonably expect that viewing biology in terms of complex feedback networks will become as essential as the role of molecular biology has become over the last few decades.

The rest of this essay will assume that the reader is aware of the debate and discussion surrounding the future ubiquitous, pervasive, networked computing continuum in which our lives will be immersed and, if not alarmed, is at least concerned as to what the scientific foundation might be for such a technology. For further discussion of the future of convergent networking, see Chapter 14 of this volume.

15.2.2 Multiscale/resolution, heterogeneity

The central multiscale issue is connecting microscopic device and component properties with the macroscopic behavior of networks and systems. The study of multiscale phenomena is not new in science. The microscopic behavior of a gas can be described in terms of the position and velocity of molecules. Ensembles of molecules can be viewed as stochastic processes with stationary distributions. Macroscopic behavior is typically described by thermodynamic quantities such as temperature and pressure, and statistical physics has been concerned with connecting all these different scales. The key feature of systems in statistical physics is that the thermodynamic properties are the consequence of the generic microscopic features, and thus sets of measure zero can be neglected.

The statistical physics view of multiscale phenomena has led to a corresponding view of complex systems that is typically quite different from what arises in engineering or biology. The complex systems studied in physics are typically homogeneous in their underlying physical properties, and complexity is most interesting when it is not put in by hand, but rather arises as a consequence of bifurcations or dynamical instabilities, which lead to 'emergent or self-organizing' phenomena on large length scales. Even when long-range correlations arise due to critical phenomena, universality justifies representing the microscopic degrees of freedom by simple, often identical, components.

The Internet is one example of a system that may superficially appear to be a candidate for the self-organizing and emergent view of complexity. It certainly appears as though new users, applications, workstations, PCs, servers, routers, and whole subnetworks can be added and the entire system naturally self-organizes into a new, robust configuration. Furthermore, once online, users act as individual agents, sending and receiving messages according to their needs. There is no centralized control, and individual computers both adapt their transmission rates to the current level of congestion and recover from network failures, all without user intervention or even awareness. It is thus tempting to imagine that Internet traffic patterns can be viewed as an emergent phenomenon from a collection of independent agents who adaptively self-organize into a complex state, balanced on the edge between order and chaos. Even the ubiquitous self-similarity of Internet statistics could be taken as the classic hallmarks of criticality.

As appealing as this picture is, the reality is that modern internets use sophisticated multilayer protocols to create the illusion of a robust and self-organizing network, despite substantial uncertainty in the user-created environment as well as the network itself. It is no accident that the Internet has such remarkable robustness properties, as the Internet protocol suite (TCP/IP) in current use was the result of decades of research into building a nationwide computer network that could survive deliberate attack. The high throughput and expandability of internets depend on these highly structured protocols, as well as the specialized hardware (servers, routers, caches, and hierarchical physical links) on which they are implemented. Yet it is an important design objective that this complexity be hidden.

The core of the Internet, the Internet Protocol (IP), presents a carefully crafted illusion of a simple (but possibly unreliable) datagram delivery service to the layer above (typically the Transmission Control Protocol, or TCP) by hiding an enormous amount of heterogeneity behind a simple, very well engineered abstraction. TCP in turn creates a carefully crafted illusion to the applications and users of a reliable and homogeneous network. The internal details are highly structured, heterogeneous, and nongeneric, creating apparent simplicity, exactly the opposite from emergent complexity.

Interestingly, and importantly, the increase in robustness, productivity, and throughput created by the enormous internal complexity of the Internet and other complex systems is accompanied by new hypersensitivities to perturbations the system was not designed to handle. Thus, although the network is robust to even large variations in traffic, or loss of routers and lines, it has become extremely sensitive to bugs in network software, underscoring the importance of software reliability and justifying the attention given to it. Computer viruses can now use the very applications that make the Internet so popular to propagate quickly and widely. The infamous Y2K bug, although not necessarily a direct consequence of network connectivity, is nevertheless the best known example of the general risks of high connectivity for high performance. There are many less well known examples, and indeed most modern large-scale network crashes can be traced to software problems, as can the failures of many systems and projects (e.g., the Ariane 5 crash or the Denver International Airport baggage handling system fiasco).

15.2.3 Robustness, reliability, high confidence

The success of the Internet is greatly due to the extreme emphasis that was placed on robustness in its design. Robustness and uncertainty management is beginning to replace information, entropy, energy, and materials as the dominant issue in complex systems of all types, including especially biology and software. It now demands the same structured and systematic mathematical and computational approach that has proven so successful with matter, energy, entropy, and information. However, just as entropy to a 17th century scientist or information theory or quantum mechanics to an 18th century one would have seemed arcane and mystical, so too do uncertainty and robustness as specific quantities to be managed in explicit and systematic ways appear to many 20th century scientists. However, one need only look at the complex systems around us to see that robustness has become *the* dominant issue.

The Boeing 777 has millions of parts, mostly rivets, but 150,000 distinct subsystems, many of which are themselves highly complex components, some with millions of subcomponents, and so on. The 777 and other modern commercial airliners are literally flying complex networks. Despite this astonishing complexity, which certainly overlaps with simple biological organisms in terms of part count, most deaths in commercial aviation are now due, not to vehicle malfunctions, but to higher level system failures, terrorist attacks, or pilot or air traffic controller error. What's important, though, is that the overwhelming proportion of the millions of parts in a modern commercial aircraft or the thousands of genes in biological organisms is there purely for robustness and uncertainty management. In both cases, the increasing complexity has produced a net improvement in robustness, provided the environment for which the system was designed doesn't change too dramatically.

Our energy, information, and transportation networks are even more complex than a Boeing 777 and have not yet reached the same level of reliability. Electric power systems are reliable enough that we generally take them for granted until a multimillion customer, multibillion dollar outage reminds us that this is naïve. Other dramatic failures in power, communication, transportation, and space systems, ecological problems, and the presence of autoimmune diseases and cancer remind us that this robustness cannot be taken for granted. Major success stories, such as the Internet, VLSI design, and the Boeing 777, have been the result of highly structured and systematic processes, with an almost obsessive attention to robustness. These experiences are suggestive and the details are often poorly understood, but they only hint at the kind of highly heterogeneous, nonlinear, dynamic, interconnected, integrated networks that are coming. A robustness perspective is particularly important in this era of exploding technological complexity and the 'better, cheaper, faster' world of virtual prototyping. And any future golden age of biology is certain to depend as much on understanding systems as on understanding their components.

The highly differential robustness of both biological and engineering systems is not accidental, but is one example of an inherent property of interconnected complex systems. By *differential* robustness we mean that insensitivity to certain uncertainties, hopefully the most likely, is obtained at the expense of increased sensitivity to other uncertainties, hopefully unlikely. Thus, designers of modern high-performance systems deliberately accept hypersensitivity to the extremely unlikely, in return for insensitivity to common uncertainties, and this trade-off drives the introduction of increasing complexity. Although the subject is in its infancy, there are even conservation laws associated with robustness and uncertainty management in complex systems that are becoming more important than more familiar limitations due to conservation of mass and energy, speed of light, quanta, entropy, information, and computation.

Complex systems engineering, and in particular, robustness and uncertainty management, is becoming the central issue throughout technology. However, it is in the biological sciences that a theory of uncertainty management promises to have even more revolutionary impact, and we argue for the creation of an entirely new discipline focused on theory and its application to *real* complex technological and biological systems. Example applications include control of cellular processes through signal transduction and cell surface receptors, and the interaction of mechanics and biochemistry in developmental dynamics and in the movement of macromolecules, cells, and tissues. Engineering applications include robustness and reliability in complex energy, information, and transportation networks. A central theme will be a theoretical foundation for the 'better, cheaper, faster' movement in engineering design.

In both biology and engineering, simple components must be put together to create complex systems, and uncertainty at the component level can have highly unpredictable consequences at the system level. Theorists in a variety of disciplines are beginning to develop sophisticated tools to address robustness and complexity, but they currently exist in a too fragmented way within narrow technical disciplines such as controls, dynamical systems, computation, communications, and statistics. One essential goal of a research program in complex systems is to create a more integrated theory of robustness.

15.2.4 Rigor

Rigor will be an essential feature of any practical theory of complex networks. The essence of rigor is having a clear picture of what is known and not known, what is proven or not proven, for what is there strong evidence, and what is conjecture. Many of the recent mainstream scientific publications in complex systems have been marked by a serious lack of rigor in all these senses. The origins of our current situation can be traced to long-standing differences in styles between scientists, mathematicians, and engineers, to which has been added the unique style of the hacker culture, which has dominated much of computing.

Mathematicians have long viewed physicists as sloppy and cavalier, while physicists retort that mathematics' emphasis on rigor would stifle physicists' creativity. Both points of view have merit, and it is natural that these two communities have developed such different styles. Physicists are seeking to explain the basic workings of the simplest aspects of the universe and ultimately have experiments to sort out good ideas from bad. The theoretical physics community rewards bold ideas that are ultimately proven correct experimentally, and there is little or no consequence for being wrong.

As mathematics has become increasingly abstract, physical intuition, let alone experiment, has been lost as a means of sorting out true from false assertions. Mathematics has thus come to rely on a theorem/proof style intended to allow the math community to have as much rigor as is possible without experimental verification. Although working mathematicians share a remarkably uniform view of what constitutes a rigorous proof, the issue is not without controversy. The arguments are not easily explained to nonexperts, however.

Engineers who deal with complex systems are often closer in spirit to mathematicians than to physicists, because the consequences of believing and implementing a bold and provocative but wrong conjecture can be disastrous. The exception to this is the 'hacker culture' that has dominated much of software development. The essence of this culture is an extreme emphasis on never letting perfection stand in the way of good enough, or the next software release, and an emphasis on features over robustness and reliability.

The 'mainstream' complex systems literature that appears in such journals as *Science* and *Nature* is largely dominated by the physicist and to some extent hacker cultures, while the more mathematical and engineering approaches are found in more specialized journals or conferences proceedings. This creates the ironic situation where a paper on, say, the Internet will appear in *Science* that could not be published in the proceedings of a conference on the Internet. The authors will argue that the so-called experts in Internetworking are conservative and resistant to new ideas, while the latter will typically argue that the research is obviously wrong.

There are positive aspects to all these cultures that must be blended properly to create a culture that is appropriate for a research community in complex networks. There is no doubt that we need the development of entirely new mathematics, and for the most part, the mathematician's style will be needed. More important, complex networks will have little that is directly accessible to either physical intuition or controlled experiments, except on their component parts. Thus one motivation for rigor mirrors that of the mathematician's – that it provides the only hope of understanding in the absence of physical insight or experimental verification. This does not mean that developing appropriate intuition is not essential as well, just that it will be more like that of the pure mathematician than the physicist.

The hacker's commitment to taking complexity head on will be valuable. If theory is to make any contribution to the practical needs of engineering and biological networks, it cannot be done in the context of purely academic examples. The systems must work in the real world. Ultimately, however, we need to have the engineer's perspective that systems must work *reliably* in the real world, not just in the lab or the demo. And finally, the physicist's desire for the simplest possible story will be needed if we are to communicate the important ideas to a broad community.

15.3 Illustrative Examples

The remainder of this chapter will discuss general issues associated with robustness and uncertainty in complex systems. The examples that follow – airbags, compact disc players, Mars Pathfinder, and Formula One racing – are used to illustrate some of the key issues in complex networks, but in simple and familiar settings. Admittedly, these are 'toy' examples when compared with the complex networks that exist and are planned, but it is exactly for that reason that they are pedagogically useful.

Often robustness is obtained at the expense of major compromises in efficiency. Digital systems generally are the most extreme example, but examples such as the Internet, VLSI design, and Mars Pathfinder are also success stories where robustness was given priority. They are also the results of highly structured and systematic processes. These can be contrasted with the abandoned upgrades to the air traffic control system and IRS software and the problems with the Denver airport baggage handling system. Although the failures of *Challenger, Galileo,* TWA 800, the Ariane 5, and the recent outages of power, telephone, and satellite networks can be traced to specific subsystems, it was the catastrophic cascading of tiny failures into system failures that is the most striking feature of these events. Complex interconnected networks will have numerous component failures, but the network must be robust enough that such failures remain isolated.

The most telling examples are often the most mundane. For example, estimates of corporate yearly costs to operate a single PC are around $24K per year. Hardware represents only $1K of this, and the low cost and high reliability of computer hardware is a consequence of highly structured and systematic design procedures. Software, networking infrastructure, and technical support cost a few thousand dollars more. Most of the total cost is associated with informal 'futzing' because of poor software robustness. This futzing is not an official part of information technology budgets, but represents the dominant cost of computing. Neither faster hardware nor new software features would affect this cost, which is the focus of most attention but is completely due to the gap between the system's behavior in a demo and its behavior in real use. This example, and the Y2K problem as well as others, also reminds us that we should not look to information technology alone as a solution to our problems of complex system robustness.

15.3.1 Complexity is driven by robustness

To help visualize the idea that complexity is driven by robustness, we can do a simple thought experiment. Take any familiar complex system, say the 777 (millions of parts), a laptop computer (billions of transistors), or a biological organism (thousands to millions of genes). List both the primary design goal of the system (e.g., to deliver passengers, to do computation, or to reproduce its genome) and the sources of uncertainty to which the system must be robust. For the 777, some uncertainties are flight timing, weather, routing, other traffic, turbulence in the boundary layer, payload size and location, uncertainty in components due to manufacturing and aging, and so on. The organism faces an even longer list of uncertainties. For the laptop, the dominant uncertainty is that the computation to be done is not fixed and known in advance. Otherwise, computers, and digital systems more generally, are unique in being perfectly repeatable.

Now imagine an idealized laboratory setting in which uncertainty is greatly reduced or eliminated. For the 777, this would be virtually impossible to actually do, but we can imagine it in principle. For the case of the idealized 777, a working vehicle could probably be built with a few thousand subsystems rather than 150,000. For the laptop and the organism, creating an idealized laboratory environment is not only possible, but wouldn't differ much from the type of laboratory experiments that standard reductionist science demands. The laptop would have exactly one simple computational problem, say, producing digits of π, and perfect sources of power and a benign environment. The organism would have a steady source of nutrients and constant environmental conditions. How many parts would be needed? The laptop only required to compute digits of π could be organized with an entirely different architecture, and if speed were an issue, a purely digital design would no longer be optimal. Digital systems accept tremendous performance penalties in exchange for robustness, and the complexity of our idealized laptop could be drastically reduced if robustness ceased to be an issue.

In biology, a free-living organism actually exists, Mycoplasma, with about 500 genes (humans have about 100,000 genes), so that gives an upper bound on required organism complexity. Researchers are working to knock out genes to lower this upper bound. It is interesting to note more generally how frequently gene knockouts that are expected to be lethal yield either no apparent phenotype or a phenotype with completely unexpected features. Often the 'no phenotype' cases are later found to have yielded an organism that lacks robustness to uncertainties that don't arise in a typical laboratory environment. Biologists often try to conceptualize this in terms of historicity, accident, and redundancy, when they really are much more subtle examples of highly evolved robust systems design.

15.3.2 Airbags

A familiar example of uncertainty management is automobile airbags. Airbags provide reduced vulnerability to some uncertainties in the environment, such as head-on high-speed collisions with drunk drivers. In return, there is both an increased cost and an increased vulnerability to component failures. For example, a sensor failure leading to spurious deployment could cause injury even in a parked car. For this reason, great care was taken in designing the crash sensors and a diagnostic unit to monitor the operational effectiveness of the airbag, so that such possibilities are minimized.

Even without component failures, airbags can make certain circumstances more dangerous, such as low-speed collisions with small passengers. There is a large net reduction in injuries and fatalities, but increased danger in certain unusual circumstances. The next generation of 'smart' airbags will include additional sensors for even greater complexity – and hopefully better robustness. It is interesting to note that experts estimate that an airbag is approximately equal in effectiveness to an additional 300 pounds of metal in the front of the car. This design trade-off clearly has to be performed at the highest system level.

Airbags illustrate an important general principle of feedback systems: providing robustness somewhere means increased sensitivity somewhere else. Indeed, it can be proven that there are mathematical formulations of 'conservation principles' that make precise this 'waterbed effect' in certain cases. Like energy and entropy, attempts to violate these robustness principles are constantly proposed and fail miserably. However, also like energy and entropy, an understanding of the principles allows us to move conserved quantities around to our advantage. These robustness principles are so important to understanding complex systems, and so completely misunderstood, that a major priority is to articulate this theory to a broader audience and expand the research effort in this area.

Informally, though, the dominant trade-offs involved in the design of airbags involve risk and cost. Greater robustness to uncertainty not only costs money, but actually increases sensitivity to component failures and certain, hopefully rare, conditions. To perform these trade-offs requires explicit models of uncertainty and an understanding of feedback. Increasing complexity means that evaluating this trade-off can be conceptually and computationally overwhelming without a sophisticated supporting technology.

15.3.3 Compact disc players

Another familiar example of an engineering system is a portable CD player, which gives high-fidelity music reproduction that is linear across large dynamic ranges in both the music volume and frequency content, despite large variations in ambient temperature and humidity and accelerations due to listener motion. A CD player connected to appropriate electronics can amplify microscopic optical variations on the CD into sound levels that can cause physiological damage to an auditorium audience, and even CD players from different manufacturers give indistinguishable performance. CD players are not typically robust with respect to, say, being crushed by a large weight or immersed in water, although presumably these could be designed for at additional expense. Our transportation, energy, information, and communication systems also exhibit similar robustness properties, and dramatic but rare failures remind us that this should not be taken for granted.

Complexity is another striking feature of engineering and biological systems. We now routinely build multimillion component systems on scales from microchips to airliners to global networks, and components themselves often have many millions of subcomponents, and so on. Biological systems are clearly extremely complex in this sense. An obvious question is whether the robustness of these 'systems of systems' is simply a direct consequence of carefully manufacturing extremely robust components at every level. Have biological systems evolved in such a way that they have finely tuned internal dynamics that are themselves robust to environmental and intra-organism uncertainties? Similarly, are the electromechanical components of the CD player manufactured in such a way that they have the same robustness properties as the entire CD player system?

The advantage for the CD designer in using 'perfect' components is obvious, as it would make the design process much easier. Imagine that we collected all the components of a CD player and considered all the possible interconnections of these components. There are a combinatorially huge number of possible designs given a set of components, and only a tiny fraction of interconnections will actually yield a working CD player, even with ideal components. Even more, an almost vanishingly small fraction (a set of measure zero in the thermodynamic limit of a large number of parts) of these potentially viable designs will work if there are nontrivial uncertainties in the components, which creates a daunting design challenge. It may then seem surprising to find that, in fact, CD players have large component uncertainties. The electromechanical parts are highly nonlinear, have limited dynamic range, and have large variation with signal amplitude and frequency, as well as temperature and vibration. Other systems have similar properties.

Thus CD players exhibit robustness to two kinds of uncertainty: those in the environment (temperature variations, vibration and movement, frequency and amplitude variations in the music) and those in the components from which they are built (variations in electromechanical components due to environment and manufacturing tolerances). It is obvious from its external behavior that a CD player has the first kind of robustness, but different functionally equivalent designs could have widely varying robustness of the second kind, and thus have widely varying costs of manufacturing. Although it is useful pedagogically to distinguish environmental and component uncertainties, systems must be robust to both, and the distinction is a little artificial.

The specific *robust design* of the interconnection is what allows the system to be much more robust than its components. Cheap parts are highly uncertain, so affordable manufacturability dictates robust design, and design of complex systems is thus effectively dominated by the trade-offs of uncertainty management. The level and nature of robustness is a critical design choice. Although the CD player is highly robust to both environmental and manufacturing uncertainties, the additional level of robustness necessary to protect against deliberate, malicious attack would be prohibitively expensive. It would typically be much cheaper to simply have extra CD players as backup, unless malicious attack was an important and common feature of the environment.

Another issue that is somewhat subtler, but turns out to be mathematically more tractable, is that of frequency and amplitude response of the CD player. Since human hearing has a limited range of sensitivity and our music has a limited range of content, in both frequency and amplitude, CD technology is tuned to those ranges. It actually performs rather badly outside those ranges, introducing large distortions that we would not normally be aware of because we can't hear them (and there is no 'music' there anyway). This is an excellent example of tuning a design to be insensitive to specific uncertainties, in this case variations in frequency and amplitude within the limited range, while accepting high sensitivity to uncertainties outside this range. Indeed, for some signals, higher fidelity reproduction would be obtained by turning off the CD player, as the error it makes is larger than the signal itself.

15.3.4 Mars Pathfinder

The Mars Pathfinder mission offers a convenient example of uncertainty management and the use of virtuality in design. The highest risk phase of the mission was the descent and landing on the surface of Mars. Without going into excessive detail, we can sketch a few of the relevant issues. This 'cartoon' will be instructive, grossly oversimplified, but not essentially wrong. For more information, see the Mars Pathfinder Web site (http://mpfwww.jpl.nasa.gov/), or particularly the Entry, Descent, and Landing (EDL) Web site (http://mpfwww.jpl.nasa.gov/mpf/edl/edl1.html).

The problem boiled down to one of uncertainty management. If the mission itself were repeated a large number of times, the outcomes (in this case, impact velocities) would likely vary by several meters per second across the missions. The systems engineer's job is to estimate this distribution accurately enough to allow cost-effective design trades to be made. (We'll ignore the actual problem of design for now, and focus on the issue of analyzing a given design.) Traditionally, uncertainty was avoided, rather than managed, through highly conservative 'stacked margins' and exhaustive physical prototyping to make sure that the impact velocities are safely within a large margin of error. This is an effective but expensive combination that is no longer an option. The 'better, cheaper, faster' imperative is primarily achieved by more careful evaluation of the cost/benefits in terms of probabilities of failures and more use of virtual prototyping to focus (but in no way eliminate the need for) physical testing. For the rationale and implications of the Jet Propulsion Laboratory's 'better, cheaper, faster' vision, see the Develop New Products (DNP) Web site (http://techinfo.jpl.nasa.gov/dnp/dnp_users_guide/ GUIDE.HTM).

A high-fidelity simulation of the dynamics of the entire descent and landing was combined with selected component tests to make overall assessments of probabilities of success. The core of the simulation consisted of commercial CAD software that essentially takes assumptions and initial conditions about the vehicle and the environment and generates a single trajectory. As an outer loop, the systems engineers wrote UNIX scripts that performed repeated Monte Carlo trials, varying parameters and initial conditions, to get an estimate of the distribution of possible outcomes (again impact velocity).

Although the Pathfinder systems engineers' approach is a step in the right direction, and they did a truly brilliant job in pulling it off, the approach has serious inadequacies (that are not their fault) that are exactly the issues which should be driving research. Currently, there are few efforts in this direction. One obvious problem is that commercial CAD packages have limited support for explicit representations of uncertainty, necessitating *ad hoc* methods such as writing UNIX script outer loops. Although there is no substitute for engineering judgment in choosing the distribution of parameters, there is little in the way of software or theoretical support for the systems engineers in this task.

Another, equally obvious, problem is that most Monte Carlo trials were wasted on benign events in order to get a statistically significant sampling to have adequate confidence in the tails of the distributions. Although there are well known statistical methods aimed at addressing this issue, none get around the issue that to avoid large numbers of samples one must have guarantees that regions in parameter space are benign, a known NP-hard problem. The high dimension of the space of uncertain yet important parameters makes exhaustive search prohibitive.

A subtler problem with this Monte Carlo approach is that many of the uncertainties in the CAD models are not parametric. For example, the uncertainty in high-frequency flexible modes of structural members, the fine-scale behavior of fabric airbags in contact with rocks and the resulting effects on tearing, the interaction of structural and acoustic modes with combustion in the thrusters, and so on, are not explicitly represented in the CAD models (the assumptions are that these effects are negligible), so their impact is difficult to evaluate. The engineers tried to reflect these in an *ad hoc* way by increasing related parameter ranges, but this is a dangerous practice. Perhaps somewhat less dangerous, but still lacking in sufficient software or theoretical support, is the replacement of chaotic fine-scale dynamics with stochastic approximations that can be directly included in the Monte Carlo simulations.

15.3.5 Complexity and Formula One racing

The technical issues underlying any discussion of complex systems are necessarily difficult to grasp, and no one example can illustrate more than a few points. Formula One automobile racing is a high-technology example with some features of complex systems of systems, but in a relatively simple setting. Formula One (F1, also called Grand Prix) is the ultimate automotive event in terms of both money and technology. It is a huge spectator sport, primarily outside the United States, as well as a testing ground for advanced automotive technology. Everything from disc brakes to fully electronic fuel injection has first been tried in F1, and then later found its way into our passenger cars. Since our passenger cars are restricted in speed and their normal operating conditions are relatively benign, we fail to notice most of the technical changes that have taken place. In F1, new technology immediately and dramatically translates into victories and is thus more visible. Billions of dollars and decades of research have been devoted to refining F1 cars, and it is a domain unequaled in fostering a single-minded pursuit of technological excellence.

In the last decade or so, something striking has happened. Passenger cars have begun to overtake F1 in technical sophistication, and it is this story that is particularly relevant to our discussion on complexity. Passenger cars have had an explosion in complexity, almost entirely in electronics, computers, and control systems. Automated and active braking, active engine and drive train control, automatic airbag deployment, and active suspensions are common, and sophisticated drive-by-wire traction, steering control, and obstacle avoidance are in development. Our cars are now safer and much more reliable and robust, enabled by a combination of computer and control technology. We pollute less, visit the mechanic less often, and survive a greater variety of dangerous conditions, although this is largely unnoticed because it has no apparent effect on day-to-day operation.

Surprisingly, these active control systems are not used in F1. They *were* introduced in a preliminary way a decade ago, with dramatic results. Again, unlike in passenger cars, new technology in F1 is immediately apparent in performance. It quickly became clear that actively controlling aerodynamics, steering, suspension, traction, and braking would have such a profound and revolutionary impact on F1 that the sport would be completely changed. Furthermore, although the use of sensors and computers is ubiquitous in F1, the addition of active feedback was viewed, quite correctly, as an entirely new and distinct technology that was both enormously powerful and poorly understood outside a narrow community. Rather than accept such a radical transformation of their sport, the ruling body of F1 simply banned all active control. Of course, the rules don't include a specific blanket ban, but in each section of the rules dealing with engines, brakes, traction, suspension, aerodynamics, structures, and so on, there is careful wording that implicitly bans active control.

Interestingly, our science and technology research budget enforces effectively the same ban as F1. The deep research issues associated with understanding and designing robust complex systems, which are likely to be among the most fundamental questions in the coming century, are relatively neglected. As in F1, there is no explicit global ban; the emphasis on components rather than systems creates an implicit ban. F1 manufacturers can and do spend enormous sums on exotic materials, extensive wind tunnel testing, and sophisticated computer-aided design, modeling, and simulation. They can and do place sensors, transponders, and video cameras on the cars, and connect these with computers for monitoring and diagnostics. They are allowed to provide actuation for power assist in steering and braking. All these are viewed as providing incremental and evolutionary effects. But they cannot use feedback for active control.

It is actually difficult for most people, even among scientists and engineers, to grasp the significance of what is banned in F1. Passive control is allowed. F1 designers can and do use a variety of surfaces and shapes to control the aerodynamics, but they are forbidden from having surfaces that move under automatic control. They can and do use all the sensors and computers and many of the actuators necessary to implement sophisticated active control. What is banned is closing the loop with active control. This is such a subtle difference that it can be difficult for F1 officials to verify. Understanding how such a subtle and often relatively inexpensive difference could make a far more revolutionary impact than much more expensive materials, structures, and aerodynamic technology requires substantial expertise not only in control engineering, but also in supporting F1 technology. The value-added technology in active control occurs at a very high level of abstraction relative to these other technologies. It would completely break the F1 'paradigm', as surely as it falls outside the usual training of most scientists and engineers.

Although we may wish that F1 had allowed active control, and allowed us to see the dramatic consequences in action, the F1 officials are not simply antitechnology. One important issue that the F1 ban on active control wisely avoids is the consequences of not just having faster cars, but ones that are responding automatically to each other and their environment on time scales much faster than drivers' reaction times. Allowing active control would ultimately result in the cars acting as a highly interconnected network. The driver would become the limiting factor, and safety concerns would demand that they be eliminated in favor of a fully automated system. The alternative would be for drivers to become mere passengers in a system they had little control over and one that could fail catastrophically due to unpredictable and 'emergent' multivehicle interactions. In any case, the spectator value of F1, which relies very much on the human element in addition to the technological, would probably suffer.

Perhaps unfortunately, society as a whole is much less prudent than F1. We are proceeding to build complex systems of systems with many of the dangers that F1 sought to avoid with a ban. It is impossible, and probably undesirable, to try to stop this trend, but we need to take more seriously the consequences.

15.4 Other Examples of Virtual Design and Uncertainty Management

A popular myth that is very ingrained in our technical culture is that the problem of uncertainty due to unmodeled dynamics is adequately handled by simply increasing the resolution of the model until all the relevant phenomena are included. Although a detailed discussion of all the flaws associated with this position would take too long, we can point out a few, at the risk of being rather abstract. The fine-scale dynamics have a large number of unknown parameters whose distributions have high variance. Examples of such parameters are those that quantify viscosity, elasticity, conductance, capacitance, permeability, and so on.

Even high-resolution models make assumptions about material properties that are uncertain, so the unmodeled dynamics may actually grow in size and complexity. The cost of higher resolution modeling would be prohibitive for very marginal return in fidelity, because, for example, the Pathfinder descent and landing is ultimately dominated by the intrinsic variability of the environment as much as by the inadequate resolution of the models. The critical issue is determining the appropriate level of fidelity for a given problem, and the Pathfinder systems engineers were again left on their own without software or theoretical support.

It may seem surprising that despite the great success of the scientific enterprise and the enormous power of CAD tools and their supporting infrastructure, the Pathfinder systems engineers, who are sophisticated and well educated, were forced to do so many *ad hoc* fixes and so much hacking. One problem is that mathematical modeling in science and engineering has traditionally been used to understand relatively simple systems in a single domain. The abstractions, assumptions, and approximations made in the modeling process were part of the domain expertise and were rarely represented explicitly.

Modern computer-based modeling grew out of this tradition. A classic example, and one of the triumphs of scientific computing, is Computational Fluid Dynamics (CFD). Here, the Navier-Stokes equations have been around for more than a century and are widely believed to reliably capture fluids on a scale below our measurement capability. Thus it is not unreasonable to view much of fluid modeling and simulation as primarily a matter of numerical solutions to PDEs and building faster computers. It is interesting to note that 'digital wind tunnels' have not replaced physical ones, despite predictions to the contrary and despite computer hardware improving at rates that have matched or exceeded all expectations. Even a cursory overview of CFD is well beyond the scope of this discussion, but there is growing evidence that the traditional focus in CFD has greatly limited its practical applicability. Particularly for the kind of shear flows that are important in aircraft design, the macroscopic aerodynamic vehicle properties are dominated by large-scale structures in the flow as well as uncertainty in the boundary conditions.

Experimentalists have made substantial progress using flow visualization to get a reasonably clear picture of the origin of vorticity in the boundary layer that is the critical feature of turbulent shear flows. Nevertheless, to quote Kline of Stanford, who was one of the major contributors to this research,

> '... the structure results now seem to provide, at long last, a reasonably complete picture of how turbulence is produced and maintained in the boundary layer and of the major eddies in the various regions of the layer. In nearly every other case in physics such increased knowledge has translated into improved models for computation. That has not been the case in turbulent boundary layers'.

Fortunately, approaches focusing more on uncertainty management and robustness may soon offer attractive practical alternatives to traditional CFD.

Another important success that has created somewhat unreasonable expectations about the ease of computer modeling is VLSI CAD, where the Mead-Conway design rules, if followed, allow the uncertainties of the physical (silicon) level and the manufacturing process to be almost completely suppressed at the functional design. (This is one of the all-time brilliant examples of a successful protocol-based uncertainty management strategy, but unfortunately this fascinating topic is far too broad to be adequately handled here.) Although this approach sacrifices performance, it is essential if designs are to be done at a fast enough pace to take advantage of the continuing advances in fabrication technology. Deep submicron designs threaten to undermine this design paradigm, because the uncertainty at the physical level cannot be completely suppressed.

The Pathfinder descent problem has little in common with CFD or VLSI CAD, except at the lowest subsystem level. The models are heterogeneous combinations of models of structures, fluids, propulsion, and electronics. No canonical PDEs or design rules are available to isolate the high-level logical functioning from the physical uncertainty. Indeed, the traditional approach to spacecraft design did approximate the Mead-Conway philosophy by building in huge conservative safety margins.

It is interesting to compare the relatively *ad hoc* approach that was taken in the Pathfinder problem with the more systematic analysis that is routinely done for the Space Shuttle reentry, which is superficially similar. The similarities are that both missions require a vehicle to use combinations of jets and aerodynamics to follow a highly nonlinear trajectory from the vacuum of space through the planet's atmosphere to land on the surface. The Space Shuttle has simplifying features, however, that allow for successful use of the primarily linear robustness analysis tools that have been developed at Caltech and Honeywell over the last 20 years. They also do repeated analysis of similar missions and so can afford to invest more effort in learning to use such tools.

More sophisticated tools could probably have made some impact on the Pathfinder program, but their application there would have been much more difficult. Indeed, a focus of current research is to extend the existing tools that have been so successful on the shuttle and other programs to make them more easily usable for problems like Pathfinder. The exact nature of such an extension, and the details of the issues involved, are unfortunately well beyond the scope of this discussion. It will require a blending of tools from robust control with bifurcation analysis from dynamical systems.

15.5 Conclusion

Other important examples exist that would need detailed expositions to begin getting a full picture, but even these simple examples suggest that in complex engineering systems, the current implicit treatment of uncertainty is inadequate. Although it is quite natural to distinguish between parametric uncertainty, noise, and unmodeled dynamics, it is also important to treat them in a unified way. Noise is typically used to describe dynamics that are not modeled in detail and are often considered part of the environment. Noise is often modeled as a stochastic process, when strictly speaking it might be more appropriate to think of it as chaotic. Unmodeled dynamics is used as a catch-all for uncertainty that is neither parametric nor noise. The problem of uncertainty modeling is discussed in more detail elsewhere.

Complex engineering systems have relied on the fact that the final design step has traditionally been to add automatic controls to an otherwise completed system, and the control engineer has had the responsibility for doing system-wide uncertainty management. Uncertainty was often treated by building in conservative and expensive safety margins. The 'better, faster, cheaper' and 'systems of systems' design paradigms that are currently being promoted are forcing a more integrated and systematic treatment of uncertainty throughout the design process.

Further Reading

Carlson, J.M. and Doyle, J. (1999) Highly optimized tolerance: a mechanism for power laws in designed systems. *Physical Review E,* **60**(2), August.

Gershenfeld, N. (1999) *When things start to think.* Holt and Co., New York.

Pool, R. (1997) *Beyond engineering: how society shapes technology.* Oxford University Press, New York.

Tenner, E. (1996) *Why things bite back: technology and the revenge of unintended consequences.* Alfred A. Knopf, New York.

16

Conclusions: Automation, Control, and Complexity

John Weyrauch

Honeywell Technology Center

'The time has come', the Walrus said, 'to speak of many things, of shoes and ships and sealing wax, of cabbages and kings' (*Through the Looking-Glass and What Alice Found There* by Lewis Carroll).

The Walrus was clearly one of the early researchers in complex systems, addressing an integrated land and sea transportation enterprise for the petrochemical business, vast secure communication networks, enterprise integration for agriculture, and the most complex of systems – national governments. In this volume we have taken the Walrus' approach to automation, control, and complexity. We have provided you a panorama of current research in complex system control and overviews of current application domains that are driving the need for safe and reliable complex systems. This final chapter is a collection of many of the key points made in earlier chapters and some thoughts on future directions and challenges.

16.1 Automation and People

Defining complexity in systems is no easy matter. Many definitions deal with physical size, number of interconnects, number of subsystem interactions, speed, and number of possible state changes. The human interaction element increases the difficulty by adding the dimension of perceived complexity. Large-scale systems that include both high levels of automation and high levels of human interaction present some of the most daunting challenges to complex system designers.

As many systems grow in size and complexity, they begin to involve teams of operators as opposed to individual operators. Human–automation interactions become much more involved in these systems because of the limitations of human-to-human communications both in speed (bandwidth) and clarity (accuracy). The bandwidth of intrateam communication and coordination is not high enough to react to rapid changes in a complex process. In many system applications (oil refinery, nuclear power plant, etc.) even occasional mistakes resulting from these limitations can be catastrophic. As stated by Cochran in Chapter 2, the system-level solution to this problem requires the development of deliberate, principled approaches to the establishment of rigorous and consistent operations cultures.

From a human operator perspective, complexity is a perceived/relative quality and has three elements – component, relational, and behavioral. As designers make systems that adapt to changes, the resulting impact to human operators is an increase in workload and/or an increase in unpredictability of the system. These negative impacts can generally be reduced in two ways: (1) by using design strategies that improve operator efficiency; and (2) by masking underlying system complexity from the operator.

A key design strategy for reducing perceived complexity is to match the conceptual framework for a system's functionality with the user's existing body of knowledge. Simply stated, make the system operate in a manner consistent with the operator's mental model of the system. This requires systems designers to possess intimate knowledge of the operator's world and to exercise extreme discipline in developing interfaces that are simple and consistent with the intended functionality of the system. If a common system design/operations model is not possible, then the system must be internally consistent with a unified operational concept. This unified, self-consistent concept allows the operator to develop a simple mental model instead of one of equal complexity to the overall system.

16.2 Sensing and Control

Modeling of complex systems is a key enabling technology for the development of effective control strategies. Research sponsored by DARPA is addressing the development of unified multimodels for use in system design and operation. These multimodels must represent system capabilities, the nature of system disturbances, goals and objectives of the system, and so on. Development of truly autonomous systems is dependent on the creation of more sophisticated models that are able to address the myriad disturbances and situations that the automation system designer was unable to predict or anticipate.

A promising new area of research in the field of control algorithms is randomized algorithms. This statistically based approach offers a systematic and consistent method for solving very hard problems. The key idea underlying the randomized algorithm approach is the development of error-controlled solutions vs. the ever more elusive error-free solutions. Further advances must be made before this approach will be practical in safety-critical applications. As Kulhavý notes in Chapter 6, specific issues remain in occasional drops in algorithm performance, convergence concerns, supervisory logic to determine initial conditions, length of simulation runs, algorithm stopping, resetting, and so on.

The analysis and modeling of biological systems provide an exciting framework for expanding the boundaries of traditional control engineering. The nature of biological systems is to evolve to provide new functions when necessary, as opposed to being designed to meet certain fixed requirements. This open-ended adaptation is an intriguing approach for controlling systems of ever increasing and changing complexity. A major stumbling block is that it is very difficult to analyze biological systems in the depth and detail necessary for direct application to autonomous systems. However, the enormous capacity of biological systems for adaptation and decision making with ambiguous and incomplete sensor information is a compelling motivation for true interdisciplinary research.

Sensing systems used to monitor and control complex systems are becoming more complex themselves. Self-monitoring, self-calibration, and self-reconfiguration are but a few examples of added functionality in the emerging field of intelligent sensors and sensor networks. New or expanded application areas such as intelligent vehicles, Free Flight for commercial aviation, active environmental monitoring and control, health care, and continuous asset management drive sensing requirements in many new directions. New concepts for sensors, local computing, and efficient and reliable networking are required. Biological sensors provide an interesting framework for understanding networks of sensors with very different performance levels, bandwidths, and adaptability.

16.3 Software and Complex Systems

Effective software architectures, efficient development processes, and cost-effective, extensible life-cycle maintenance are critical issues for complex systems. A key to dealing with software complexity is the development and disciplined application of methods, tools, and processes to handle software design and life-cycle. Current research is focused on virtual machines, software architectures that cover entire families of applications, process automation provided by model-based tools, and highly disciplined software processes that foster collaboration in the development process.

Significant research in the systems and software communities is centered on the development of agent-based software architectures. Agents provide a promising framework for partitioning problems into manageable pieces and for integrating those pieces in a flexible and reliable way. Advantages include reduced development time, increased reliability, reduced cost, modular development, and reconfigurability. Disadvantages center around verification and validation of overall system performance for the countless possible interactions between various agents.

Significant research emphasis is being placed on improving the life-cycle cost and longevity of complex systems through the use of advanced diagnostic and prognostic techniques. In this context, our focus in this book has been on machinery diagnostics and prognostics based on all available sensor data in a system. The approach described by Hadden *et al.* in Chapter 11 emphasizes finding consistent correlations between different sensors for specific wear and failure conditions. This approach is built on using multiple algorithms and combining the results to obtain higher detection and identification sensitivity. A key issue involves resolving the conflicting conclusions that sometimes occur. Higher level decision functions can be used effectively to resolve these conflicts; however, these decision functions lead to the major issue of system validation since they are nondeterministic in nature. A further complication in system validation is that some failure modes are rare, and therefore wear and failure signatures are very difficult to define accurately.

16.4 Complexity Management and Networks

Demands of society for new and expanded services and capabilities with reduced impact on the environment and reduced energy usage are driving the need for systems of higher and higher levels of complexity. In addition to these ever increasing demands for performance, society is becoming both increasingly risk averse as well as increasingly reliant on these systems. These growing demands result in more stringent safety and availability requirements. The emerging air traffic system described by Green and Jackson in Chapter 12 is characterized by multivehicle, multiperspective automation and optimization with multiple stakeholders and high safety requirements. Similarly, national power generation, transmission, and distribution systems represent some of the most complex systems in use today. Complex adaptive systems offer great promise for autonomous, safe, and optimized operation. Adaptive controls can be applied to distributed control of the power grid in many ways: self-calibrating, self-diagnostic sensors; locally autonomous controllers; networking local controllers to achieve global optimization and failure recovery; and containment of faults using local intelligent controllers. The Internet is a key enabler of integrated control and optimization of this large geographically distributed system. Power grid control is also motivating research to develop the mathematical foundations and robustness of complex networked systems. Efforts are aimed at designing, analyzing, and validating robust distributed control; network interactions at the individual component and group levels; fault containment; and network-level self-stabilization, self-optimization, and self-healing.

In Chapter 15, Doyle makes the compelling argument that our understanding of uncertainty in complex engineering systems is inadequate. Automatic controls have traditionally been used for system-wide uncertainty management. Adequate safety has often been achieved by building in conservative safety margins. New design paradigms are forcing a more integrated and systematic treatment of uncertainty, but the string of recent failures in space exploration dramatically demonstrates that much still needs to be done.

16.5 Future Challenges

My interest in complex systems began in the mid-1970s while I was working on the development of the Space Shuttle orbiter entry through landing guidance, navigation, and control system. In the 1970s, the Space Shuttle design pushed aerospace systems to a dramatically higher level of complexity. The software was written in a higher order language for the first time, and the number of subsystems and level of redundancy were much higher than for any preceding system. The orbiter also represented the first operational use of cathode-ray tube displays. Developing the concepts for operator roles and information content on the displays plowed new ground in knowledge management for flight systems. The design community's response to all these new challenges was to apply the tried and true design techniques of the day. This approach led to many design errors, since basic assumptions underlying these techniques were being violated by the much more complex nature of the problem. This shortcoming was realized at the time and was overcome by one of the most exhaustive verification and validation efforts in the history of systems engineering. Fairly simple guidance and control algorithm designs were literally checked and iterated constantly over a six-year period. Even this effort proved insufficient, and many design errors were discovered in the early flights of the orbiter. Fortunately none were catastrophic! Many had that potential, but they were discovered and corrected by analyzing unexpected behavior in relatively benign flight conditions. Again, the saving grace was exhaustive analysis.

The systems and applications described in this book are many orders of magnitude more complex than the Space Shuttle. Research in several disciplines has resulted in many new design techniques and processes that greatly improve the reliability of new designs.

However, as described in several chapters, incidents still occur in all of these systems. I believe strongly that a new approach must be taken to complex system design. Our current time-sequenced, highly partitioned, design, development, and maintenance approaches are inadequate to develop safe, reliable, available systems of ever increasing complexity. Modularizing system elements to reduce complexity is very appealing but will often lead to unpredictable interactions during extreme conditions that can result in catastrophic failure of the system. To this end, much work remains to be done in verification and validation of agent-based systems. Similarly, the concept of the human operator as the final safety net for system operations is no longer reasonable. Many examples exist in aviation, power generation, and petrochemical processing where the operator directly drove the system to an unsafe state because of a lack of knowledge about design assumptions, deficiencies, and undocumented, unallowed operator actions.

A revolution in multidisciplinary collaboration is the key to developing new design approaches that will work. Our current problem is that our education system emphasizes specialization. It is very difficult for teams with diverse backgrounds (engineering, biology, psychology, physics, chemistry, computer science, etc.) to develop radical new concepts because they have a hard time appreciating the often subtle advantages one perspective may have over another. Individual multidisciplinary training is a key enabler for making substantial progress in developing a revolutionary new approach to systems engineering. Emerging fields such as bioengineering and genetic engineering will speed this process along, as will a community-wide emphasis on lifelong learning. In conjunction with this new look at education, we must re-evaluate risk/reward models used to evaluate designs and operational concepts. Emphasis should be placed on fault containment versus fault elimination, bounded performance versus consistent deterministic performance (allow for a bad but not catastrophic day), and simple universal response techniques for failure situations where the specific nature of the failure is unknown (equivalent to the vomit response mechanism in mammals). The current thrust in human-centered design of systems must be quickly adopted throughout the design and development community. As outlined in several chapters of this book, designers must have a strong understanding of operator requirements, perspectives, and issues. Designers must anticipate unintended operator actions and make the system robust to these actions without overly constraining the operator's interactions with the system.

We hope you have enjoyed the survey of research and applications we have provided in this text. The field of complex systems is in its infancy, and we look with eager anticipation to future breakthroughs in systems engineering that will enable functionality and reliability that is beyond our current imagination.

Current Affiliations and Addresses of Contributors

Massoud Amin, D.Sc.
Electric Power Research Institute (EPRI)
3412 Hillview Ave.
Palo Alto, CA 94304-1395
U.S.A.
mamin@epri.com
http://rodin.wustl.edu/~massoud/
 amin.html

Bonnie Holte Bennett, Ph.D.
Knowledge Partners of Minnesota
9 Salem Lane, Suite 100
Saint Paul, MN 55118-4700
U.S.A.
bbennett@kpmi.com
http://www.kpmi.com/

Peter Bergstrom
Honeywell Technology Center
3660 Technology Drive
Minneapolis, MN 55418
U.S.A.
peter.a.bergstrom@honeywell.com
http://www.htc.honeywell.com/

Ulrich Bonne, Ph.D.
Honeywell Technology Center
12001 State Highway 55
Plymouth, MN 55441
U.S.A.
ulrich.bonne@honeywell.com
http://www.htc.honeywell.com/

Peter Bullemer, Ph.D.
Honeywell Technology Center
3660 Technology Drive
Minneapolis, MN 55418
U.S.A.
peter.bullemer@honeywell.com
http://www.htc.honeywell.com/

Edward L. Cochran, Ph.D.
Honeywell Technology Center
3660 Technology Drive
Minneapolis, MN 55418
U.S.A.
ted.cochran@honeywell.com
http://www.htc.honeywell.com/

John Doyle, Ph.D.
Mail Code 107-81
Control and Dynamical Systems
California Institute of Technology
Pasadena, CA 91125
U.S.A.
doyle@cds.caltech.edu
http://www.cds.caltech.edu/~doyle/

Steven M. Green
MS 210-10
NASA Ames Research Center
Moffett Field, CA 94035-1000
U.S.A.
sgreen@mail.arc.nasa.gov
http://www.ctas.arc.nasa.gov/

George D. Hadden, Ph.D.
Honeywell Technology Center
3660 Technology Drive
Minneapolis, MN 55418
U.S.A.
george.d.hadden@honeywell.com
http://www.htc.honeywell.com/

Joseph W. Jackson, Ph.D.
Commercial Electronic Systems
Honeywell
21111 N. 19th Avenue
Phoenix, AZ 85036-1111
U.S.A.
joe.jackson@honeywell.com

*Jonathan Krueger**
Guidant Corporation
4100 Hamline Ave N
St. Paul, MN 55112
U.S.A.
Jonathan.Krueger@guidant.com
http://www.guidant.com/

Rudolf Kulhavý, Ph.D.
Honeywell Technology Center
Pod vodárenskou věží 4
CZ - 182 08 Prague
Czech Republic
rudolf.kulhavy@honeywell.com
http://www.htc.honeywell.cz/

Chris Miller, Ph.D.
Honeywell Technology Center
3660 Technology Drive
Minneapolis, MN 55418
U.S.A.
chris.miller@honeywell.com
http://www.htc.honeywell.com/

Blaise Morton, Ph.D.*
DRW Investments
Suite 900
311 Wacker Drive South
Chicago, IL 60606
U.S.A.
bmorton@drwtrading.com

Victor Riley, Ph.D.
Honeywell Technology Center
3660 Technology Drive
Minneapolis, MN 55418
U.S.A.
victor.riley@honeywell.com
http://www.htc.honeywell.com/

Tariq Samad, Ph.D.
Honeywell Technology Center
3660 Technology Drive
Minneapolis, MN 55418
U.S.A.
tariq.samad@honeywell.com
http://www.htc.honeywell.com/people/
   tariq_samad

Ricardo Sanz, Ph.D.
Universidad Politécnica de Madrid
Departamento de Automática
Jose Gutierrez Abascal, 2
E-28006 Madrid
SPAIN
sanz@disam.etsii.upm.es
http://www.disam.upm.es/~sanz

George J. Vachtsevanos, Ph.D.
School of Electrical and Computer
   Engineering
The Georgia Institute of Technology
Atlanta, GA 30332-0250
U.S.A.
gjv@ee.gatech.edu
http://users.ece.gatech.edu/~gjv/

*J. Kruger and B. Morton were employed at Honeywell Technology Center when they contributed to this book.

*Joe Van Dyke**
Systems Analysis and Software
  Engineering
9665 Timberlane Place
Bainbridge Island, WA 98110
U.S.A.
joevandyke@predictusa.com

John Weyrauch
Honeywell Technology Center
3660 Technology Drive
Minneapolis, MN 55418
U.S.A.
john.r.weyrauch@honeywell.com
http://www.htc.honeywell.com/

A. Martin Wildberger, Ph.D.
Strategic Science & Technology
Electric Power Research Institute (EPRI)
5406 Victory Court
Fairfield, CA 94533-9730
U.S.A.
mwildber@epri.com

J. David Zook, Ph.D.*
Dresden Associates, Inc.
2450 Dresden Lane
Golden Valley, MN 55422
U.S.A.
dzll@msn.com

*J. Van Dyke was employed at PredictDLI and J.D. Zook was employed at Honeywell Technology Center when they contributed to this book.

Index of Names

Aarsten, A., 184, 189
Aarts, E.H.L., 106, 114
Abbott, T.S., 63, 73, 225, 239
Abelson, H., 181, 189
Agre, P., 183
Aho, A., 161, 168
Ahrens, J., 166, 168
Alamán, X., 189
Alarcón, I., 190
Alarcón, M., 177, 189
Alexander, C., 167–68
Allen, P., 144, 149
Almeida, L., 189
Amin, M., 14, 216, 263, 273, 279, 285–86, 309
Amit, Y., 104, 113
Ancevic, M., 6, 15
Aracil, R., 189
Arbuckle, P.D., 240
Ashford, R., 237, 239
Audet, S.A., 131, 134, 150
Augustin, L., 169
Avrunin, G., 164, 168
Ayres, R.U., 148–49

Bagwell, D., 240
Bak, P., 12, 15, 251, 260
Balaguer, C., 189
Bang, E.S., 227, 239
Barrientos, A., 173, 189
Barron, A.R., 87, 95, 141, 149
Barzilai, A.M., 150
Bastin, G., 114
Bayes, T., 100
Bejder, H., 189
Bennett, B.H., 191, 194, 213, 309
Bergstrom, P., 191, 213, 309
Bernardo, J.M., 114
Bernoulli, J., 100
Bialek, J.W., 286

Billings, C.E., 50, 53, 57, 225, 239
Boehm, B.W., 166, 169
Bonasso, R.P., 183, 189
Bonne, U., 14, 76, 131, 139, 143, 149, 309
Brehmer, B., 57
Bright, J.R., 50, 51, 57
Bristow, J., 192, 213
Brooks, F., 153, 169
Brugali, D., 189
Bruno, G., 161, 169
Bryan, D., 169
Buede, D.M., 5, 15
Bugajski, D., 94
Bullemer, P., 14, 17, 19, 309
Bullock, T.H., 125, 129
Bumby, J.R., 286
Busch, D., 213
Buschman, F., 184, 189
Butler, A., 124, 129

Carlsen, K., 268, 286
Carlson, J.M., 301
Carlson, L., 238–39
Carroll, L., 303
Cebrowski, A.K., 280, 285
Cerny, V., 105, 113
Charnock, J., 240
Chen, L.-S., 110, 113
Chen, M.-H., 104, 114
Cheng, J., 213
Chew, R.G., 223, 239
Christensen, S., 169
Cladis, P.E., 249, 260
Clements, P.C., 182, 191
Cochran, E.L., 14, 17, 19, 304, 309
Cofer, D., 95
Cohen, S.G., 169
Constantine, L., 119, 130
Corbett, J., 168
Cowan, G.A., 249, 260

Crandall, B., 7, 15
Cringely, R., 63, 73
Cunningham, W., 167, 169
Czarnik, A.W., 145–46, 149–50

Davis, J., 120, 130
Davis, T.J., 233, 239
Dayan, P., 255, 261
de Antonio, A., 190
de Pablo, E., 189
DeGroot, D.V., 114
Demazeau, Y., 183, 190
DeMers, B., 57, 73
Demmou, H., 169
den Braven, W., 240
Desvergne, J.P., 145–46, 149–50
Dillon, L., 168–69
Doble, J., 167, 169
Dobson, I., 261
Doll, J.D., 114
Doyle, J.C., 10, 14–15, 216, 274, 287, 301, 306, 309
Dy-Liacco, T.E., 268, 285

Echauz, J., 207, 213
Edwards, T.G., 194, 213
Efron, B., 101, 113
Erzberger, H., 233, 239

Fang, T.C., 240
Fanger, P.O., 140, 149
Farinella, P., 286
Fennick, J.H., 9, 15
Feynman, R., 24
Fink, L.H., 268, 286
Firby, R.J., 189
Fischer, K., 183, 189
Fogel, D.B., 11, 15
Fogel, L.J., 107, 113, 254, 260
Fontaine, L., 189
Forrest, S., 261
Fouad, A.A., 267, 286
Frank, R., 131, 134, 150
Franklin, S., 180, 189
Friedman, D., 252, 260
Friedman, H.L., 114
Frolow, I., 218, 240
Fuster, J., 120–21, 125, 129

Galán, R., 190

Gambao, E., 189
Garcia-Ortiz, A., 273, 285–86
Garstka, J.H., 280, 285
Gat, E., 189
Geisser, S., 113
Gelatt, Jr., C.D., 113
Gelfand, A.E., 101–2, 104, 109, 113–14
Gell-Mann, M., 12, 15, 249, 260
Geman, D., 104, 110, 113
Geman, S., 104, 110, 113
Gershenfeld, N., 301
Gevers, M., 114
Geyer, C.J., 113
Gibson, P., 213
Gintis, H., 252, 260
Glass, B.J., 232, 239
Godbole, D., 95
Gödel, K., 127–28, 130
Goka, T., 240
Goldberg, Rube, 38
Goldman, R., 57
Gómez, P., 189
Goodstein, L., 57
Gordon, N.J., 110, 113
Gould, S.J., 11, 15
Grabbe, S., 240
Grace, M.P., 234, 240
Grade, J.D., 150
Graesser, A., 180, 189
Green, S.M., 14, 215, 217, 234–35, 237, 239–40, 306, 309
Grenander, U., 113
Grest, G.S., 106, 114
Grigsby, D., 240
Gutierrez-Osuna, R., 150

Hadden, G., 14, 152, 191, 194, 209, 213, 305, 310
Hall, A., 164, 169
Haraldsdottir, A., 230, 232, 240
Harp, S.A., 123, 129, 256, 260, 286
Hastings, W.K., 102, 113
Hayes-Roth, B., 183, 189
Haykin, S., 122, 129
Heermann, D.W., 104, 113
Heimerman, K.T., 237–38, 240
Hess, J.A., 169
Heydt, G.T., 271, 286
Hinton, D., 235, 240
Ho, Y.C., 109, 113

Hodos, W., 124, 129
Holland, J.H., 122, 130, 254, 261
Horridge, G.A., 125, 129
Huhns, M., 183

Ishikawa, S., 168

Jackson, J.W., 14, 215, 217, 234, 240, 306, 310
Jalali, S.G., 246, 261
Jennings, N.R., 181–83, 188–90
Jessel, T., 130
Jiménez, A., 190
Johnson, S., 161, 168
Joshi, A., 260–61

Kahn, A., 243
Kahne, S., 218, 240
Kandel, A., 123, 130
Kandel, E., 116, 122, 129–30
Kang, H., 205, 207, 213
Kang, K.C., 155, 169
Kenny, T.W., 150
Kersten, G.E., 184, 189
Kim, I., 213
Kinny, D., 190
Kirkpatrick, S., 105–6, 113
Kline, S., 300
Knickmeyer, J., 273
Knill, D., 130
Koch, C., 120, 130
Koenig, W., 89, 95
Kortenkamp, D., 189
Koza, J.R., 107, 113
Kramer, K., 213
Krause, J., 89, 95
Kroot, P., 143, 150
Krueger, J.W., 14, 151, 153, 161, 164, 169, 310
Kulhavý, R., 14, 75, 97, 304, 310
Kundur, P., 267, 286
Kutty, G., 169

Laarhoven, P.J.M. van, 106, 114
Landry, S., 11, 16
Laplace, P.S., 100
Lasseter, R.H., 261
Latour, B., 13, 15
Lau, T.W.E., 109, 113
Laudeman, I.V., 238, 240
Lea, D., 184, 190

Lettvin, J.Y., 115, 130
Leveson, N., 154, 169
Levi, K., 57
Levine, D., 169
Lewis, B., 169
Lewis, H.W., 8, 15
Lewis, R., 33
Lewis, S., 213
Lindley, D.V., 114
Liu, C.-C., 271
Liu, C.H., 143, 150
Luckham, D., 160, 169

Ma, M., 260, 261
Machowski, J., 267, 286
Maier, M., 9, 16
Mann, W., 169
Margolus, N., 252, 261
Margulis, L., 147, 150
Maslen, E.H., 7, 15
Matía, F., 190
Maturana, H.R., 130
McCarthy, G., 285
McCulloch, W.S., 122, 130
Mead, C., 118, 130
Mecham, M., 68, 73
Mehra, A., 188, 190
Melliar-Smith, P., 169
Meltzer, D., 260
Menga, G., 189
Menon, S., 213
Metropolis, N., 102–6, 108, 109, 114
Meunier, R., 189
Mezaros, G., 167, 169
Middelhoek, S., 131, 134, 150
Miller, C., 14, 17, 35, 55, 57, 310
Miller, D., 189
Miller, M.I., 114
Mills, S., 8, 15
Minsky, M., 189–90
Misiak, C., 57, 73
Misra, P., 82, 95
Mitchell, M., 252, 261
Moncelet, G., 164, 169
Morcom, R., 62, 73
Morton, B., 14, 76, 115, 286, 310
Moser, L., 164, 169
Mufti, M., 205, 213
Mumford, D., 120, 130
Mungee, S., 169
Musliner, D., 95

Nagel, D.C., 73
Nagel, E., 127, 130
Nagel, K., 252, 261
Nagle, H.T., 146
Nelson, K.S., 209, 213
Newman, J.R., 130
Nicolas, G., 12, 15
Nielsen, P.E., 185, 192
Nimmo, I, 33
Nolan, M.S., 218–19, 240
Nordin, P., 189
Norman, D., 60, 73
Noronha, S.J., 184, 189
Norris, G., 7, 15
Novak, W.E., 169

Occello, M, 183, 190
Opperman, R., 40, 57
Owens, A.J., 113, 260

Palffy-Muhoray, P., 249, 260
Paludetto, M., 169
Parth, F.R., 10, 15
Partridge, A., 150
Pejtersen, A., 57
Penrose, R., 128, 130
Perkins, S., 146, 150
Perrow, C., 8, 16, 37, 48, 57
Perry, T.S., 218, 240
Peterson, A., 169
Piccioni, M., 113
Pierce, J.R., 138, 150
Pines, D., 260
Pitts, W., 122, 130
Polson, P., 67, 73
Pool, R., 13, 16, 301
Porras, J., 169
Poston, R., 165, 169
Prieto-Diaz, R., 155, 169
Prigogine, I., 12, 15
Prywes, N., 166, 168
Puente, E.A., 190

Ramakrishna, Y., 169
Randelman, R.E., 106, 114
Rasmussen, S., 252, 261
Rassmussen, J., 50, 54, 57
Rechenberg, I., 254, 261
Rechtin, E., 9, 16
Reynolds, J.K., 150
Rhodes, L., 239

Richards, D., 130
Riley, V., 14, 18, 36, 46, 57, 59, 71, 73, 310
Ripley, B.D., 101, 114
Rockstad, H.K., 150
Rodríguez, P., 189
Rohnert, H., 189
Roland, P., 120, 129–30
Rosenbluth, A.W., 114
Rosenbluth, M.N., 114
Rosenschein, S., 183
Rossi, A., 276, 286
Rossky, P.J., 104, 114
Rubin, D.B., 101, 114
Rust, J., 110, 114

Sagan, D., 147, 150
Sage, A.P., 9, 16
Sahasrabudhe, V., 188, 190
Sahlin, N., 57
Saiedian, H., 164, 169
Salmond, D.J., 113
Saltaren, R., 189
Samad, T., 1, 14, 76–77, 95, 115, 123, 129, 131, 191, 310
Sanz, R., 14, 151, 171, 183–85, 188–90, 310
Sarter, N., 64, 67–69, 73
Schiffman, S.S., 150
Schmalz, B., 57, 73
Schmeiser, B., 104, 114
Schmidt, D., 160, 169
Schoess, J., 213
Schwartz, J., 130
Schwefel, H.-P., 254, 261
Searle, J., 128, 130
Segarra, M.J., 190
Sekigawa, E., 68, 73
Selic, B., 164, 169
Shannon, C., 138, 150
Sheremetov, L.B., 183, 190
Sherry, L., 67, 73
Sheth, K.S., 240
Silver, B.L., 11, 16
Silverstein, M., 168
Singh, M.P., 183, 260–61
Slack, M., 189
Smirnov, A.V., 183, 190
Smith, A., 243, 250
Smith, A.F.M., 101–2, 104, 109, 113–14
Snyder, D.L., 114
Solomon, S., 131, 150
Sommerlad, P., 189

Soros, G., 251, 261
Sperandio, J., 51–52, 57
Sreenivas, R.S., 113
Sridhar, B., 238, 240
Srivastava, A., 114
Stal, M., 189
Stout, T., 153, 169
Suga, N., 146, 150
Sussman, G.J., 181, 189
Sussman, J., 181, 189
Swager, T.M., 146, 150
Swenson, H.N., 233, 240
Sycara, K., 183, 189

Taylor, R., 160, 169
Teller, A.H., 114
Teller, E., 114
Tenner, E., 50, 57, 301
Thompson, R.P., 243
Tierney, L., 101–2, 114
Toffoli, T., 252, 261
Tommila, T., 190
Travis, J., 145, 150
Tzafestas, S., 190

Vachtsevanos, G.J., 191, 195, 205, 207, 213, 310
Vakili, P., 113
Valsecchi, G.B., 286
van der Ziel, A., 136, 150
Van Dyke, J., 191, 213, 311
Vapnik, V., 87, 95

Vecchi, M.P., 113
Venta, O., 190
Vera, J., 169
Vestal, S., 169
Vicente, K., 51–52, 57
Vidyasagar, M., 110, 114
Vivona, R.A., 240

Walsh, M.J., 113, 260
Wang, C., 183, 190
Wang, H., 183, 190
Watkins, C.J.C.H., 255, 261
Weinberg, S., 11, 16, 84, 95
Weiner, E.L., 68, 73
Weyrauch, J., 15, 303, 311
Wildberger, A.M., 14, 215, 241, 246, 252, 261, 277, 279, 285–86, 311
Williams, D.H., 240
Williams, D.W., 235, 240
Williams, T.J., 153, 169
Wilson, E.O., 11, 16
Wittig, T., 183, 190
Wolfenbuttel, R.F., 131, 150
Woods, D., 64, 67–69, 73
Wooldridge, M., 183, 189, 190
Wootton, J., 285–86

Yourdan, E., 119, 130

Zipf, G.K., 12, 16
Zook, J.D., 14, 76, 131, 311

Subject Index

Abnormal Situation Management (ASM), 33, 194
Abstraction Hierarchy, 54
accidents, 8–9, 16, 20, 24, 29, 37, 42, 50, 53, 57, 67–68, 173, 272, 289, 294; *see also* failures
Active FAST (AFAST), 233
actuators, 1, 5–7, 76, 78, 84, 116, 118, 127, 131, 134, 142–43, 153, 173, 178, 192, 194, 209, 226, 246–47, 254, 256, 269, 278–79, 298
adaptation, 40, 57, 78, 87, 100, 118, 129–30, 166, 182, 189, 215, 241–42, 247–50, 253–61, 265–69, 277–80, 285–86, 304, 306
  definition of, 250
adaptiveness, 18, 39, 40–42, 56
Advanced Traffic Management System (ATMS), 272–73
Aeronautical Telecommunication Network (ATN), 232
agents, 14, 41, 44, 46, 50–54, 151, 171, 180–90, 215, 241–42, 247–61, 269, 277–81, 286, 289, 305, 307
  adaptation in, 250, 253–56, 259, 279–80, 286
  communication between, 187
  control applications, 186
  definition of, 181
  modeling applications, 250–53
  properties of, 182
air traffic management and control, 3, 9, 14, 18–21, 28, 30, 47, 51, 56–57, 66, 71–72, 81, 164, 215–18, 227, 239–40, 290, 293
  air traffic growth, 230
  airspace capacity, 218–19, 235
  communication in, 218, 237
  cost of inefficiencies, 230
  separation, 48, 52, 67, 217–22, 227, 229, 232, 235–37
Airbus, 53, 67–68, 226

aircraft, 1–2, 6–9, 17–21, 29, 40, 45–55, 59, 64–70, 76–79, 82, 86–90, 94, 115, 148, 204, 211, 215–27, 231–32, 235–40, 253, 290, 293, 295, 300
  Airbus A300, 68, 73
  Airbus A320, 53, 67–68
  Boeing 737, 19, 68
  Boeing 777, 6–9, 70, 225, 239, 290, 293
  McDonnell Douglas MD-80, 68
Aircraft Communication and Reporting System (ACARS), 226
Airline Operational Control (AOC) center, 223–28, 231, 236
airports, 6, 15, 19, 25, 49, 52, 218–25, 228, 231–32, 235–40, 293
  Munich II international airport, 6
alarms, 2, 47, 53, 152, 163, 172, 193
algorithms, 5, 11, 76, 79, 82, 87, 90–94, 97, 101–14, 119–20, 143, 151–52, 159, 179, 192–95, 198–204, 212–15, 226, 239, 242, 247, 253–56, 259, 270, 278–79, 304–5; *see also under* control
analog computing, 118, 122, 127–28
Analog Devices, 144
architecture description languages (ADLs), 157–61
ARCHON, 186–87, 190
Ariane 5, 290, 293
Arizona State University, 282, 285–86
Army Research Office (ARO), 265
Arrival Sequencing Program (ASP), 222
artificial intelligence, 88, 100, 128, 152, 183, 189, 213, 246, 248, 253, 260, 265, 273
artificial life, 252, 277
artificial noses, 146
asset management, 76, 149, 186, 305
ATC System Command Center (ATCSCC), 221–22
ATC Towers, 220–21

Automatic Dependent Surveillance (ADS), 226–27, 232, 235
Automatic Dependent Surveillance Broadcast (ADS-B), 232, 237
automobiles, 4–7, 12, 21, 24, 37–38, 48, 50, 62, 76, 148, 216, 218, 247, 252, 294, 297–99
  airbags, 4–5, 15, 148, 216, 293–98
  antilock brakes, 4, 50
  seatbelts, 4–5
  sensors, 137
autonomous vehicles, 75, 78, 89–94
autonomy, 2, 6, 14, 55, 75–79, 83–84, 87, 89, 94, 116, 127–29, 152, 171, 174, 181–83, 189–90, 204, 213, 248, 250, 257–60, 279–80, 286, 304, 306
avionics, 6, 57, 73, 190, 218, 223–27

bandwidth, 60, 63–64, 69, 106, 138, 232, 303
banking, 216, 243, 263, 272, 281
baseball, 117
bats, 146
behaviorism, 118
belief functions, 207
best practices, 33, 111
biochemistry, 145, 291
bioengineering, 307
biology, 10–14, 76, 86, 115–24, 127–29, 133–35, 140, 145–48, 216, 250–54, 273, 279, 287–95, 304–7
biomedical devices, 6, 137
biomimetics, 266
biotechnology, 266
blender, 60
Boeing, 6–9, 68, 70, 225–26, 290
Boston University, 282, 286
brains, 117–30; *see also* central nervous system; neurons
  efficiency of, 118
  imaging, 120, 129
British Midlands, 19, 26
buildings, 1–2, 6, 15, 20, 32, 70, 77, 94, 102, 117–20, 137, 147, 150, 167, 171, 173, 176, 180–82, 185, 189, 193, 212, 233, 243, 252, 288–89, 300–301, 306

C++, 198, 200, 259
California Institute of Technology (Caltech), 274, 281, 283, 287, 301, 309
California Power Exchange, 244
Capability Maturity Model (CMM), 167–69

Carnegie Mellon University (CMU), 281, 283, 286
Carrier, 211
cascade effects, 154, 165, 219, 224, 264–65, 270, 274–77, 280–84
cellular automata, 277
cement plants, 173, 178
Center-TRACON Automation System (CTAS), 233
central nervous system (CNS), 76, 116–29
  algorithmic processing in, 120
  cerebral cortex, 117, 120
  invertebrate architecture, 125–26
  sensory and motor pathways, 120
  vertebrate architecture, 76, 124–25
centrifugal chillers, 152, 196
certification, 136, 142, 165, 225, 237
chaos, 11–12, 27, 38, 43, 168, 175, 236, 250–51, 261, 288–89, 297, 301
charge coupled device (CCD), 132, 135, 144–45
chemical processes, 8, 20, 47, 80, 88, 98–99, 152, 172–73, 187
  batch, 20, 88, 106–7, 190, 249
chemistry, 1, 11, 62, 79, 82, 106, 164, 250, 307
Chinese room thought experiment, 128
Church-Turing hypothesis, 128
Civil Aeronautics Board, 243
Cockpit Control Language, 18, 71–73
cognitive science, 2, 128
collaboration, 17, 21, 25, 31–33, 168, 182, 230, 305
Collaborative Decision Making (CDM), 232–34
COM (Component Object Model), 194–200
commercial aviation, 3, 6, 12, 28, 32, 57, 59, 70, 73, 78, 148, 217, 230, 239, 290, 305–7
commercialization, 5, 12, 143
Commonwealth Edison, 265, 282
communication in teams, 22, 29–32
  declarations, 29
  offers, 30
  promises, 30
  requests, 30
communication networks, 97, 105, 153, 174, 192, 281, 303
communication, navigation, and surveillance (CNS), 231

Subject Index

communications, 2, 20, 25, 28–32, 171, 204, 218–19, 224–26, 232, 237, 247, 256, 285, 288, 291, 303
compact disc players, 216, 293–96
complex adaptive systems (CASs), 14, 215, 250–52, 255, 260, 264, 279, 306
  flocking in birds, 252
  freeway traffic, 252
complex interactive networks, 216, 263–66, 271, 277
  examples, 267
Complex Interactive Networks/Systems Initiative (CIN/SI), 216, 263–66, 276, 280–81, 284–85
complexity
  behavioral, 36–41, 48–49, 56
  cognitive, 36
  component, 36–37, 47, 56, 60–62, 71–72
  definition of, 35, 303
  functional, 36, 59–60, 67
  interactive, 37
  of system, 18–20, 36, 42–43, 50, 151, 177, 196, 219, 304
  relational, 36–40, 47–48, 52, 56
  types, in control systems, 151, 175
complexity management, 1–3, 8–15, 21, 35, 39, 42, 47–53, 76, 79, 85, 94, 115–16, 143, 151–52, 176, 215–17, 305
computational fluid dynamics, 300
computer science, 59–60, 79, 88, 110, 167, 171, 250, 273, 307
computer viruses, 290
computer-aided design (CAD), 6, 9–10, 15, 106, 179, 297–300
conceptual frameworks, 18, 60–64, 69, 72, 304
Conflict Alert, 222
consciousness, 76, 127–28
conservation laws, 82, 291, 294
control
  adaptive, 87, 100
  algorithms, 1–3, 77–80, 87–88, 137, 152, 161, 172, 192, 259, 304, 306
  distributed, 172–75, 189, 215, 242, 247–49, 255–56, 259, 264–65, 269, 279–80, 284–86, 306
  feedback, 77–80, 83, 97, 191, 247, 298
  feedforward, 80
  loops, 47, 78–80, 115, 134
  model-predictive, 76, 79, 110, 180
  nonlinear, 175

  passivity-based, 177
  plantwide, 172, 187
  proportional-integral-derivative (PID), 247, 256
  robust, 76, 87, 110, 114, 175, 180, 264, 301
  stability, 48, 83–86, 178
Control Display Unit, 64–65, 69–73
control science and engineering, 14, 75, 79, 83, 152, 171, 176–77, 183, 188, 273, 299, 304
control systems, 1–7, 14, 17, 46, 53, 76–81, 84–88, 92, 94, 115–17, 129–40, 144–57, 160, 163, 168, 171–83, 186–90, 212, 216, 236, 242, 248, 256, 259–60, 265, 269, 281, 285, 298, 306
Controller/Pilot Data Link Communications, 227, 232
CORBA, 160, 182–83, 188
Cornell University, 243, 281, 283, 286
cost efficiency/reduction, 6, 9, 51, 112, 137, 144, 149, 165, 173–74, 188, 218–20, 245, 296, 305
cultural issues, 14, 20, 24, 28, 62, 64, 68, 189

data concentrators, 195–98, 203–5, 209–11
data fusion, 200
data mining, 272, 278
databases, 71–72, 82, 111, 129, 180, 184, 200, 203–5, 225, 232, 272
DCOM (Distributed COM), 194–98
decision support, 112, 187, 212, 222–24, 231–33, 238, 277
Defense Advanced Research Projects Agency (DARPA), 94, 304
Dempster-Shafer theory, 195, 201, 207, 213
Denver International Airport, 290
deregulation, 14, 215, 241–46, 249, 264, 267–70, 279, 284–85
design patterns, 156, 167, 183–84, 190
  pattern languages, 167–69, 184
  Portland Pattern Repository, 167–69
diagnostics, 5–6, 27, 30, 77, 100, 129, 133, 137, 149, 152, 172, 193–213, 278, 282, 286, 294, 298, 305; *see also* detection and identification of *under* faults
differential equations, 82, 98, 104, 175–77, 250, 277, 300
dispatchers
  airline operations, 218, 223–24, 227–28
  power systems, 9, 244, 267–68

displays, 2, 6, 22, 70, 122, 133, 196, 200, 218–19, 222, 231–35, 253, 306
Distance-Measuring Equipment (DME), 219
Distributed Air-Ground Traffic Management (DAG-TM), 236
DIXIT, 185
domain analysis, 151, 155–56, 161, 168–69, 186
domain engineering, 183–85
domain knowledge, 1, 77–78, 86
DoME, 161, 169
DOS (operating system), 59–63, 69
dynamic programming, 110
dynamical systems, 12, 109, 171, 248, 278, 291, 301
   discrete event, 109, 249, 282
   hybrid, 88, 277

e-commerce, 261, 269, 288
economics, 4, 12, 52, 114, 167, 191–92, 213, 217–19, 223, 228–31, 238, 246, 250–51, 256, 260, 263–68, 272, 273–74, 277–79, 285
   econometrics, 98, 100
education, 12, 37, 50, 98, 307
electric enterprise, 241–42, 245–49, 259, 279, 286
electric power, 12–14, 151, 215–16, 241–46, 249, 256, 263–71, 274, 279, 284–86, 290
   brokers, 241, 243
   co-generation, 176
   commodity versus service, 243
   distribution, 241, 244, 267
   generation, 7, 80, 174, 241, 244, 267, 306–7
   growth in U.S. demand and capacity, 269
   hydroelectric, 174, 242, 257
   load forecasting for, 259, 268
   markets and exchanges for, 244
   nuclear, 8, 13, 242, 304
   outages in U.S., 274
   power grids, 215, 242, 259–60, 267, 270, 281, 284, 306
   real-time pricing in, 258
   thermal, 175, 178
   transmission, 186, 241, 244–45, 267, 269
   wheeling, 242, 245
   wind, 174
Electric Power Research Institute (EPRI), 216, 241–42, 246, 256, 260–61, 263, 265, 279–86, 309, 311

emissions, 5, 80, 139, 141
En route Descent Advisor (EDA), 233
energy efficiency, 7, 118, 139, 246, 305
engine control, 5, 137
Enhanced Traffic Management System (ETMS), 223, 237
enterprise integration, 188, 303
environmental impact, 4, 49, 98, 137, 146, 148, 173, 305; *see also* emissions; pollution
ergonomics, 42, 45
ESPRIT, 174, 186
Estonia ferry, 8
European Commission, 189
European Science Foundation, 189
evolution, 11, 22, 82, 98, 105, 107, 113, 122, 129, 145–48, 168, 188, 216, 231, 240, 249, 251, 254, 260–61, 264, 267, 272, 279, 286
evolutionary computing, 11, 76, 92, 107–8, 253, 255
evolutionary programming, 107–8, 254–55
evolutionary strategies, 254–55
Excel, 198
exhaustive analysis, 15, 164, 297, 306
Expedite Departure (EDP), 233
expert systems, 100, 123, 130, 177–80, 187, 192, 198, 201

failure mode effects analysis (FMEA), 192, 194
failures
   cascading, 264–65, 270, 274–77, 280–84
   catastrophic, 8–10, 19–20, 118, 137, 212, 279, 282, 293, 299, 304–7
   prediction of, 152, 192, 199–202, 209, 212, 278
   systems failures, 17, 20–21, 290, 293
fatigue, 42–46
Fault Current Limiter (FCL), 269
faults
   containment of, 306–7
   detection and identification of, 100, 152, 187–88, 192–93, 213
fault tolerance, 84, 175
feature extraction, 205–6
Feature-Oriented Domain Analysis, 155, 169
Federal Aviation Administration (FAA), 217–23, 231–33, 239–40
Federal Aviation Regulations, 227, 239
Federal Energy Regulatory Commission (FERC), 245, 261
   rules 888 and 889, 245

Subject Index

feedback, 77–80, 83, 97, 120–25, 134, 147, 149, 165, 191, 195, 247, 251, 278, 288, 294–95, 298
financial markets and systems, 12, 216, 244, 252, 263, 265, 272, 281, 288
Fisk University, 282, 286
fitness functions, 108, 123, 249, 251, 254, 279
Flexible AC Transmission Systems (FACTS), 246, 256, 259–60, 269
flight control, 1, 78–84, 88, 113, 145, 226
flight management systems, 18, 59, 64–73, 78, 115, 225–26, 232, 237, 239
formal methods, 164, 169
Formula One, 216, 293, 297
Fourier transform/FFT, 92, 192, 194, 204
Free Flight, 3, 14, 215, 218, 231–40, 305
Future Air Navigation (FANS-1/A), 226
future-proof, 176
fuzzy logic and control, 36, 79, 83, 123, 130, 176–78, 185–87, 190, 192, 195, 204–7, 213

game theory, 248, 278
genetic algorithms, 11, 107, 122, 129, 148, 177–78, 253–55, 259–61, 279, 310
genetic engineering, 307
genetic programming, 178, 279
genomics, 145, 266
George Washington University, 281, 286
Georgia Institute of Technology (Georgia Tech), 191, 195, 212, 310
Global Positioning System (GPS), 226, 269, 271
Gödel's proof, 127–28
governments, 4–5, 12, 137, 216, 218, 231, 240–45, 265, 267, 280, 303
Ground Proximity Warning System (GPWS), 53
guidance, 65–68, 71–72, 213, 225–26, 232–33, 236, 239–40, 273, 306

Harvard University, 282–83, 286
health care, 76, 149, 305
heart pacemakers, 7, 137
heuristics, 75, 81–83, 201
hierarchies, 29, 36, 63, 85–86, 104, 109, 119–20, 158, 188–89, 209, 227, 267–68, 279, 289
High Voltage Direct Current (HVDC) interconnections, 246
HIPO method, 119
home automation, 156, 161–62

Honeywell, 71, 142, 157, 211–12, 301
  Commercial Electronic Systems, 217, 310
  Home and Building Control, 150
  Honeywell Round thermostat, 134
  Honeywell Technology Center (HTC), 1, 17, 19, 35–36, 53, 59, 77, 89, 97, 115, 131, 153, 169, 191, 194, 211, 256, 303, 309–11
  Total Plant System, 153
human error, 20, 22, 28, 45, 47, 231, 256, 290
human factors, 2, 18, 22, 35, 42, 45, 50, 64, 70, 73, 231, 236, 238
Human Genome Project, 266
human-centered design, 18, 36, 41, 57, 70, 73, 231, 307
hybrid, 88, 104, 176, 267, 277, 284

ICa, 152, 183–90
IEEE Transactions on Automatic Control, 178
image processing, 100, 104, 106, 110
immune system, 116, 277
independent power producers (IPPs), 243
Independent System Operator (ISO), 244, 285
information theory, 12, 69, 138, 150, 290
infrastructures, 14, 148, 189, 216, 263–67, 274–77, 280–85
instrument flight rules (IFR), 227, 231
Instrument Landing System (ILS), 219
integrated circuits, 9, 38, 98, 106, 118, 132, 147, 153, 246, 287, 290, 293, 295, 300
intellectual property, 143, 166
intelligent control, 178, 189–90, 247, 256, 260, 306
intelligent systems, 120, 189–90
Intelligent Transportation Systems (ITS), 272
Interconnected Operations Services (IOS), 245
Intermodal Surface Transportation Efficiency Act (ISTEA), 272
International Space Station, 195
International Standards Organization (ISO), 140, 150, 168
Internet, 50, 160, 181, 192, 244, 263, 266, 269–71, 288–93, 306
Internet Protocol (IP), 289
Iowa State University, 282, 285–86
Iridium, 276

Java, 160, 181, 200
Jet Propulsion Laboratory (JPL), 297
Johnson noise, 136

knowledge backplane, 53–55
knowledge fusion, 152, 194–95, 200–204, 212
Knowledge Partners of Minnesota, 191, 309
knowledge, as socially constructed, 25–26, 31

learning, 45, 57–62, 67, 76, 95, 114, 118, 120, 123, 126, 148, 177–78, 186, 242, 245, 247, 250, 255, 259, 261, 277, 301, 307
linguistics, 17, 30–31

machine learning, 254–55
    Q-learning, 255, 259
    supervised, 255
    unsupervised, 255
machinery, 14, 21, 195–96, 199, 203–5, 212–13, 305
Machinery Condition Assessment, 194, 204
Machinery Prognostic and Diagnostic System (MPROS), 194–204, 211
maintenance, 1–2, 20, 27, 76–77, 115, 149, 152–53, 162, 167, 174, 179, 186, 192–95, 198, 201, 204, 212–13, 223–24, 267–68, 280, 305, 307
    condition-based, 152, 194–95, 213
    preventive, 6, 193
    scheduled, 137, 148, 152
Markov chains, 101–2, 105–6, 113–14
Mars Pathfinder, 216, 293, 296
Massachusetts Institute of Technology (MIT), 281–82, 286
mathematics, 62, 154, 216, 273, 292
McDonnell Douglas, 68
Mead-Conway design rules, 300
Medtronic, 137, 150
megacities, 266
mental models, 1, 5, 8–9, 14, 17–18, 46, 53, 60–61, 75–94, 97–99, 104, 109–12, 119, 122, 124, 127, 129, 141, 152, 161–65, 168–69, 176–78, 180–82, 187, 199, 212, 250, 252, 255–56, 259–61, 265, 276–81, 295, 297, 299–300, 304, 307
MetaH, 157–60, 164, 169
metaphors, 18, 46, 59–64, 70–73, 164, 215, 260
    desktop, 63, 70–71
meteorology, 224, 228, 237
metrics, 7, 168, 237–38
microprocessors, 117–18, 144, 151, 153, 256, 279
Microsoft, 196, 198, 200

Microsoft Research, 110
microstructures, 132, 143, 246
    micromachining, 132, 147, 150
miniaturization, 6–7, 137
missiles, 19, 45, 90
models, 1, 5, 8–9, 14, 17–18, 46, 53–54, 59–61, 75–99, 104, 109–12, 119, 122–29, 141, 150–52, 161–69, 176–82, 187, 199, 203, 212, 250–52, 255–56, 259–61, 265, 276–81, 295–300, 304–7
models in control engineering
    adaptation of, 75, 87
    applications, 79–80
    discrete event, 282
    empirical, 80
    first-principles, 79, 82, 88, 98
    heuristic, 81–82
    linear, 80–84, 100
    linearized, 76, 88, 176–77
    local, 81, 84–85
    nonlinear, 79, 83–84, 88, 281
    regression, 79, 86, 139
    state-space, 79
    steady-state, 83
    stochastic, 82, 98
    time series, 192
    types of, 81
modes, 20, 63, 67, 84, 88, 158–61, 195
    aircraft, 53, 64, 67–68, 71, 90, 226
    brain, 119
    failure, 136, 192–94, 202, 205–11, 256, 268, 305
    power system, 267–68
Monte Carlo, 76, 98–102, 106, 113–14, 297
multimodels, 14, 77, 80, 94–95, 111, 304
multiresolution, 75, 88–93, 207, 216, 276
multiscale, 88, 216, 263–64, 267–68, 284, 287–89
multivehicle systems, 94, 217, 299, 306
Mycoplasma, 294

nanotechnology, 7, 266
NASA, 24, 194, 217, 231, 236, 239–40, 309
    Ames Research Center, 217, 309
National Airspace System, 215–23, 227–32, 235–39
National Energy Policy Act, 243, 245
National Institute of Standards and Technology (NIST), 33
National Public Radio, 275
natural selection, 11, 113, 118, 147

Subject Index

Nature, 292
Naval Research Laboratory, 194, 212
Navier-Stokes equations, 300
navigation, 6, 65–72, 79, 148, 218–19, 221, 225–27, 232, 236–37, 240
 navaids, 218–19, 227, 232
negotiation, 152, 182, 184, 230, 240, 245, 278
networks, 14, 21, 54, 76, 88, 110, 122–23, 141, 147, 150–51, 215–16, 246, 252–54, 261–75, 278–93, 295, 305
 convergence of, 216, 288
 protocols, 202–4, 232, 244, 266, 287–89, 300
neural networks, 79, 88, 106, 110, 122–23, 129, 141, 148, 177–78, 187, 195, 205–7, 213, 254–55
neurons, 117–22, 125–26, 145, 255
 loss in humans, 118
New York Mercantile Exchange (NYMEX), 243
Newsweek, 275
North American Electric Reliability Council (NERC), 244–45, 261, 274
NP-hard, 105, 108, 297

Object Management Group (OMG), 164
Object-Oriented Ship Model (OOSM), 194, 196, 199–203
Office of Naval Research (ONR), 194–95, 213
Official Airline Guide (OAG), 224
oil refineries, 1–3, 6–9, 21, 25, 33, 40, 47–49, 54, 79, 94, 115, 193, 304
olfaction, 76, 125, 145
Open Access Same-time Information System (OASIS), 245, 269
operations research, 88
operators, 1, 3, 6, 9, 14, 17–18, 21–27, 30, 36, 41–49, 52–55, 70, 77, 108, 115, 123, 131, 134, 137, 174, 178–79, 191–93, 212, 216–18, 231, 254–56, 264, 267, 281, 284–86, 303–4; *see also* teams
 beliefs of, 19–20, 23–30
 operator capacity, 18, 42–51
optimization, 3, 11, 75, 80–82, 87, 92–97, 102, 105–14, 122, 172, 182, 218, 223–24, 230, 239, 254, 260–61, 264–65, 273, 277–78, 284, 306
 combinatorial, 105–7
 ordinal, 108–9, 282
 plant-/enterprise-wide, 3, 78, 80

route, 91–92, 225
 stochastic, 108
orient-evaluate-act model, 22–25

PageNet, 275
parallel processing, 92, 209, 254, 259–60
 in brains, 117–18
Passive Final Approach Spacing Tool (P-FAST), 232–33
pattern recognition, 122–23
pattern theory, 120
perceived complexity, 14, 17–18, 22, 32, 35–42, 48–52, 55–56, 59–63, 72, 303–4
performance monitoring, 81, 186
personal computers, 59, 62–64, 70, 100, 196–97, 203–4, 224, 289, 293
 corporate yearly costs, 293
Petri nets, 164
pharmaceuticals, 5, 149
Phillips, 27
physics, 7, 11–12, 79, 82, 100, 110, 113–14, 130, 133, 139, 149, 164, 249–52, 257, 266, 279, 289, 292, 300, 307
PID control, 247, 256
pilots, 9, 18, 26, 53, 57, 64–73, 78, 90, 115, 218, 221, 224–28, 231–32, 236, 240, 290
pipelines, 216, 263, 280, 284
Piper Alpha oil platform, 27–29
planning, 27–28, 31, 55, 89, 120, 125, 168, 187, 219, 223–28, 231, 235–37, 240, 268, 273, 278
Plant Reference Model, 54
pollution, 5, 148, 173, 242, 272, 298; *see also* emissions
population, poverty, and pollution, 266, 284
power laws, 11–12, 281, 286, 301
pragmatics, 31
PredictDLI, 191, 195–96, 204–5, 211–12, 311
Predicted Mean Vote (PMV), 140
Problem Analysis Resolution and Ranking (PARR), 234
procedures, 13, 17, 21, 25, 28–32, 45, 48, 52, 67, 71–72, 77, 81, 110, 193, 204, 215, 218–19, 222–25, 230–39, 254, 293
process control, 14, 145, 149–53, 169, 172, 186
 statistical process control, 98
process maturity frameworks, 151
process plants, 17, 28, 151, 172; *see also* chemical processes; oil refineries
productization, 143

prognostics, 6, 77, 137, 152, 193–205, 209–13, 305; *see also* prediction of *under* failures
  prognostic vector, 201–2
project management, 151, 165, 168, 171
psychology, 11, 42, 73, 307
Purdue University, 282–83, 285

quantum gravity, 128
quantum mechanics, 84, 98, 288–90

randomized algorithms, 14, 75, 92, 97, 100, 108–10, 113–14, 304
randomness, 11–14, 75, 97–100, 107–10, 113–14, 250, 254, 304
Rapid Application Development (RAD), 165–66, 169
reductionism, 249, 288, 294
regulations, 4–5, 173, 187, 211, 223, 242
reliability, 9, 15, 30, 76, 131–32, 136–37, 148–49, 159, 164, 188, 200, 213, 231–32, 235, 242–45, 261, 265, 269–70, 284–85, 288–93, 305–7
Rensselaer Polytechnic Institute (RPI), 281
requirements, 9, 20, 26, 31–33, 46, 49–50, 55–57, 66–67, 70, 77–78, 84–85, 91–92, 119, 153–56, 165–68, 172, 179–80, 184–88, 212, 223–27, 240, 243, 253, 256, 304–7; *see also* specifications
  analysis, 155–56, 166
  engineering, 186
rigor, 287–88, 291
ripple effects, *see* cascade effects
risk, 8–9, 15, 28, 57, 152, 155, 174, 184, 187–88, 228, 231, 235, 245, 269–70, 275, 278, 295–96, 299, 306–7
RiskMan, 185–88
robotics, 151, 173–74, 189, 279–80
robustness, 148–49, 215–16, 231, 235, 256, 265, 270, 284, 287–96, 300–301, 306
rotorcraft, 55

SafeFlight-21, 232
safety, 1, 4–5, 8–9, 15, 20, 29, 50–52, 57, 73, 76, 81, 84, 87, 136–39, 142, 148, 159, 164, 169, 172–74, 187–88, 191–92, 212, 217–19, 222, 225–37, 242, 269, 272, 279, 299–301, 306–7
  fail-safe, 142
  intrinsic, 76, 136

sampling, 98, 101–5, 108–9, 113–14, 193, 196–97, 205, 247, 297
  Gibbs sampler, 104, 109–10, 113
  importance sampling, 101, 110
  Langevin sampler, 104, 110
  Metropolis sampler, 102–9, 114
  rejection sampling, 101–2
Santa Fe Institute, 12, 249
Satellite Communication System (SATCOM), 226, 232
satellites, 20, 91, 216, 226–27, 232, 263, 269, 270–75, 280, 284–86, 293
  Galaxy-IV, 275
  Galileo, 293
scheduling
  airline, 217–19, 228
  applications, 106
  ATM switches, 97
  electric power, 244, 267–68
  in biology, 125
  maintenance, 152, 192, 203, 268
  of computation, 157–59, 164, 203
  project management, 166
Science, 292
self-healing, 216, 265–66, 269, 274–77, 280–85, 306
self-organized criticality, 12, 251–52
sensors, 1–7, 14, 54, 75–80, 84, 90, 99, 102–3, 116–17, 120–21, 127, 131–53, 156, 162, 176, 188–89, 192–93, 196–209, 212, 226, 246–47, 254–56, 259–60, 265–66, 269, 273–74, 278–83, 286, 294, 298, 304–6
  analog, 134
  arrays of, 139, 144–45
  biological, 76, 116, 132–35, 145–46
  calibration and self-calibration in, 134–39, 246, 260, 305–6
  chemical, 76, 133, 136, 145–46
  comfort, 140
  compensation and self-compensation in, 131, 138–39, 142
  cross-sensitivity in, 76, 136–39, 145
  definition of, 131–32
  failures of, 102–3, 294
  in electric power systems, 246
  inferential, 76, 79–80, 131, 140–41
  integration, 144
  intelligent/smart, 133, 139, 150, 256, 278, 305
  manufacturing of, 143
  mass flow, 132, 143

olfactory, 76, 116, 139, 145
power dissipation in, 137–38
self-checking, 76, 142
stability of, 133, 138, 142
ships, 14, 17–19, 27–28, 115, 152, 194–99, 203–5, 211–12, 303
   U.S.N.S. Mercy, 195
   U.S.S. Vincennes, 19, 27
signal processing, 76, 83, 122, 133, 137–41, 144–47, 161, 192, 196
signal-to-noise ratio, 76, 87, 132, 136–38
simulated annealing, 105–8, 113–14
simulation, 9–11, 76, 98–100, 104–5, 108–14, 120, 162–64, 177, 190, 249, 252–65, 272–84, 297–300, 304
Simulator for Electric Power Industry Agents (SEPIA), 256–60
situation awareness, 223, 232–33, 236
Smalltalk, 182
sociobiology, 11
software
   abstraction strategies, 151
   architectures, 85, 151–52, 155–56, 159–63, 168, 189, 192, 305
   bugs, 151, 155, 200, 290
   configuration management, 168, 239
   dependencies, 151, 154, 159, 165
   excess complexity, 151–53
   interfaces, 159
   languages, 161, 164, 176, 306
   object oriented, 94, 152, 156, 180–82, 199, 249, 253, 257, 279, 287
   process, 165
   reuse, 151, 159, 166, 168
   tools and metatools, 151, 161–62, 167–68
software development, 14, 151, 154, 162–69, 188, 292
   document-based, 162–63
   model-based, 161, 163
software engineering, 2, 151, 155, 166, 169–71, 176, 248
   component-based, 180, 189
Software Engineering Institute, 167
space debris, 276
Space Shuttle, 15, 49, 209, 276, 301, 306
   Challenger, 24, 36, 49, 293
Space Station *Freedom*, 195, 209
Space Station *Mir*, 276
space structures, 82
Special Use Airspace (SUA), 79, 219, 231

specifications, 9–10, 116, 135–38, 143, 150, 154, 158–61, 164–69, 180, 183, 278–79; *see also* requirements
spiral model, 9, 24, 146, 155, 165–66, 169
stability, 83–84, 133, 136, 142, 178, 191, 225–26, 259, 270, 282
stability augmentation, 225–26, 268
state estimation, 80, 101, 109, 113
state-based feature recognition (SBFR), 194–95, 203–4, 209–13
statistical physics, 102–4, 110, 289
statistics, 8–9, 12, 24, 76, 82–83, 86, 95, 98–110, 113–14, 138–41, 205, 289, 291, 297
   Bayesian, 76, 100–4, 109–10, 113–14, 130, 201–3
structure control, 188
superconductors, 259, 266, 269
Surface Management System (SMS), 233
Surface Movement Advisor (SMA), 232–33
system complexity, 18–20, 36, 42–43, 50, 151, 177, 196, 219, 304
system design, 17, 21, 27, 36, 39, 49, 77–79, 94, 144, 164, 269, 303–4, 307
system health management, 2, 6, 14, 81, 152, 191–94
system identification, 100
Systems Analysis and Software Engineering, 311
systems engineering, 9–10, 15–16, 57, 168, 171, 175, 179, 235, 291, 296, 306–7
systems failures, 17, 20–21, 290, 293

task models, 54
TCP/IP, 289
teams, 6, 9, 17, 21–22, 25–33, 45, 53–54, 176, 204, 239, 256, 303, 307; *see also* communication in teams
   authority, 26, 29, 31–32
   shared understanding in, 15, 26–27
technoscience, 13
telecommunications, 164, 243, 247, 263, 270, 273, 275, 280–81, 284
Tennessee Valley Authority (TVA), 265, 282
Terminal Radar Approach Control (TRACON), 220–22, 232–33
Texas A&M University, 281, 286
thermodynamics, 138, 289, 295
Total Quality Management (TQM), 168
Traffic Collision Avoidance System (TCAS), 53, 70, 228–29, 232
Traffic Flow Management (TFM), 217

Traffic Management Advisor (TMA), 232–34, 240
Traffic Management Coordinator (TMC), 222–23, 228–30, 233, 237–38
Traffic Situation Display (TSD), 222–24
training, 9, 17, 28, 32, 39, 41, 44–46, 50, 52, 60–64, 67, 70, 73, 79, 123, 208, 223, 253–56, 299, 307
transducers, 133–39
Transfer Capability Evaluation (TRACE), 269
Transmission Control Protocol (TCP), 289
transportation systems, 12, 97, 215–16, 247, 263, 272–73, 281, 285–86, 290–91, 303
capacity of, 7
trend vectors, 222, 229
turbulence, 148, 223, 237, 293, 300
Turing equivalence, 122, 127
TWA 800, 293

uncertainty, 14, 33, 75, 86–87, 99–100, 108–9, 112, 138, 140, 152, 175–78, 182, 205, 216, 236, 278, 287–301, 306
Underwriters Laboratory (UL), 136
Unified Modeling Language (UML), 164, 169
Unified Power Flow Controller (UPFC), 269
Universidad Politécnica de Madrid, 171, 310
University of California–Berkeley, 281, 286
University of California–Los Angeles (UCLA), 281, 286
University of California–Santa Barbara, 281, 286
University of Illinois, 281, 286
University of Massachusetts–Amherst, 282, 286
University of Minnesota, 256, 281, 286
University of Tennessee, 282, 286
University of Washington, 271, 282–83, 285
University of Wisconsin, 282, 286
UNIX, 297
unpredictability, 11, 18, 21, 24, 38–49, 55–56, 63, 89, 176, 188, 291, 299, 304, 307
usability, 59, 63–64, 69–70
U.S. Department of Defense, 195, 216, 263, 265, 280–81, 284–85
U.S. Department of Energy, 245
U.S. Department of Transportation, 272
user interfaces, 14, 17–18, 22, 36, 39, 42–46, 51–56, 59–64, 160, 163, 166, 180, 187, 198–99, 257, 304
graphical, 60–64, 69, 94, 161, 164, 200, 257

playbook, 54
tasking, 18, 55–56
User Request Evaluation Tool (URET), 232–34, 240
User-Preferred Trajectories (UPTs), 235
U.S. National Science Foundation, 246
U.S. Navy, 33, 152, 204
U.S. Presidential Commission on Critical Infrastructure Protection, 216, 264

verification and validation, 2, 5, 9, 79, 102, 112, 159, 165, 185–86, 211, 278, 292, 305, 307
Very-high-frequency Omni-range Receiver (VOR), 219
VHF radio, 226
Virginia Tech, 282, 285
virtual engineering, 9–10, 15, 179, 291, 296–97, 299
virtual machines (VMs), 151, 159–62, 165, 305
Visual Basic, 182, 198, 200
vomit response, 307

wake vortices, 227, 235
Washington State University, 281, 286
Washington University–St. Louis, 282, 286
waterfall model, 9, 165–66
wavelet neural networks, 194–95, 203, 207–8
wavelets, 89–92, 192, 207–8, 213
waypoints, 66–67, 71–72, 78, 83, 225
weather, 72, 90–92, 115, 148, 217–19, 223–25, 228, 230, 236–37, 293
Western Systems Coordinating Council (WSCC), 244
Wide Area Measurement System (WAMS), 246, 269, 282
Windows (operating system), 63–64, 196–198, 257
wireless, 137, 144, 270
WM Engineering, 212
workforce reductions, 4, 17, 50, 139, 174, 192, 213
workload, 18, 21–22, 39–52, 55–57, 219, 221, 225–27, 232, 237–38, 304
World Wide Web, 94, 131, 215, 260

Y2K bug, 275, 290, 293
York, 211–12